Signal From Noise, Information From Data

Signal From Noise, Information From Data

William A. Luker Sr.

and

Bill Luker Jr.

Graphics by Gerald D. Luker

To order additional copies of this book, contact:
Xlibris Corporation
1-888-795-4274
www.Xlibris.com
Orders@Xlibris.com
61651

CONTENTS

Preface 7

Part I: The Descriptive Statistical Domain

Introduction 17
Chapter 1: About Numbers, Variables, and Measurement............................19
Chapter 2: Tables, Pictures, and Central Tendency.......................................31
Chapter 3: Variation and Normality 50
Chapter 4: Other Useful Descriptors..74

Part II: Inferential Statistics

Introduction 87
Chapter 5: Probability and Sampling 91
Chapter 6: Sampling Distributions, Standard Errors, and Estimation 106
Chapter 7: Two Alternatives: Hypothesis Testing and Statistical
    Significance 119
Chapter 8: Hypothesis Testing: More than Two Alternatives.........................135

Part III: Analysis

Introduction 153
Chapter 9: Relationships and Causation 155
Chapter 10: Measuring Relationships: Covariance
    and the Coefficient of Correlation 166
Chapter 11: Bivariate Regression: Ordinary Least Squares (OLS) 186
Chapter 12: Univariate Multiple Regression 201
Chapter 13: Assumptions of the OLS Regression Model...............................217
Chapter 14: Curvilinearity in OLS Regression 230
Chapter 15: Missing Data and Outliers..244
Chapter 16: Autocorrelation, Heteroscedasticity, Multicollinearity,
    Model Specification, and Normality...270

Part IV: Special Issues in Multiple Regression

Introduction 289
Chapter 17: Nominal Variable Coding 291
Chapter 18: Interactions 308
Chapter 19: Proportions, Counts, Ranks, and
    Dichotomous Dependent Variables 327
Chapter 20: Causal Inference with Structural and
    Reduced Form Equations 343

Part V: Time Series Analysis

Introduction 365
Chapter 21: Forecasting with Classical Decomposition 367
Chapter 22: Forecasting with Moving Averages 385
Chapter 23: ARIMA Models 401
Chapter 24: Causal Inference with Time Series 420

Part VI: Multivariate Analysis

Introduction 439
Chapter 25: Factor and Principal Components Analysis 441
Chapter 26: Canonical Correlation and
    Jacob Cohen's Set Correlation 470

Index 489
References 499

P R E F A C E

Statistics define this book's theme. More precisely, the book focuses on how to make decisions using the science and tools of statistics.

In setting out to define what we mean, let us first get the eight-hundred-pound grizzly bear out of the corner of the room and introduce him: Statistics mean numbers, and numbers lead inevitably to some rigorous, systematic thinking, and decisions based on that thinking. As an old professor of ours used to say, rigor does not mean "rigor mortis."

While a seasoned manager's gut feelings often provide more accurate information in guiding decisions than carefully prepared reports, wouldn't it be nice (for her and you) to know that she backed her gut feelings with statistically valid information about the future when she and you and the rest of the team execute her next decision? Therefore, we designed the book for the "quant jock" in all of us. Yes, numbers proliferate, but we use them with a decision-making approach that relies on information extracted from data.

We know that many people have concerns about their math skills. We, however, appeal to a simple maxim that holds true no matter how poor you think your math skills—or worse, how bad you think you are, intrinsically, at doing math. That maxim is as follows: do not believe everything you think, or everything other people say, especially about your own shortcomings in the world of numbers.

We do not think you have poor math skills; we certainly do not think you are somehow intrinsically bad at math. You probably have fallen prey to bad teaching and invidious comparison. Certainly, some people have a greater facility than others, which does not mean you cannot understand and use statistics.

On occasion, we will use statistical equations. When we do, we follow that equation with a verbal explanation. For example, we offer a familiar equation for the arithmetic mean or average value of some set of numbers.

$$M = \frac{\Sigma X_i}{N}$$

In simple language, this equation translates as follows. M (or whatever symbol we choose to use as a symbol for the average) equals the sum (Σ), the total, of all the observations (X_i) in the group divided by the number (N) of items in the group.

For the set of numbers 13, 22, 25, 33, and 35, N = 5 and the sum (the total) of the numbers, ΣX_i = 130. Thus, the average equals

$$M = \frac{\Sigma X_i}{N} = \frac{130}{5} = 26$$

Was that hard? Please remember that statistics uses applied math, not pure math. You will see what we mean later in this preface.

In any event, we promise you will come away from this book with a newfound respect for your ability to make and comprehend decisions based on statistics.

STATISTICS DEFINED

One use of the word *statistics* refers to a collection of numbers (some factual, some not) like those usually found in an almanac. For example, the *Statistical Abstract of the United States* contains a wide variety of measurements, ranks, and counts, such as annual rainfall in all the states, ranking of college football teams, temperatures by seasons, and the proportions of minority ethnic groups in cities, towns, and counties. In this context, defining statistics as a plural noun equates it to raw data. Therefore, our first definition uses the word *statistics* as a plural noun, a collection of numbers.

Our second definition refers to an integrated body of procedures and techniques called the statistical method used to help us choose among courses of action. The method of statistical inquiry pervades every field, including, but not limited to, education, business, and government. As a singular noun, we define statistics as a tool for making effective decisions.

Often, large enterprises or organizations possess a wealth of data (statistics as a plural noun), but very little useful information. We read or hear about the nightmare of eight-hundred-page PowerPoint presentations, showing every conceivable dashboard metric. We imagine generals and business executives confronted with an incomprehensible mass of data, backing slowly out of the room and bolting for the door in terror. The statistical method—in its broadest sense—extracts useful information from data. Statistical methods feed on data, like a whale swimming through an ocean full of krill.

Unlike what comes out of the back end of a whale, however, statistical methods produce

- o descriptive measures like means (averages) or pictures like histograms or tables like a frequency distribution that represent general characteristics of a population (universe) or sample,
- o inferences from samples to populations, and
- o analyses of relationships among measures.

Thus, the statistical method has three domains: description, inference, and analysis. We will tackle each of these arenas in the chapters that follow.

OTHER FEATURES OF THE BOOK

Now that you have some sense of the book's bearings, let us look at some of its other features.

First, as we said, our book does not feature mathematics. Occasionally, the language of statistics demands mathematical notation. In those cases, we provide the verbal background for understanding. We note that we use neither calculus nor matrix algebra, relying instead on examples, pictures, and analogies.

Second, our discussions give prominence to applications. What do we mean? Mathematical statisticians develop the mathematical theory behind the methods we use, constantly inventing new ways to estimate the strength of ever more complicated relationships in the worlds of the natural and social sciences. Their fundamental work opens the door for applications used by applied data analysts.

After all, someone must apply the tools, or they have no use, and therefore no meaning, except as an interesting intellectual challenge. Everywhere we encounter data that require description, inference, and analysis to make the data useful for decision making. To extract information contained in data, applied analysts strive to bridge the gulf between the theoretical and practical worlds from which mathematical statisticians are often unfortunately segregated.

Third, unity characterizes our approach. At the risk of repetition, our book moves from descriptive statistics, measuring the general characteristics of data, to sampling inference to analysis, always showing the links among these methods. In other words, statistics represents a unified system, tied by a golden thread, not a grab bag of isolated procedures. A search for signal and dampening noise specifies that golden thread. When you see a set of raw numbers not treated statistically, you see noise. All the elements of the method—description, inference, analysis—aim toward enhancing signal and minimizing noise. Our book helps develop an understanding of unity because that insight counteracts mindless mechanical applications of methods, enabling hard thinking about appropriate approaches to statistical problem solving.

Our book places heavy emphasis on to the heart of modern statistical practice, analytic statistics.[1] Of the three domains we cover in our unified approach, analytic statistics is as follows:

- The most difficult to master
- The most exciting, relevant, and useful; and
- The source of endless debate on how to conduct our lives, both as persons and as a society

Here, we uncover *associations* (relationships) and describe their strength. Using relationships, we make *predictions* or *forecasts* (two different kinds of prognostications, as you will learn) and inferences about *causation* (causes and effects). While we have much more to say about analysis subsequently, we note that causal inference forms the basis for interventions to make conditions better. Taking the process

---

[1.] This statement does not disparage description and sample inference as they are useful and critical on their own.

of teaching reading as an example, only when we understand the causal forces acting on children's ability to read can we intervene to create the most effective methods for teaching reading.

Analytic statistics, more than the heart of modern statistics, represents its public face. Because it comes into play in countless instances every day, the main reason for reading this book involves learning its ways and means.

Every time you ask a subordinate (or a manager asks you) for the cause of a certain failure or success, you are asking—whether you know it or not—for conclusions or recommendations derived from analytic statistics. Economic, demographic, and public health forecasts flow from analytical statistical studies. Each time the news media report associations among events with a particular outcome (smoking and lung cancer being the most obvious example), newspersons relate the results of analytical statistical studies. Moreover, in an identical vein, every time we make predictions about the results of actions that as persons, families, or a society may take, these predictions come from analytical statistical studies.

In short, all the things we really care about in statistics lie within this sphere. Not surprisingly, in this domain, we find its greatest strengths, weaknesses, uses, and abuses. We will sort out these apparent contradictions with you, the reader, as best we can.

A note on statistical software

Recently, we analyzed a set of questionnaire responses collected at a conference of managers. When queried about the turnaround time for the analysis, our answer was "about forty-five seconds." The person acted stunned, as if confronted by an inexplicably magical act. Fifty years ago, we would have understood his reaction. Our answer, however, lay outside calculator, pencil, and paper technology. The correct statistical procedure was theoretically well understood in the 1960s, but almost impossible to execute with more than a handful of observations. As recently as a quarter century ago, in the mid-1980s, the procedure in question remained almost exclusively in the hands of academic and managerial scientists with large businesses and universities with mainframe computers equipped with expensive specialty software. Even with this technology, limited access and long waiting lines produced turnaround times of weeks. Or even months.

Now, inexpensive powerful personal computing hardware has revolutionized the field. Statistical studies dominate the public mind precisely because statistical software packages (ranging from free to less than $1,000) running on cheap personal computers (less than $500) have made many statistical procedures accessible to almost any person. Indeed, the so-called information explosion, fed at least partly by an exploding increase in published statistical studies, has led some critics to call it a triumph of noise over signal.

This revolution opened doors to understanding and practices impossible before. It has also enabled distortion and quackery to occur with unprecedented frequency. Our book tries to counter those forces.

In our book's examples, we use several software packages. Our recommendations center on those we think are the most useful, accurate, easy to use, and affordable. Some are even free.

Commercial Packages

If we had to choose only one general-purpose package, our choice is *Number Crunching Statistical System (NCSS)*.[2] Developed by Dr. Jerry Hintze of Utah State University, we cannot say enough good things about him and this program. This program—powerful, menu driven, and easy to use—comes with an outstanding help system, complete documentation of its routines, wonderful graphics, and complete output. If you have limited resources and, as a result, can only afford a single program, choose NCSS.

For ease of use, choose *Statistix* (SX).[3] While you may not fully realize the value of ease, after reading our book, you will. SX has good graphics, but its limited range of techniques makes it less useful than NCSS. SX makes only a passing nod to multivariate statistics.

Both menu and command driven, *SPSS*[4] (Statistical Package for the Social Sciences) denotes another outstanding choice. By command driven, we mean that you can write commands for routines not fully

2. Hintze, J. (2007), *NCSS, Pass, and Gess*. Kaysville, Utah, www.NCSS.com.

3. Analytical Software, PO Box 12185, Tallahassee FL 32317-2185, www.*statistix*.com.

4. 2009 SPSS Inc., SPSS Inc. Headquarters, 233 S. Wacker Drive, 11th floor, Chicago, Illinois, 60606.

accessible by the automated menu. Years in the making, for many people, SPSS still represents the standard.

Another package we like, SYSTAT[5] also uses menus and command lines. This command feature makes it a little harder to learn than NCSS and SX. The extra learning costs should not dissuade you. SYSTAT was created in 1982 as the brainchild of the statistical genius Dr. Leland Wilkinson while a professor at the University of Illinois, Chicago. His was among the first statistical analytical packages to run on a microcomputer and, as a result, became a leader in scientific field research.

For a package more difficult to learn and use than any of those previously mentioned, we recommend Bill Greene's Limdep,[6] a package originally written for memory-limited (64k) microcomputers. Nothing short of genius, Greene, as a graduate student, efficiently programmed those limited early computers to do incredible things. Limdep's primary function estimates econometric models, particularly those models using nominal data.

Free Packages

The relatively high cost of the programs listed so far to persons (although affordable for organizations of any size) requires a listing of free, downloadable packages from the Net. OpenStat,[7] the most comprehensive of the free packages, embodies almost everything you need, especially if you cannot personally afford any of the commercial packages we recommend. OpenStat, which is easily downloadable, has an excellent manual explaining the concepts and methods and will do practically all the calculations we use in this book.

We highly recommend an abbreviated version of Systat, MyStat.[8] Given its zero cost, your toolbox should certainly include it. MyStat also has

---

[5].   Systat Software, Inc., 225 W Washington St., Suite 425, Chicago, Illinois, 60606 USA

[6].   Econometric Software, Inc., 15 Gloria Place, Plainview, New York, 11803 USA.

[7].   William G. Miller, openstat@msn.com.

[8].   Systat Software, Inc., 225 W Washington St., Suite 425, Chicago, Illinois, 60606, USA.

a fantastic help system, invaluable to all researchers, novices, and professionals. We note some clearly spelled-out limitations on its use.

Our third choice of free programs, *Instat*,[9] produced by the Statistical Service Centre at the University of Reading, Reading, England, has general procedures useful in teaching statistical ideas, yet has the power to assist research in any discipline requiring data analysis. You can download the full version of Instat and use it for noncommercial purposes. The program has no copy protection, time limit, or data-size restriction other than Instat's inherent size limitations.

Gretl,[10] our fourth choice among free programs, has some methods that the other three do not, but users report some difficulty in mastering. Even so, we urge you to try it because it can function as an excellent supplement.

Finally, we emphasize, that no single statistical package can satisfy every preference or need. Nor do we argue that our choices correspond to best. Other people recommend different programs, usually because they have special needs or see the world differently. We use these recommended programs by virtue of their statistical power, ease of use, and strong tutorial guidance across a range of problems. We note that the developers of these programs update their packages periodically to add routines requested by their users and new applications arising out of mathematical statistics.

The Internet as a Source of Data, Programs, and Explanations

We would be remiss if we did not alert you to the Internet's almost-infinite supply of information about statistics, useful programs, data, and explanations of procedures. Explanations in the Net will vary from mathematically sophisticated to simple metaphors to verbal analogies. To access these resources, just type into your browser the information you need, letting the browser do the search. We will call your attention to useful sites. One example is an electronic statistics

---

9.   www.ssc.rdg.ac.uk/software/instat/instat.html.

10. Gnu Regression, Econometrics and Time-series, Allin Cottrell, Department of Economics, Wake Forest university, Ricardo "Jack" Lucchetti Dipartimento di Economia, Università Politecnica delle Marche.

textbook, *Statsoft*. This text is comprehensive, easy to use and a great complement to this book.

REFERENCES

In writing this book, we covered a wide range of statistical methods and techniques from simple description to complex multivariate analysis. How much of this requires detailed documentation? We have come down on the side of light. As required, we cite specific quotations or paraphrases. Much of the material is well known and can be found anywhere in the Net. We reference material from which we "borrowed a poesy of other men's flowers."[11] We also make special reference to work that influenced our picture of statistics in a special way. For example, we frequently cite the work of Jacob Cohen because of his unifying vision. Before Cohen, many practicing disciplinarians in psychology, sociology, political science, and pedagogy did not seem to understand the general linear model and, as a result, did not see multiple regression as a general data analytical system with analysis of variance and chi-square as sun-sets of regression. They tended to treat statistics as a grab bag of methods, without developing its unity. While Cohen was not alone in in appreciating the general linear model, his work, along with the development of microcomputer technology, made the analytical breakthrough possible.

At the end of the book ,we have listed references with the coded abbreviations used in our footnotes.

A FINAL NOTE

As appendices to some of the chapters, we have provided data sets that correspond to material in the chapter. You can copy these data sets into your computer using your scanner and read them into a spreadsheet. Our intent is that you use these data sets to work through the ideas discussed in the chapter. Osmosis does not work. We learn by doing.

[11.]   Montaigne

The Descriptive Statistical Domain

Introduction

Part 1 covers the domain of descriptive statistics. Descriptive statistics is not simply the first domain of the statistical method, but its foundation. All around, we see myriads of descriptive statistical examples, like leaves swirling around on a windy autumn day. However, like so many other things, we are generally unaware of their significance.

Sports provide a wide range of examples: Major League Baseball; the NBA and NFL and college football, basketball, volleyball, and soccer. And as school-year starts again with its own endless rounds of statistics—grades, percentile ranks in performance, and of course school athletic scores inspire endless discussions by tens of millions of people from all walks of life.

Ironically, when we ask most people whether they know anything about statistics, they will vociferously deny all knowledge of the subject, as if statistics would saddle them with some incurable social disease, like chronic halitosis or the dreaded stink foot. But to paraphrase Shakespeare, they protest too much.

Indeed, with the greatest fluency and erudition, these self-styled statistical ignoramuses rattle on about grade-point averages, batting averages (percentages), on-base percentages, quarterback ratings (ranks), and all manner of other descriptive statistics about teams and players. Some of these indicators are, mathematically speaking, more difficult to derive and interpret than the simpler measures to which you will be exposed in this book.

What's more, these same people use these statistics to do precisely what statistical analysts do all the time: make *comparisons* (in this case, of the play of their favorites, or most despised, other players and competitive teams) and *predictions* about future performance. In addition, these comparisons extend description to analysis.

We mentioned earlier that too many people believe they lack mathematical ability. That probably has something to do with the huge gulf just illustrated between their actual and perceived knowledge of basic descriptive statistics. Our argument here, as in the earlier and more general case, is that you already know more about descriptive statistics than you think; and if these examples don't prove it, the rest of the chapters in part 1 will.

Descriptive statistics is a domain where we make useful comparisons between various sets of events or groups of persons. While our professional training cautions us to refrain from predictions based solely on descriptions, in some cases, these descriptions form the basis for predictions. In addition, you will get better at the prediction game when we tackle the rest of Part I, and the inferential and analytic domains in Parts II and III.

So in part 1, we show you the more formal background for the methods and measures *that most people use daily* to discuss general characteristics of groups of numbers (say, *all* major league baseball players or samples of that larger group (players on *one* major league team).

We will also show you how to use these descriptive tools for accurate and statistically valid comparisons, which provide the basis for more informed deductions about future behavior. We give you important insights into their miss-use. Finally, we show you how descriptive statistics indeed provides the foundation on which we build and use the tools of the inferential and the analytic domains. This is the first step in extracting information from data or, as we put it, signal from noise.

CHAPTER 1

ABOUT NUMBERS, VARIABLES, AND MEASUREMENT

Statistics, numerical data, represent the raw materials of the statistical method. This chapter discusses how we create these numbers, their purposes, their forms, and their limits. Because numbers take different forms, we must often use different statistical techniques to describe and analyze them.

Definitions

Our discussion requires three definitions: observations, variables, and data sets. Variables signify quantities that can assume any set of values and, in our context, define the thing measured. For example, if we measure the weight of one hundred persons, these weights will vary. Observations exemplify individual items in the variable, and data sets consist of collections of variables used in doing statistical work. Table 1.1 amplifies our definitions.

Table 1.1 A Data Set

Name	Weight (in pounds)	Age (in years)
Jones	180	41
Smith	165	22
Ellsworth	202	37
King	178	29

Table 1.1 shows a small excerpt from a spreadsheet containing the names, weights, and ages of one hundred people. Any single name, weight, or age corresponds to an *observation*. These observations define the unit of analysis or the general level of phenomena treated

statistically (persons, groups, countries, corporate brands). Our example marks observations in bold. We marked variables with italics. We call them variables, as opposed to constants, because they can vary without losing their defining characteristic(s). The entire collection of observations and variables form a *data set*.

Some General Considerations in Measurement

The Nature of Measurements

Some measurements are simple, like counting the number of cars in a parking lot, straightforwardly tapping into the variable of interest. We also measure lengths or heights in inches with a *yardstick*.

But even the simplest measures, e.g., a weight scale, are a response to a need. In other words, using some kind of conceptual scheme as a guide, we construct measurements, or variables, to solve problems. This is true of doctors and the problem of obesity—a "constructed" concept that we know gives rise to many chronic diseases—and is true also of economists and accountants, who constructed the variables known as gross domestic product and unemployment rates in response to the problem of ending the Great Depression of the 1930s (and the Great Recession of the 2000s!)

While you may find this idea counterintuitive, we do not simply find variables; we construct them. So do not ask if our measurements capture the *true essence* of a concept. That question has no answer. We ask, instead, about a measurement's usefulness; we ask if our measurements provide the information needed to solve the problems confronting us.

Reliability, Precision, and Validity

The reliability of a measure lies in its consistency and repeatability. Suppose your doctor asks you to measure your weight ten times in a row, with no time for drinking, eating, or exercising in between. The first weighing shows 180.10, the next 180.07, the next 180.08, and on to the last measurement, all very close to 180 pounds. Even though there is some small variation—because no matter how precise the instrument, all measurements vary—for our purposes, the scale produces consistent and, therefore, reliable measures of weight. But if one weighing registers 180 pounds, the next 350 pounds, and the next 400 pounds, the scale is inconsistent and produces unreliable measures of weight. A completely unreliable measure has no value because it measures nothing.

Precision refers to the number of decimal points required by a measurement. In our example of the bathroom scale, we rounded the weight measurements to hundredths of a pound. If your health problems required more precision, perhaps to thousandths of a pound, you would immediately understand that precision is relative to the use you want to make or the use we want to make of the measurement.

Validity is the degree to which a device or test actually measures the variable of interest. We illustrate this notion, also relative to the purpose of taking a measurement, with the belief of health professionals that your weight, along with other measures, reflects the general level of your health. Thus, the bathroom scale should function as such an indicator. In other words, validity relates to some pattern of behavior or condition external to the measurement. If a measure relates to nothing but itself, we call it circular, invalid. If it relates to some pattern of behavior or condition not in our interest, we also call it invalid. For example, if we wanted to measure introversion and our scale measured extroversion, for our purposes, the measure is invalid.

A completely reliable measure can be valid on some variable but invalid on another because it is not the variable you need or want to measure. If a variable has zero reliability, it measures nothing and is, by definition, invalid.

In the everyday world, the extensive use of complex variables like teaching effectiveness makes knowledge of reliability and validity crucial. You should always seek to know as much as you can about the reliability and validity of variables. Later on, we will show you how to estimate reliability and calculate validity.

Continuous and Discrete data

Continuous data are measured to a theoretically infinite precision. In measuring body weight, the numbers in each observation can potentially extend to infinity—as many decimal places, say, as you could fit on a page or two or ten—or as much precision as the measuring instrument allows.

By contrast, discrete data have only a finite number of decimal places between data points. For example, if a variable represents persons, we can record observations of one person or two persons or fifty persons; we cannot record one-and-a-half persons or one-and-a-third persons.

Ways of Forming Quantitative Variables

In creating quantitative variables, we use four methods: measuring, ordering, counting, and categorizing.

Measuring

Measuring takes two forms: cardinal (ratio) and interval.

Cardinal Variables

Cardinal variables, also called ratio variables, have two characteristics. First, the distance on a scale from the number 1 to the number 2 measures the same distance between the number 2 and the number 3. An example is a variable constructed or created by a twelve-inch ruler. The distance between 0 and 1 demarcates the same distance as between 1 and 2 and 3 and 4, and so forth. On some pencil and paper tests, we call the score on the test cardinal if the distance between a score of 24 and 26 measures the same difference as between 31 and 33. So we say that cardinal data, or cardinally measured and created variables, have equal-scale intervals.

Second, cardinal data have a true zero. Zero means that the variable has none of the measured quantity. These two characteristics, equal-scale intervals and true zero, allow, among other things, addition, subtraction, division and multiplication.

These features also permit ratio statements like 4 divided by 2 gives 2, 2 divided by 4 gives 0.50, 4 is twice as big as 2, and two is half as large as 4. Money in bank accounts is another useful way to think of cardinal data. If I have $50.00 in my account and you have $100.00, I have half as much money as you. We note that many statistical techniques and interpretations apply only to cardinal data.

Interval Variables

Interval data share equal scale differences with cardinal data, but interval data have no true zero. The effect of having no true zero makes ratio comparisons either impossible or theory constrained. Temperature provides an example of an interval scale. While some temperature scales, measured either in Kelvin or degrees, have an absolute zero, where all molecular motion stops, at least theoretically, no one has ever observed absolute zero, and physicists don't think they will.

So temperature has equal scale distances, but we are not exactly sure what that means. It surely means that 110 degrees Fahrenheit expresses more hotness than 100 degrees, but how much hotter? Does an increase from 90 degrees to 100 degrees express the same increase in "hotness" as an increase from 100 to 110? Whether ratio comparisons are possible in the face of this uncertainty depends on the theory on which the measurement was based and its purpose. What this means is that some interval scales are treated as cardinal and some as ordinal. In classifying variables as ordinal or cardinal, we tend to respect the practices of different academic disciplines. In psychology, sociology, and social psychology, to name but three, analysts construct interval scales to measure things like introversion and alienation, and they allow ratio comparisons of these observations. Their claim to do this is based on the theory and purpose of the measuring scale.

Ordering

Ordinal data are rankings based on some quality. Examples of ordinal data are as different as the weekly ranking of football or basketball teams and survey research that aim to uncover interpersonal differences in attitudes about a wide variety of subjects.

Things that have been ranked have more or less of some quality, but they do not specify the amount because they have neither equal-scale intervals nor true zeros. Ordinally ranked data represent relative positions within a group.

Referring again to the example of the weekly football polls, the distance between a ranking of first and second may be much greater or lesser than the distance between second and third. This characteristic prevents subjecting ordinal data to addition, subtraction, multiplication, and division and is the source of much of the so-called controversy about which team is better at any point in any given season. Relative rankings also give rise to the unsurprising intransitivity of team rankings when a lower-ranked team loses to a higher-ranked team one week, but defeats a higher-ranked team the next after the higher-ranked has fallen to an even lower-ranked team the previous week. As we all say, go figure. The lesson here is that you must treat ordinal data with different statistical techniques than you would cardinal data.

The rating scale is the second example we mentioned. A marketing firm may ask customers to rate a car rental service for cleanliness using the following scale:

4 Outstanding   3 Excellent   2 Fair   1 Poor 0

One customer may rank the service as excellent; another customer might rank the same service as fair. We understand that there is some difference between the two, but just exactly what is it? The standards used for excellent and fair remain irretrievable; the standards might be the same or very different.

Some marketers argue that we might transform ranked data into intervals because of the equality of successive points. We remain skeptical of such an argument preferring to treat rating scales as ranked data. Again, this conundrum points to the need for different statistical techniques than those employed in describing and analyzing cardinal data.

Counting

The term counting literally means counting. For example, if we count the number of special needs children in a school, school 1 has twenty-five, school 2, thirty-three. Variables produced by simple counting often require special statistical treatment, a notion explained in a subsequent chapter.

Categorizing (Nominal Data)

Nominal data are qualities or categories to which you can assign arbitrary numbers, assuming no hierarchy or ranking. These qualities can be distinguishing features of individuals, such as marital status (married, single, divorced, widowed), religion (Protestant, Catholic, Jewish, Muslim, other), or gender (male, female); and for the most part, they are mutually exclusive (e.g., people are either male or female). We can assign numbers such as 1 and 0 to these qualities, thereby differentiating *male* from *female* and bringing qualitative characteristics like gender into the domain of quantitative description, inference, and analysis.

Some Special Cases

Percentages

Percentage, defined as "parts per one hundred," reflects a special case of nominal coding. When we say the word *percent*, we mean "per one hundred." In other words, a percentage provides a way of expressing a number as a fraction of one hundred. We can express percentages using the % sign, where "45%" reads as "forty-five percent" and equals 45/100 or as a decimal fraction 0.45.

We use percentages to compare relative size of one number with another number where the first number is a part of the second. In a classroom of forty pupils, if ten are Hispanic, we can compare the number of Hispanics to the total by dividing ten by forty. The result is 0.25 or 25%. The statistical treatment of percentages also requires different statistical techniques than those used with cardinal and interval data because they are constrained (limited) to values between 0 and 100.

Rates of Change

Rates express relative change over time. They are cardinal data because they have a true zero and equal-scale intervals. An example is the change in the price of a car between 2007 and 2008. Subtract the 2007 price from the 2008 price and divide the difference by the 2007 price. If $20,000 is the 2007 price and $22,000 the 2008 price, the $2,000 increase divided by $20,000 equals 0.10 or a 10% rate of change.

If the price declined by $2,000, we calculate the rate of decline in a similar way or -$2,000, -0.10 or -10%.

Ratios

Ratios express comparisons. We call the variable compared to another variable or variables the base. We calculate ratios by dividing the other variable(s) by the base. Thus, the ratio variable provides a relative comparison of one or more variables to a base variable. An example illustrates the idea.

Let us say that deaths attributable to auto accidents in 2009 in a hypothetical community were 50. The total deaths in this community for the same year were 50,000. If we want to compare deaths attributable to auto accidents to total deaths (the base), we divide 50 by 50,000 yielding a ratio of 0.001. This ratio tells us that one in one thousand deaths in this community resulted from auto accidents. Thus, ratios embody a variation of percentage variables.

Percentiles, Deciles, and Quartiles

Percentiles are the percentages of observations in the group that are less than that value. A percentile, then, represents the rank of that number compared to others. Often, we express certain scales as percentile ranks because they have no direct interpretation in their raw form. If a student makes a raw score of 300 on an entrance examination, that score cannot be understood unless it is compared to the scores of the other students who took the exam. If the test makers tell the student that her score was greater than 70% of the scores of other test takers, her percentile rank is 70.

The highest rank is the ninety-ninth percentile. Using the test score example, no person's score can be in the one-hundredth percentile because the highest score would have to rank higher than the highest score. Similarly, the lowest score is in the first percentile. No score can be in the zero percentile rank because the lowest score would have to be lower than the lowest score.

The formula for calculating a percentile rank is PR
= .... $(L_N \div N) \times 100$ (Equation 1.1)

Where

> PR = percentile rank,
> L_N = lowest number = the number of observations lower than the item,
> and
> N = the total number of observations.

So find the number of observations with values less than the observation to be transformed to a percentile rank, divide that number by the total number of observations (this yields a decimal fraction), and multiply by 100.

Suppose we have 11 scores on a test: 73, 26, 31, 88, 77, 91, 63, 51, 44, 47, and 50. If you were the person who scored 63, what is your percentile rank?

To find out, first arrange the scores from highest to lowest, as in

$$
\begin{array}{c}
91 \\
88 \\
77 \\
73 \\
63 \\
51 \\
50 \\
47 \\
44 \\
31 \\
26
\end{array}
$$

Now, count the number of scores less than 63 and divide that number, 6, by the number of scores 11. This yields a decimal fraction 0.5454. Multiplying by 100 yields 54.54. Rounding 54.54 to the nearest tenth produces a percentile rank of 54.5, meaning that 54.5% of the persons taking the test made a score lower than yours.[12]

---

12.   Occasionally we calculate the percentile rank to show the number equal to or lower instead of just lower. SX makes that calculation. In that case, we count the number of cases equal to or less than and divide by N + 1. In our example, 7 divided by 12 yields a *percentile rank equal to 58.3*. Thus we can say that 58.3% of the persons measured had a score equal to or less than your score.

Using the same data and logic, if we gave you a percentile rank of 54.5 and asked you to find the test score equivalent to that percentile rank, our calculation is as follows:

$$Pr = \frac{L_N}{N} \text{ or}$$

$$0.5454 = \frac{L_N}{11} \text{ or}$$

$$11 \times 0.545 = L_N \text{ or}$$
$$L_N = 5.995 \text{ rounded to } 6$$

Thus, the score equivalent to a percentile rank of 54.5 is the score one higher than LN, 7, or 63.

Similarly calculated are deciles (deca means "ten") and quartiles (quart means "a fourth"). For example, the 1st decile includes the 1st to 9.9th percentile. The 1st quartile includes the first to the 24.9th percentile.

Cross-sectional data

Cross-sectional data are measured at a single-time point and can be in any of the forms discussed above except rates of change. We create cross-sectional data by collecting many items (persons, firms) at the same moment in time. The U.S. census, purporting to take a snapshot of a host of variables during a single year, is an example of cross-sectional data. So we can think of cross sectional data as data collected by a snapshot using a still camera.

Time Series

A time series collects measurements over several periods usually in sequence—hours, days, months, and years. Because time series have internal dynamics—in economics called trends, cycles, and seasons—they present special statistical challenges. An example of a time series is GDP over the last fifty years. Time series analysis is an extraordinarily fruitful arena for accurate and effective forecasting and causal analysis. We will have much more to say about these opportunities in part 5.

Panel Data

Combining cross-sectional and time series data creates panel data. When time is long and cross-sectional observations small, we call the data "time series, cross-sectional." When the period is short and the cross sections large, we call these data "repeated measures." In either case, we not only make observations across time, but each time point has a cross section. Panel data require special statistical treatment.

Summary

The critical ideas discussed in this chapter are the following:

1. Data sets, variables, and observations

    Data sets are collections of variables (columns in the data) and observations (rows in the data). Variables represent the things measured across our sample of observations. Observations define the unit of analysis or the level that our measurements take (individuals, groups, countries, corporate brands.).

2. Simple and Constructed Variables

    Measurements may be as simple as counting the number of something (horses, cars, people) or as complex as GDP, unemployment rates, alienation). Whether variables are simple or complex, we base them on some conceptual scheme, and we always construct them.

3   Continuous and discrete measurements

    Continuous data measure to any precision (number of decimal points) permitted by our measuring instrument. Discrete data have only a finite number of decimal places between data points

4. Reliability and Validity

    Reliability is the degree to which repeated measurements yield almost the *same* results. Validity of a measurement refers to the degree to which the measurement represents what we want it to measure. A measure can be reliable but not valid; no measure can be unreliable and valid.

5. Ways of forming quantitative variables

We form quantitative data by measuring, ordering, counting, and categorizing. Measured data may be cardinal or interval. Cardinal data have equal-scale intervals and true zeros. Interval data have equal-scale intervals but no true zero. We call ordered variables ranked data. We form count variables by counting, e.g., the number of cars in a parking lot. Categorizing data means assigning arbitrary numbers to qualities, for example, religion.

6. Percentages, rates of change, and percentiles

Percentages are special cases of nominal data, in which *percent* means "parts per one hundred." Rates involve changes over time; we treat these data as cardinal. Percentile ranks measure relative positions in a variable and are treated as rank data.

7. Ratios

We use ratios, special cases of percentages, for comparisons of one or more variables to some base variable.

8. Cross Sectional, time series, and panel data

Cross-sections are data collected at a single moment in time. Time series are data collected at the same unit of analysis at different points in time. Panel data combines cross sections with time series.

CHAPTER 2

TABLES, PICTURES, AND CENTRAL TENDENCY

Description, the first dimension of the statistical method, represents a body of techniques and calculations used to describe a population or sample. A statistical population marks out the total of all the observations relevant to our statistical interest. If our interest concerned the political attitudes of the voters in the city of Dallas, Texas, the relevant statistical population would encompass all the eligible voters residing in the city of Dallas.

Samples represent a smaller part of statistical populations. In our example, a single voting precinct denotes a sample. Whether data represents a population or sample depends on research interests, feasibility, and resources. In the previous example, if we limited our research interest to a single precinct, the sample now becomes the population. Most books use uppercase symbols to represent descriptions of populations and lowercase to represent descriptions of samples. For ease of reading, we decided to use only uppercase symbols throughout the book. For some calculations, we use different formulae in describing samples, and in such cases, we so indicate. Figure 2.1 helps understand the difference between a population and a sample.

Figure 2.1 Sample and Population

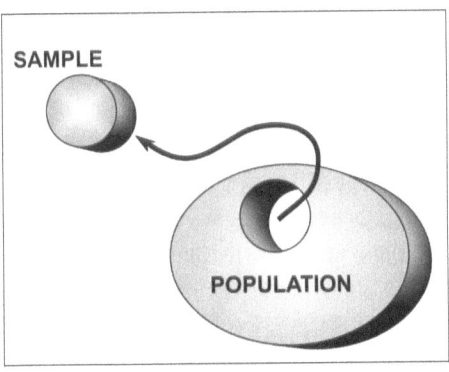

Frequency Distributions

Our first descriptor, a frequency distribution, uses tables and pictures (figures) that organize unstructured variables into comprehensible wholes. Organizing data into frequency distributions loses detail, but gains a sense of the whole.

Table 2.1, below, shows the ungrouped scores of math aptitude of one hundred applicants for *StatSys, Inc.*, a hypothetical firm, given between January and March 2008.

Table 2.1 Math Aptitude Scores of 100 Job Applicants for StatSys, Inc., Jan-Mar '03

81	82	79	83	84	94	82	85
76	88	87	73	73	68	77	84
79	79	76	80	81	58	83	81
76	77	68	71	80	72	72	79
74	82	71	80	85	53	71	80
57	72	78	80	68	89	87	87
73	80	82	81	81	83	80	84
75	63	75	87	78	70	76	54
75	75	81	78	73	80	89	79
75	76	89	87	74	89	64	76
68	80	84	75	86	82	73	57
75	65	81	79	85	77	75	79
80	68	73	84				

When anyone sees this table, they are likely to say, "This table means nothing." What they mean is that the table contains no information (it's all noise) that would help them understand StatSys's potential employee pool.

Creating a frequency distribution

One response to interpretive frustration constructs a frequency distribution of the scores. Here is how we create a frequency distribution. First, create a set of mutually exclusive classes and tabulate the number (frequency) of observations falling into each class. We rarely see frequency distributions with less than six or more than sixteen classes because less than six loses too much detail; more than sixteen loses the big picture.[13] H. A. Sturges[14] provided a starting rule:

$$N_C = 1 + (3.3 \times Log_{10} N)$$ (Equation 2.1)

The equation reads as follows: a good starting place for estimating the number classes in a frequency distribution is to take the log (base 10) of the data set size, N, multiply that by the constant 3.3 and add the constant 1. In our hypothetical data set, N = 100. The logarithm (base 10) of 100 is 2. Multiplying the log of 100, 2, by the constant, 3.3 gives 6.6. Adding one gives 7.6 and rounding gives eight classes as our starting point.

The second step determines the class interval. First, locate the highest and lowest observations in the data set, 94 and 53, and subtract the lowest from the highest. This calculation yields the crude range, 41. Dividing the crude range 41 by the number of classes, 8, yields about 5, the beginning class interval.

Third, construct the classes by selecting an observation 1 less than the lowest value in the variable, 52. This value is the lower limit of the first class. Then add the class interval 5 to give the upper limit of the first class 57. Thus, the first class is as follows:

52 but less than 57

An observation equal to or greater than 52 but less than 57 goes in the first class.[15] This wording makes the classes mutually exclusive. Then we

---

13.   CC, pp. 59-161. We cite Croxton and Cowden because their work is such a valuable source for descriptive statistics.

14.   Sturges, H. A. "The choice of a class interval." *Journal of the American Statistical Association*, 1926, pp. 65-66. There are other rules, for example, the Rice rule (NC= 2 x $N^{0.333}$) or the cubed root of N multiplied by 2 is a good starting point for the number of classes. The cubed root of 100 equals about 4.64. Two times 4.64 yields about 9.28. Rounded to the nearest unit gives 9, one class more than Sturgis. We urge you to *play* with the construction of frequency distributions using Rice or Sturgis as a starting point until you achieve a picture that is a *satisfactory* approximation of the variable's configuration.

15.   Using this language assures that the classes are mutually exclusive because it leaves no doubt as to which class an observation belongs.

construct the remainder of the classes around this scheme. Fifty-seven but less than 62 defines the second class and so forth. Finally, we count the number of observations in each class, putting in each class the count or frequency of the scores.

Notice that we used the sign < to mean "and less than." The last class uses "and above" to accommodate eight classes. We also constructed three additional columns showing class percentages, and cumulative percentages for interpretation and comparison with other data sets. Table 2.2 shows the completed frequency distribution.

Table 2.2 Frequency Distribution of Math Aptitude Scores

Classes	Frequency	Cumulative Frequency	Percent (%)	Cumulative Percent (%)
52 and less than 57	4	4	4	4
57 and less than 62	1	5	1	5
62 and less than 67	3	8	3	8
67 and less than 72	12	20	12	20
72 and less than 77	25	45	25	45
77 and less than 82	32	77	32	77
82 and less than 87	17	94	17	94
87 and Over	6	100		100
Total		100		100%

What new information (over the raw data) do we glean from a frequency distribution? First, we see a slightly unbalanced (asymmetrical) variable, in this case, a tail trailing off to the lower scores with the majority of cases concentrated in the range of 72 to 87. We call an unbalanced distribution skewed; in this example, we have left skewed data. Second, our data apparently have no gaps, a virtue. Third, the data have a single peak (also a virtue). Fourth, we now see what proportion of the scores fall above or below a given level. For example, 45% of the pool of prospective employees has scores less than 77. Now, we have reduced noise and picked up a signal.

The Graphic Portrayal of Frequency Distributions

To obtain another perspective, we can construct a picture of the variable, a bar chart called a histogram, shown in figure 2.2.

Figure 2.2 Histogram of Math Scores for the Firm StatSys

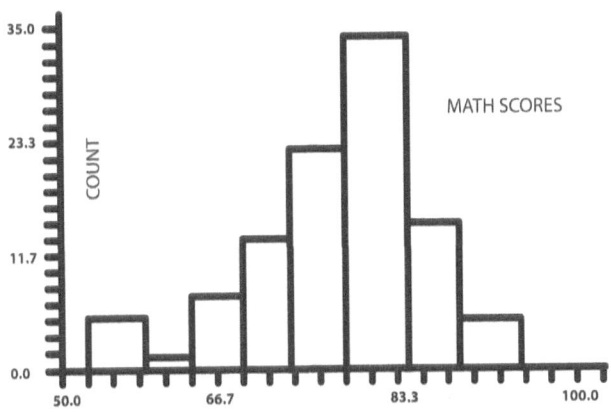

On the x-axis (the horizontal line at the bottom), we created a scale representing class intervals. On the y-axis (the vertical line on the left), we constructed another scale representing the number of scores (frequencies). From the midpoint of the first-class interval, we erected a bar showing the number of scores in that class. Then we erected more bars representing all classes. This picture enables us to see more clearly the single peak, no gaps, and the left asymmetry.

We can also construct a frequency polygon, a smoothed line chart, and superimpose it on the histogram. Figure 2.3 shows the polygon on top of the histogram. We constructed the line chart using the same method as the histogram using lines instead of bars. The lines in this graph smooth the picture.

Figure 2.3 Polygon Overlaid on Histogram for StatSys Math Scores

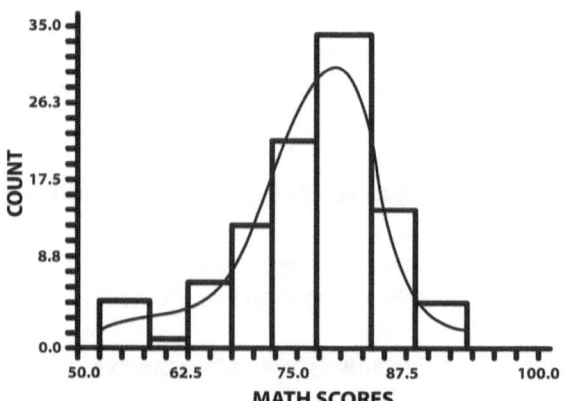

The polygon, shown in figure 2.3, is more idealized, giving less detail but a more general sense than the histogram.

Quantitative Descriptors

Numerical (quantitative) descriptors focus on four dimensions of a variable: central tendency (averages), dispersion (variation), skewness (asymmetry), and kurtosis (middleness). We call descriptive measures of populations, parameters, descriptive measures of samples, statistics.[16] Following the logic of our book, we borrow a term from Kurnow, Ottman, and Glasner[17] and call these descriptors decision parameters because they aid in rational decision making.[18]

Central tendency captures the inclination of observations to cluster around some central value. Dispersion measures variation from a central value. Skewness measures asymmetry. Kurtosis measures thick or thin shoulders.

---

16. Here is another use of the word *statistics*. Now we have three.
17. KGO.
18. The great classical economist Frank Knight distinguished *RISK* from *UNCERTAINTY*. Risk is a condition in which we can identify alternatives and assign probabilities to them. Uncertainty is a condition in which we do not know or cannot estimate probabilities. Statistical methods, including description, transform uncertainty to estimates of risk through identified alternatives and probabilities, thus making decisions more rational.

Central tendency (averages)

Central tendency, the more technical term for averages, is the degree to which observations cluster around a single value—i.e., an average is a single number that best represents a variable. If, for example, we have a population of annual income data for a particular geographic area, we can describe the data with a single number that effectively represents the series. In this chapter, we discuss four measures of central tendency: the arithmetic mean, median, mode, and midrange. We discuss two other measures, the geometric and harmonic means, in chapter 4.

The Arithmetic Mean

The arithmetic mean personifies the measure that most people recognize and call an *average*. Throughout our book, we refer to the arithmetic mean as simply the "mean," while referring to other averages by their full names.

We define the arithmetic mean of a population (or sample) as the sum of all the observations in the population (or sample) divided by the number of observations. Equation 2.2 formalizes this definition.

$$M = \frac{\Sigma X_i}{N}$$
(Equation 2.2)

The symbol M, standing for the mean, is the uppercase Greek letter *Mu*. The symbol Σ, the uppercase Greek letter sigma, defines an instruction to sum all the observations in the variable. Uppercase X_i represents each individual observation in the population. Uppercase N is the symbol for all the observations in the population or sample. To illustrate, assume we have a variable consisting of the annual income (in thousands of dollars, 2005) of five corporate executives.

129   148   120   112   151

The arithmetic mean is as follows:

$$M = \frac{151 + 129 + 148 + 120 + 112}{5} = \frac{660}{5} = 132 \text{ or } \$132,000$$

The sum of all the incomes, 660, divided by the number of observations, 5, equals 132, the arithmetic mean annual income (in thousands) of the five executives. How do we interpret this measure?

First, we interpret the mean as the value $132,000 that represents the series better than any other number. We can let the number stand-alone, compare it to salaries of other sets of executives in other industries, or compare the salaries of the same group of executives over time.

A second interpretation defines the arithmetic mean as the measure that "fits the data best," where the phrase "best fit" has profound connotations. Consider table 2.3 below:

Table 2.3 The Sun of Deviations Squared Equals a Minimum

	Salaries in Thousands of Dollars	Algebraic Deviations from the Mean	Deviations Squared
	151	19	361
	129	-3	9
	148	16	256
	120	-12	144
	112	-20	400
Sums	660	0	1170

Arithmetic
Mean = 132

In Table 2.3, we constructed the second column by subtracting the mean from each of the observations X_i - M_x. When subtracting algebraically, 132 from 151, we change the sign of the subtractor and add, giving 19. Subtracting 132 from 129 gives the difference -3. The algebraic sum of these differences (column two) equals zero. We also call the sum of this column divided by N, the first moment about the mean, m_1, always zero. The mean is the only value in the variable from which the sum of the deviations equals zero, one criterion for "best fit."

Because the sum of D_i = zero, the sum of D squared (D^2), column 3, must be a minimum, smaller than the sum of D^2 from any other value in the variable. What makes this minimum important? As a representative number, the mean is a signal. An observation's deviation from the mean represents noise. Therefore, the sum of D squared (ΣD^2) measures error (noise). To put the issue another way, the calculation of the mean adds information over that provided by the frequency distribution, strengthening signal. The mean represents our minimum knowledge. Because the mean minimizes ΣD^2, it minimizes error or noise and this makes it the measure that "fits the data best." We also call the ΣD^2 SS or sum of squares and we call ΣD^2 divided by N the second moment around the mean or m_2. While the notion of best fit may still have for our readers an elusive quality, our discussion of variation in chapter 3, elaborates and expands its implications.

Two other characteristics of the mean provide additional insight. First, if we multiply $N \times M$, the number of items in the series multiplied by the mean, the result is the total of the series.

$$N \times M = \text{Total (Aggregate)} \qquad \text{(Equation 2.3)}$$

$$5 \times 132 = 660$$

Another characteristic of the mean is that it is not robust, i.e., it is influenced by extreme values. As an example, consider once again our example of five executive salaries. Suppose the salaries were:

129   148   120   151   312

We call the last salary, $312,000, an "outlier," an unusual observation (we will have more to say about outliers later). In this example, the outlier *pulls up* the mean from $132,000 to $172,000. You can see that extreme values or outliers on the high or low end of a variable heavily influence the mean. We use the words "not robust" when outliers exercise large influence on a descriptive measure. While one hundred and seventy two thousand dollars meets our "best fit" criterion, it distorts our variable and, depending on our purposes, we might want to look at an alternative measure. If our purpose was to show a value representing the aggregate of the series, the mean is our choice. However, if our purpose was to produce a more representative number, we might want find a more robust measure.

Returning to our math scores, the sum of all the scores equals 7,730. Dividing 7,730 by 100, the number of applicants, yields the mean value of 77.3. The mean 77.3 seems to capture the nature of the math scores for this group of applicants. Please note something else: in the absence of any other information, the mean specifies our best guess of the value of any individual score.

We can check the mean's centrality by looking again at figure 2.2.

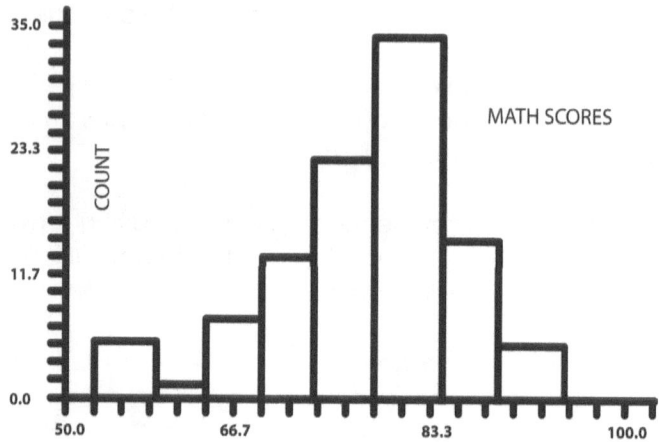

The mean value 77.3 falls at the midpoint of the bar with the highest frequency. While the lower scores on the left tail of the distribution pull the mean down slightly, it is still strongly representative and the value that "fits the data best." Among the measures of central tendency discussed in this chapter, the mean marks out the most useful measure of central tendency. For instance, from the means of several groups of observations (assuming constant N among the groups), we can find the mean of the combined groups by adding their means and dividing by their numbert. In addition, the mean, as compared to other measures, plays a central role in inferential and analytical statistics.

If asked for information about the general quality of applicants, you could report the mean, 77.3, as a summative representative number of math aptitudes.

The Proportion as an Arithmetic Mean of Nominal Data

Suppose we have a variable composed of 10 people, 7 females, and 3 males. If we assign the numbers 1 to females and 0 to males, we create the following hypothetical variable. Assigning the numbers 1 and 0 implies no hierarchy, and as explained in chapter 1, we call these data nominal.

Molly	1	Brenda	1
Jane	1	Geneva	1
Mary	1	Frank	0
Phyllis	1	John	0
Julie	1	Robert	0

Now suppose we calculate the mean of the data using the numbers (nominal codes) assigned to females and males. The sum of the ones and zeros equal seven. The number 7 divided by N, 10 equals 0.7, a decimal fraction. Converted to a percentage, the mean equals 0.70 or 70%. Of course, if we coded boys 1 and girls 0, the proportion would be 0.30 or 30%. No matter how we code boys and girls, this variable consists of 70% female, 30% male or 30% male and 70% female. Changing the coding changes nothing. Therefore, perhaps surprisingly, a percentage is a mean of nominally coded qualitative data. The importance of this idea centers on our ability to convert qualitative data to nominal codes treating them as useful statistical numbers.

In our StatSys variable, we could use the same coding scheme (scores equal to or greater than 80 coded one; scores less than 80 coded zero) and calculate the percentage of applicants with scores equal to or greater than 80, the minimal math aptitude score, for StatSys to consider applicants for a job. Obviously, percentages play an important role in descriptive statistics.

The Median

The median is the midpoint in a series of numbers; half the data values are above the median, and half are below. Using the midpoint definition, we can also say that the median falls at the 50th percentile. Instead of calculating the median, we *locate* it by arranging our data set from lowest to highest values. The observation in the middle of the series is the median. Variables with an even number of observations produce two medians. With two medians, we take the mean of the two. The symbol we use for the median is *Md*.

Medians have two interesting characteristics. First, medians are robust because the magnitude of extreme values does not influence them. To illustrate, we will again use the executive salaries containing an extreme value.

129   148   120   312   151

Putting the data in an array, 120, 129, 148, 151, and 312, the middle observation defines the median, 148. Only the presence of another observation, not its magnitude, influences the median. Thus, if our goal is to produce a measure of central tendency, uninfluenced by outliers, we might choose the median because it is robust.

The median's second characteristic is that it minimizes non-algebraic differences. To understand what this means, look at table 2.4.

Table 2.4 Absolute Deviations from Median Equal a Minimum

Salaries in Thousands of Dollars	Absolute Deviations from the Median
112	|22|
120	|9|
129	|0|
143	|14|
151	|22|
Absolute Sum = 67 = Minimum	|67|

Median = 129

We created the column headed "Absolute Deviations from the Median" by calculating the difference, between each observation and the median disregarding signs (the vertical lines on each side of the deviations means that we have disregarded signs), between the median and each observation in the variable and summing those absolute differences. That sum , is smaller than the absolute sum from any other observation in the variable.

One useful application of this characteristic is in locating a position that minimizes absolute distances. Suppose we asked you to locate a distribution center that would minimize the total travel distance of trucks to five retail outlets. Assume that we define the distances between retail stores by points on a line: $A = 4$, $B = 12$, $C = 18$, $D = 40$, and $E = 58$. Figure 2.5 portrays these outlets.

Figure 2.4 Median Minimizes Absolute Distance

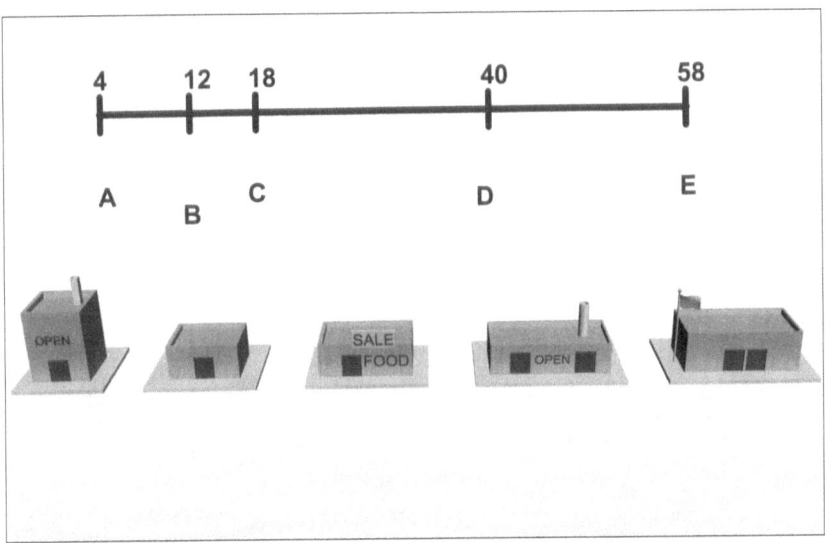

The meaning of these points is the distance between the stores. Store A is 54 miles from store E; B is 46 miles from E, etc. If our goal were to locate the distribution center at a point that minimizes the total travel distance for delivery trucks, we would locate the distribution center at store C, the median, 18 because the median minimizes the absolute travel distance.

The median is also a useful measure of central tendency when a variable is not cardinal as, for example, with percentages or percentile ranks.

The midrange[19]

A variation of the median is the midrange, the sum of the lowest and highest observation in the series divided by two.

$$MR = \frac{H+L}{2}$$

(Equation 2.4)

We illustrate the midrange's usefulness with the variable, $A = 0$, $B = 8$, $C = 14$, $D = 36$, and $E = 54$, representing homes at various distances.

Figure 2.5 shows these family distances.

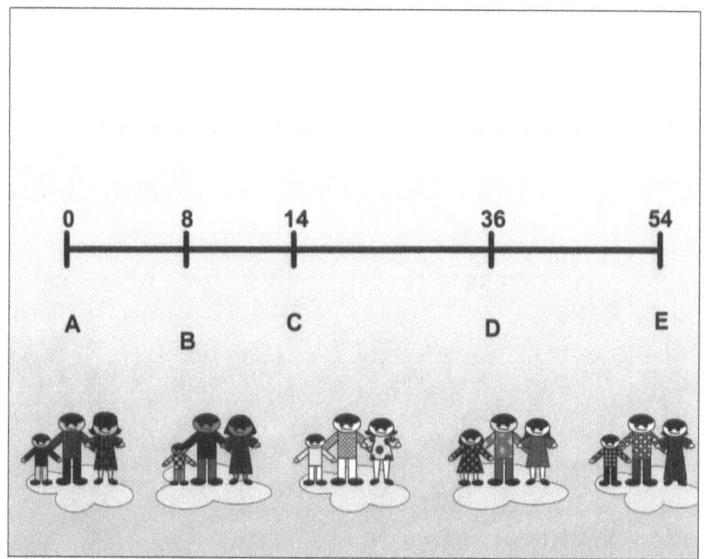

Our problem is deciding the location of a walk-in medical clinic. Now, instead of minimizing the absolute travel distance between houses, we want to minimize the travel distance to the clinic for any family. The range of our data is 54, the distance between home A and home E. One half of 54 is 27, the midrange.

---

19.   We are indebted to Joseph Malkevitch, Department of Mathematics York College (CUNY) Jamaica, New York, 11451 for the examples that use the midrange in location decisions, posted on his Web site (*malkevitch@york. cuny.edu*).

Locating the clinic at the midrange 27 means that no family has to travel more than 27 units (yards, miles, etc.) to get to the clinic.

Returning to our math scores, we construct an array from the smallest to largest observation and identify the median. The median lies between the 50th and 51st observations, both of which are 79. Thus, the median of our scores is 79, 1.7 points greater than the mean, not much. The unbalance to the left of the distribution pulls the mean down in the direction of the skew but the magnitude of the lower values at the distribution's left does not affect the median.

An important point: In a perfectly symmetrical distribution, both sides (defined by a line drawn from the peak of the distribution to its base) mirror each other; the mean, median, and mode (see below) are the same. We strongly suggest that when you report the mean, you also always report the median and mode because, if these values are different, their differences tell a story about skewness.

The Mode

The mode, Mo, is the observation occurring most often in a variable, i.e., its highest frequency. In our array of the scores data set, the mode is 80, greater than the mean and the median. This is true because neither the magnitude nor existence of skewed observations affect the mode. We can confirm this assertion by looking again at figure 2.2.

The modal value 80 is to the right of the median. In most skewed distributions, the median will fall about one-third of the way between the mean and the mode.[20] There are many situations, however, in which the mode does not exist. In addition, we cannot use the mode effectively in other calculations. Once again, in a symmetrical distribution the mean, median, and mode are the same. Therefore, the question of which measure (the mean, median, or mode) best represents a series is relevant only if a distribution is skewed. If a mode exists, it is always a good idea to report the mode along with the median and mean.

---

20. This relationship between mode, median, and mean does not always hold. For example, in a right skewed distribution, with some discrete data, the median may be larger than the mean. In addition, outliers may easily change this basic relationship, one reason why we need to be especially sensitive to their effect.

Summary

The critical ideas discussed in this chapter are the following:

1. Frequency distributions

    Frequency distributions organize and summarize unstructured variables. While some specificity of a variable is lost, portraying variables as frequency distributions allows the user to see "patterns" in the data, and understand how observations are distributed.

2. Histograms and polygons

    Histograms, a type of bar chart, show the distribution of observations from low to high in a variable. Polygons show the same information connecting levels (e.g., tops of the bars in the histogram) with lines providing another visual view of the variable.

3. Central tendency

    Central tendency measures how the individual observations cluster around a single value. Measures of central tendency include the arithmetic mean, median, midrange, and mode.

4. Arithmetic mean

    The arithmetic mean is the sum of all observations in the population (or sample) divided by the number of observations. We commonly refer to the mean as the average. Multiplying the arithmetic mean by the number of observations yields the total for the sample. Extreme cases in variables (i.e., outliers) *pull up* or *pull down* the mean. When we represent data as zero or one, the proportion of ones equals the arithmetic mean of the nominal variable.

5. First and Second Moments

    The first and second moments introduce the notion of "best fit" by showing that m_1 is always zero and m_2 is minimal. Thus, the mean fits the variable best because it minimizes variation (noise).

6. Median

The median locates the midpoint in a series of numbers. Half of the data are below the point, and half above. The median is also the 50th percentile of the distribution. You locate it by ranking the data high to low and finding the point halfway between the high and low values of the variable. The median, by minimizing the sum of the absolute deviations, can play an important role in location decisions. Robust in the face of outliers, the median can often portray representativeness better than the mean.

7. Midrange

The midrange, calculated by finding the sum of high and low values of a variable and dividing by 2, plays a valuable role in location decisions.

8. Mode

The mode is the observation occurring most often, or with the highest frequency. In a symmetrical distribution, the mode, median, and mean are equal. With an asymmetrical, the values of the three measures are different.

9. Signal and noise

The mean (as a representative number) is a signal. The degree to which the observations deviate from the mean is the extent to which the mean is not representative (noise). Noise represents the statistical problem: explaining and reducing noise and enhancing signal.

Appendix

Chapter Two

The table below is our first data set. The first column of our data set is an identification number. The data are in ASCII delimited, space separated format so they can be easily scanned and read into almost any spreadsheet or statistical package.

Your tasks are the following:

1. Using your imagination, define the observation (person, firm, school, etc.) and give names to the five variables in the data set consistent with the data used in your organization.

2. Using hand calculations, construct frequency distributions and calculate the four measures of central tendency discussed in this chapter for at least one of the four cardinal variables.

3. Use your statistical package to compare your hand calculations with the output of your program. Again, using your statistical package construct frequency distributions and calculate the four measures of central tendency discussed in this chapter for all the variables in the data set.

4. Write a brief essay explaining the meaning of your effort.

ID	VAR1	VAR2	VAR3	VAR4	VAR5
1	63.71	65.09	70.07	189.40	0
2	39.58	46.75	100.24	151.25	0
3	44.98	51.94	81.38	200.12	0
4	33.36	69.36	98.72	132.05	0
5	39.02	64.39	105.46	114.35	1
6	41.68	59.77	82.65	137.28	0
7	54.44	60.80	106.22	149.10	1
8	49.63	50.99	74.71	196.52	0
9	53.39	56.67	122.13	172.65	0
10	63.56	57.97	83.02	175.86	0
11	47.83	52.69	81.49	140.27	1
12	63.40	43.42	118.23	115.43	1
13	44.88	99.60	77.80	118.36	0
14	53.50	68.58	110.13	175.94	0
15	48.36	62.72	67.36	163.15	0
16	51.07	66.25	111.70	174.29	1
17	47.31	76.65	83.53	165.07	1
18	51.38	50.7	73.36	160.63	0
19	59.08	52.49	112.46	156.92	0
20	40.64	62.23	86.75	140.63	0

21	61.10	63.94	77.45	218.00	1
22	45.15	56.00	91.85	138.22	0
23	49.66	55.73	66.72	137.54	0
24	52.20	45.31	116.26	157.42	1
25	42.29	52.68	117.06	150.35	1
26	40.64	80.10	91.76	157.45	1
27	51.01	47.35	99.01	155.87	0
28	56.04	66.51	99.40	139.16	0
29	51.72	57.26	78.49	165.27	0
30	45.93	63.84	81.83	115.68	0
31	43.02	49.76	101.65	165.58	0
32	55.63	54.93	113.52	182.56	1
33	40.77	51.00	123.61	185.87	0
34	42.50	59.49	121.93	157.52	0
35	63.47	67.20	74.73	167.52	0
36	39.31	56.79	120.99	131.88	1
37	40.15	62.35	92.24	210.53	0
38	58.60	53.27	93.36	125.23	0
39	44.25	47.51	121.71	137.91	1
40	43.31	64.44	90.61	194.42	1
41	42.85	64.15	84.46	129.12	0
42	49.42	54.37	71.44	171.17	0
43	59.97	48.66	106.81	151.52	1
44	57.53	65.81	110.79	183.87	1
45	48.55	48.39	109.46	141.88	1
46	52.31	53.15	83.81	160.97	0
47	42.51	59.25	107.43	186.05	1
48	37.85	55.46	88.37	109.30	1
49	49.24	64.48	83.54	122.96	0
50	47.68	78.86	57.36	214.79	1

CHAPTER 3

VARIATION AND NORMALITY

In chapter 2, we explained that central tendency measures how individual observations in a variable tend to cluster around a central value, a representative number. Variation, sometimes called dispersion, demarks the second arena of description measuring the degree to which individual observations fall away from that central value. We also refer to variation as the spread of a variable. If you discover that a population has a mean of 50, in the absence of any other information, the value 50 constitutes your best guess of any individual observation in the series. Dispersion tells you how much error is in that best guess.

The word *variable* suggests that all data vary. Persons vary in weight and intelligence. Businesses vary in size and efficiency. If all observations in a variable are identical, there is no need for statistics; no variation means that we do not have a statistical problem. As we shall see, understanding variation: measuring it and determining its cause defines analysis. Variation and central tendency, then, lie at the heart of the statistical method including inference and analysis.

Variation Measured

We measure variation in at least six different ways: the standard deviation (S), the variance (S^2), the sum of squares (SS) or the total variation (TV), the coefficient of variation (CV), and the crude range (CR). We take each of these in turn showing how we calculate them and how they are useful.

The Standard Deviation

The standard deviation represents the principal measure of variation in the descriptive realm. We use S as the symbol for the standard deviation and calculate it using the following equation.

$$S = \sqrt{\frac{\Sigma D^2}{N}}$$ (Equation 2.1)

This equation instructs us to calculate the deviations from the mean, square and sum the deviations, divide the summed squared deviations by the number of observations in the variable, and take the square root of that quotient. Table 3.1 illustrates this calculation.

Table 3.1 Calculation of the Standard Deviation

Salaries in Thousands of Dollars	Algebraic Deviations from the Arithmetic Mean (D)	Deviations Squared (D²)
Column 1	Column 2	Column 3
142	15	225
129	2	4
132	5	25
120	-7	49
112	-15	225
Sums	0	528

Arithmetic Mean = 127
Sum of D^2 = 528

The table consists of the annual salaries of five executives rounded to thousands. We begin the calculation by algebraically subtracting the mean 127 from each individual salary in the variable. Algebraically subtracting 127 from 142 tells us to change the sign of the subtractor 127 and add that to the number from which it is subtracted 142. Therefore,

-127 added to 142 gives 15. Following that scheme, 127 subtracted from 129 equals 2. Continuing through the last three items in the series, we calculate the deviations 5, -7, and -15. As we learned in chapter 2, the sum of these deviations equals 0. Obviously, we cannot use the sum of D as a measure of variation because it always equals 0. Because we need a sum greater than 0 in column 3, we square the values in column 2. The sum of the squared deviations equals 528. Dividing the number 528 by N, the size of the variable (5) yields the quotient 105.6. We call this number the second moment, m_2. The square root of the number 105.6 equals the standard deviation, approximately 10.28. We see that we base the measurement of the standard deviation on the second moment.

How do we interpret the standard deviation? In chapter 2, we introduced the notion of signal and noise, with the mean as signal and variation from the mean as noise. With this discussion of the standard deviation and the second moment, we now have measures of noise. In this context, we want to emphasize that the fundamental goal of the statistical method is to amplify and strengthen signal and minimize noise. To broaden the interpretation of the standard deviation, we take a side trip to introduce the normal distribution.

The Normal Distribution

Normal, in this context, does not refer to a norm or value judgment. Here, we use the word *normal* as a label of a variable's distribution defined by a mathematical equation.[21] For our purpose, the normal distribution has two characteristics: perfect symmetry, commonly referred to as a bell curve, and a special relationship between the standard deviation and the arithmetic mean.

In other words, perfect balance delineates the necessary, but not sufficient condition for normality. The sufficient condition focuses on the relationship between the mean and standard deviation. We say a variable approximates normality if adding ± one standard deviation to the mean that produces a range that takes in about 68% of the observations in the variable, adding ± 1.96 standard deviations to the mean, this range encompasses about 95.00% of the observations

---

[21]. For those who are interested this is the equation defining the normal distribution: $f(x) = \dfrac{1}{(SDx) * (2\Pi)} e^{-.5 \frac{x - Mx}{SDx}}$.

in the variable, and adding ± 2.57 standard deviations to the mean includes about 99% of the observations in the series.[22]

Figure 3.1 graphically shows the relationship between the mean and the the standard deviation in a normal distribution.

Figure 3.1 The Normal Curve

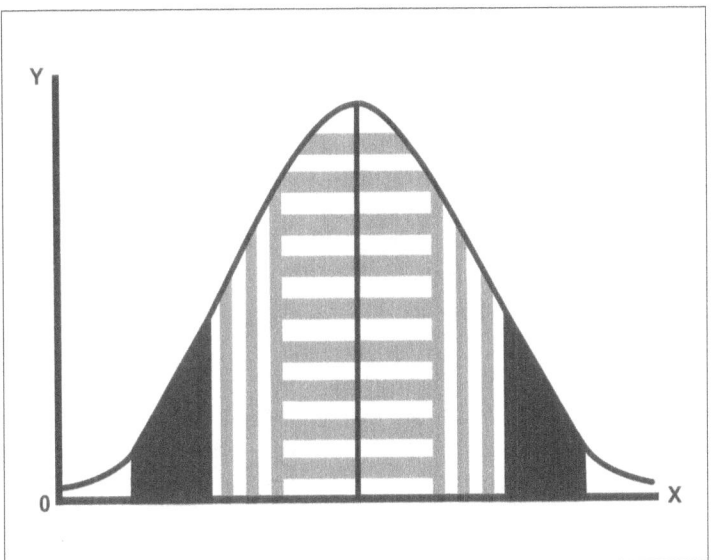

The area depicting horizontal lines figure 3.1 shows that adding ± one (1) standard deviation to the mean encompasses about 68% of the cases in the series. The vertical part of the figure plus the horizontal area, ± 1.96 standard deviations from the mean encompassing about 95% of the observations in the variable. The dark areas plus the horizontal and vertical, encompass plus and minus 2.57 standard deviations from the mean encompassing about 95% of the observations in the variable. The white portions of the figure lie outside the range ± 2.57. We say the curve in figure 3.1 is asymptotic to (getting closer and closer to but never touches) the x-axis suggesting that no matter how many ± standard deviations we add, we never encompass all the observations in the series.

---

[22] You can find on-line a number of free programs that will calculate the percentage of observations falling between ± any range of standard deviations in a normal distribution. One of these programs is HyperStat Online.

Not all normal data look like this example. Some are flatter, some steeper, and, in the real world, some may even list a little to the right or left. Nevertheless, all normal curves look bell-shaped and all approximate this unique relationship between the mean and standard deviation. Thus, normality represents a theoretical idealization, never actually realized in the world of measurement. The distributions we actually encounter only approximate normality.

What forces cause a variable's distribution to approximate normality? A distribution will approximate normality when many small, independent effects influence each observation in a variable. To put the issue another way, multiple forces acting randomly with equal force on each observation in a variable, will cause approximate normality. Alternatively, the opposite, a single (or a few) force(s) acting nonrandomly on a variable will cause skewness in a distribution. We emphasize again that we never have a perfectly normal variable, only sample approximations.

Before the advent of the microcomputer and Internet technologies, textbooks contained tables showing percentages falling between ranges of ± standard deviations, e.g., one-half or one-third a standard deviation from the mean. Although we have not included such a table, you can easily download any one of many free programs from the Net that will calculate these percentages for you. The Web sites, *SurfStat and HyperStat Online*, have easy to use statistical tables.[23]

While the mean gives a single number that best represents a variable, the standard deviation tells how the observations in the variable deviate or fall away from the mean or the extent to which there is error or noise in the mean. If given no other information except the mean, and asked to make a best guess of a single observation's value (called a point estimate) in a variable, the mean offers your best guess. If we know the mean and standard deviation of an approximate normal variable, when asked to estimate the value of any single observation, you would say, as before, the mean. But now, with knowledge of the standard deviation and the distribution's normality, we would qualify our point-guess with a statement of confidence, an interval estimate, by saying that we

---

[23]   Dear, Keith and Robert Brennan, *SurfStat Statistical Tables*, the Australian National University.

are 99+% confident that the our best guess falls somewhere between ± three standard deviations from the mean. Knowledge of the mean and standard deviation gives us both signal (the mean) and noise (the standard deviation). Therefore, truly, the standard deviation is an error or noise measurement, a measure of our ignorance.

In our math score data set, the mean equals about 77.3 and the standard deviation approximately 7.8. This means that, if our variable's distribution approximates normality, about 68% of the scores will fall between 69.5 and 85.1, approximately 95% will fall between 61.7 and 92.9, and practically all will fall between 53.9 and 110.8. Check it out.

Other Useful Interpretations of the Standard Deviation

The Camp-Meidell Inequality

In the face of non-normality, what can we say about the relationship between the mean and standard deviation? Even in asymmetric distributions, we can estimate error using Camp-Meidell's inequality.[24] As long as a distribution has only one mode and the difference between the mode and the mean does not exceed the standard deviation, Camp Meidell found

$$\text{Camp-Meidell Inequality} = 1 - \frac{1}{2.25 K^2} \qquad \text{(Equation 3.5)}$$

In this equation, K equals the number of standard deviations from the mean. The equation tells us that the percentage of observations falling between ± 1 standard deviations from the mean will equal

$1 - (\frac{1}{2.25 \times 1^2})$ or $1 - \frac{1}{2.25} = 1 - 0.44 \approx 56\%$ (about 56%); between, ± 2

standard deviations from the mean, at least $1 - \frac{1}{2.25 \times 2^2}$ or $1 - \frac{1}{9}$

$= 1 - 0.11 \approx 89\%$ of the observations in the series. In that same series, plus and minus three standard deviations from the mean will include

at least one $- \frac{1}{2.25 \times 3^2} =$ or $1 - \frac{1}{20} \approx 95\%$ of the observations.

---

24. CC page 221.

Chebycheff's Inequality

If a distribution does not conform to the Camp-Meidell rules, we can apply Chebycheff's inequality.[25] Chebycheff found that, in any data set, no matter what its distribution, bi-modal, badly skewed, whatever,

$$\text{Chebycheff's Inequality} = 1 - (\frac{1}{K^2})$$
(Equation 3.6)

will yield the percentage of observations that fall between $\pm K$ standard deviations. In any distribution, then, ± 2 standard deviations from the mean will encompass at least $1 - (\frac{1}{2^2}) = 1 - 0.25$ or 75% of the observations in the series and ± 3 standard deviations will include at least $1 - \frac{1}{3^2}$ or 89% of the observations.

Therefore, if we know the mean and standard deviation, even if we do not know the variable's distribution, we can still calculate the percentage of observations falling between $\pm K$ standard deviations no matter how the variable is distributed. Now we see the necessity of always reporting the standard deviation when we report a mean.

The Coefficient of Variation (CV)

A useful addition to the standard deviation, an absolute measure of dispersion, is the coefficient of variation, CV, a relative measure. We define CV as the standard deviation divided by its mean or

$$CV = \frac{S}{M}$$
(equation 3.7)

This calculation makes the standard deviation, S, relative to the mean allowing the comparison of variation between two or more variables with different units of measure.

The Variance (S^2)

The variance, the standard deviation squared, also called the second moment, m_2 has major uses in the province of statistical inference

---

25.   Op cit

and analysis. Hence, we will postpone any extended discussion of its usefulness until we address that topic. We repeat the equation for the variance below.

$$S^2 = m_2 = \frac{\Sigma D^2}{N}$$ (Equation 3.8)

The Sum of Squares (SS) or Total Variation (TV)

The sum of squares (SS), also called the total variation (TV), is the sum of the squared deviations from the mean.

As a reminder, the equation for SS is

$$SS = TV = \Sigma D^2$$ (Equation 3.9)

The SS or total variation lies at the core of analytical statistics; we will have much more to say about Total Variance in that context. Still, a short explanation adds flavor to our discussion.

Total Variation or sum of squares defines the STATISTICAL PROBLEM. Total variation represents our lack of knowledge. The goal of the statistical method is not just to describe, although that can be interesting and valuable on its own; limiting our statistical inquiry to description accepts the world as it is. Explaining the cause of variation directs interventions that change the world and makes things better. For example, the reading abilities of children vary. The goal of analysis is to understand WHY reading ability varies and what we can do (how we can intervene) to increase reading ability. We take up the discussion of analysis in part 3. Early statistical applications methods focused almost entirely on description.

The Crude Range (CR)

The crude range, simplest of the measures of variation, is the difference between the largest and smallest observations in a variable. CR can be valuable as a rough tool in estimating the standard deviation. If we can assume approximate normality, one-sixth of the crude range will yield a rough estimate of the standard deviation.

The Standard Deviation and Variance of Nominal Data

Another interesting insight relates to the standard deviation and variance of nominal data. In chapter 2, we explained a percentage as a special kind of mean of nominal data coded zero and one. Therefore, if a nominal variable has a mean, it also has a total variation, a variance, and a standard deviation. Consider the variable in Table 3.2, a single nominal variable with boys and girls assigned numbers zero and one, introduced earlier.

Table 3.2 A Gender Variable Nominally Coded

Names	Gender	Names	Gender
Molly	1	Brenda	1
Jane	1	Geneva	1
Mary	1	Frank	0
Phyllis	1	John	0
Julie	1	Robert	0

Now we build another table, 3.3 with deviations from the mean and the squared deviations.

Table 3.3 A Gender Variable Nominally Coded (in columns)

Names	Girls = 1; Boys = 0	Deviations	Deviations Squared
(Column 1)	(Column 2)	(Column 3)	(Column 4)
Molly	1	0.3	0.09
Jane	1	0.3	0.09
Mary	1	0.3	0.09
Phyllis	1	0.3	0.09
Julie	1	0.3	0.09
Brenda	1	0.3	0.09
Geneva	1	0.3	0.09
Frank	0	-0.7	0.49
John	0	-0.7	0.49

| Robert | 0 | -0.7 | 0.49 |
| Sums | 7 | 0 | 2.1 |

The mean of the nominal variable is the sum of column 2 divided by ten the total number of observations or 0.7. Subtracting the mean from each observation gives column 3, the deviations from the mean. Not surprisingly, column 3 sums to zero. Column 4 squares and sums the deviations in column 3, 2.1. The value 2.1 represents the Total Variation (SS) for the variable, gender. Dividing the value 2.1 by 10 yields the variance (S^2) of gender, 0.21. The square root of the variance of gender yields its standard deviation (S), 0.46.

How do we interpret these calculations? A short cut may help. In the calculations above, we can, without proof, easily calculate the variance, S^2, of gender by multiplying the proportion of females (P) times the proportion males (Q), $P \times Q$ or $0.7 \times 0.3 = 0.21$, the variance. We designate the first proportion P and the second proportion Q. Hence the variance (S^2) of a nominally coded variable (1, 0) equals $P \times Q$. So the maximum variation for a nominal variable coded one, zero equals $0.5 \times 0.5 = 0.25$ giving S, the square root of the variance. Thus, the square root of 0.25, 0.50 equals the standard deviation of gender. Think about it. If we had five girls and five boys for the nominal variable, the variance is maximal. If we had nine girls and one boy, the variance is minimal. If we had all 10 girls, variance equals zero.

We will return to this notion in our discussion of statistical inference.

Measuring Departures from Normality

One of the assumptions of many statistical analyses is that variables are normally distributed. Because we rarely deal with population data, i.e., our data are usually samples from a larger population, it is virtually impossible for a finite sample of data to be normally distributed. It is possible, however, that we obtained the sample data from a theoretically normal population. Thus, we use tests of normality to provide some level of confidence that this sample may have come from a theoretically normally distributed population. We address this question in detail in part 2. In this section, we discuss the basic methods for calculating departures from normality.

Skewness (SK)

Skewness or asymmetry measures a variable's imbalance. Because all normal distributions have perfect symmetry, skewed distributions, by definition, cannot be normal.

The Third Moment (m_3) as a Measure of Skewness

The first moment, m_1, provides the basis for measuring central tendency and the second moment, m_2, for measuring dispersion. You will not be surprised, then, to learn that the third moment, m_3, provides the base equation for measuring skewness. Table 3.6 below shows the calculation:

$$m_3 = \frac{\sum D^3}{N}$$ (Equation 3.10)

Table 3.6 shows the calculation of m_3 in a left skewed variable.

Table 3.6 A left skewed distribution

	Salaries in Thousands of Dollars	Algebraic Deviations from the Arithmetic Mean	Deviations Squared D2	Deviations Cubed D3
	142	15	225	3375
	129	2	4	8
	132	5	25	125
	120	-7	49	-343
	112	-15	225	-3375
Sums	635	0	528	-210
Arithmetic Mean	127			

The negative sum of the column D³ in table 3.6 (cubing restores the signs) and m_3, $\frac{210}{5}$ = -42, indicates an unbalanced distribution (skewed) to the left. That is, the tail stretches left and pulls the mean down and to the left of the median. The mass of the distribution piles-up on the right. Figure 3.2 shows a left skewed distribution.

Figure 3.2 A Left Skewed Distribution

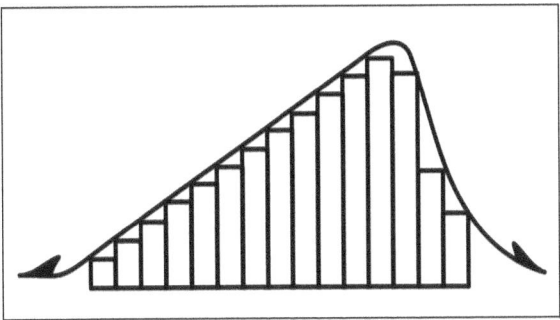

When m_3 equals zero, a variable is symmetrical because the negatives and positives cancel each other. Table 3.7 shows the calculation of m_3 in a balanced distribution.

Table 3.7 A Balanced Distribution

Salaries in Thousands of Dollars	Algebraic Deviations from the Arithmetic Mean (D)	Deviations Squared (D^2)	Deviations Cubed (D^3)
10	-20	400	-8000
20	-10	100	-1000
30	0	0	0
40	10	100	1000
50	20	400	8000
Sums 150	0	1000	0

Arithmetic Mean = 30 $\Sigma D^3 = 0$

This variable is perfectly balanced because the ΣD^3 is zero and, of course, the third moment, m_3 is zero.

A positive m_3 shows right skewness.

The data in table 3.8 shows the calculation of m_3 in a right skewed variable. In this case, an outlier causes the skewness, making the right tail longer, and pulling their mean to the right of the median with the mass to the left.

Table 3.8 A Right Skewed Distribution

Salaries in Thousands of Dollars	Algebraic Deviations from the Arithmetic Mean (D)	Deviations Squared (D2)	Deviations Cubed (D3)
10	-30	900	-27000
20	-20	400	-8000
30	-10	100	-1000
40	0	0	0
50	10	100	1000
90	50	2500	125000
Sums 240	0	4000	90000

We see that the outlier, 90, makes the ΣD^3 and m_3 positive skewing the distribution to the right. Figure 3.4 is a picture of a right skewed distribution.

Figure 3.4 A Right Skewed Distribution

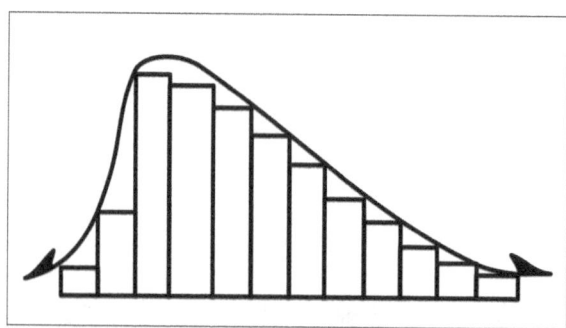

The Coefficient of Skewness ($\sqrt{\beta_1}$)

The third moment, m_3, an absolute measure of skewness, expresses the presence of skewness, not its degree. To obtain the degree of skewness we need a standardized measure or one that holds constant the unit of measure. To standardize the measure, we divide the third moment, m_3 by a measure of variation, m_2, the second moment. We calculate the coefficient of skewness, using the equation shown below.

$$\sqrt{\beta_1} = \frac{m_3}{m_2^{1.5}} \qquad \text{(Equation 3.11)[26]}$$

This equation instructs us to divide the third moment by the second moment raised to the 1.5 power. In other words, $\sqrt{\beta_1}$ measures skewness relative to its variance. In the case of the data in table 3.8, the third moment is 15,000; the second moment is 666.67. Raising m_2 to the 1.5 power yields 17313. Therefore, $\sqrt{\beta_1} = \frac{15000}{17313} = 0.871434$.

Values greater or lesser than +/- 0.50 indicate variables that are not typical of normal distributions.

The usefulness of knowing a variable's skewness should be clear. It is one-step in determining normality. Good descriptive practice always reports a skewness coefficient. Many of the statistical programs we recommend use $\sqrt{\beta_1}$ or some variation to measure skewness.[27]

Kurtosis (Ku)

Even if a distribution is balanced, it may not be normal. Symmetry is the necessary, but not sufficient condition for normality; the sufficient condition is mesokurtosis. We define kurtosis as "middleness/tailness."

---

[26] CC pp. 230-232

[27] Virtually all statistical programs use a variation of the ratio m_3/m_2 skewness relative to variation. One of our recommended programs, NCSS, uses a measure of skewness called g_1 and interprets it for you.

You will not be surprised to learn that the base calculation for measuring kurtosis is the fourth moment.

The Fourth Moment

In order to explain this notion, we require a brief digression.

Mesokurtosis and Normality

A symmetrical, mesokurtic variable is normal. Observe figure 3.3 below.

Figure 3.3 A Symmetrical, Mesokurtic Normal Distribution

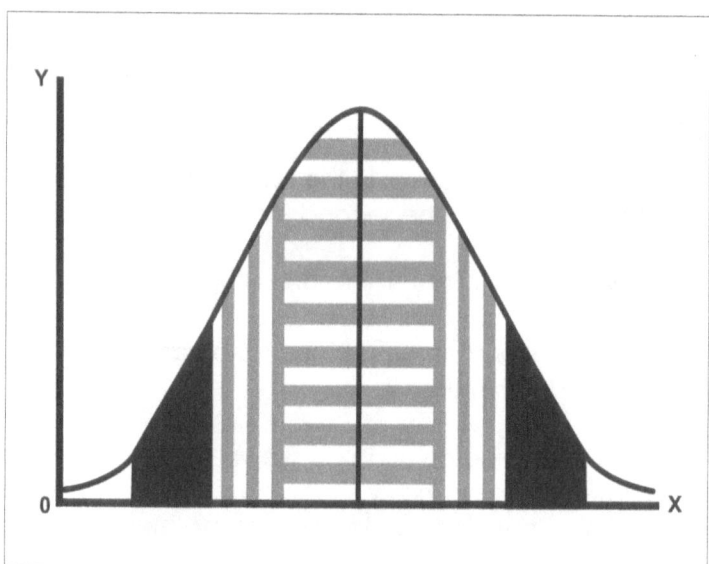

In figure 3.3, drawn as a symmetrical, mesokurtic, normal distribution. With respect to middleness/tailness, the normal curve is mesokurtic.

Platykurtosis

Now assume that in the normally distributed variable in figure 3.3 we transfer observations from below Q_1 and above Q_3 into the space between Q_1 and Q_3, figure 3.4.

Figure 3.4 Loading from Tails to the Middle

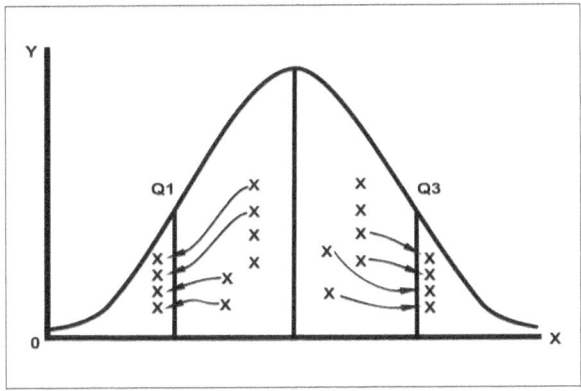

If we evenly load enough observations into the space between the first and third quartiles, the distribution is still symmetrical but not normal because it has a heavy middle with light tails. We call this condition platykurtic.

Leptokurtosis

If we return to the original state of normality and start loading observations from the middle space between Q_1 and Q_3 to the two ends as shown in figure 3.5, the variable, while still symmetrical, is not normal and will have heavy tails and a light middle. We call this condition leptokurtic.

Figure 3.5 Loading from the Middle to the Tail

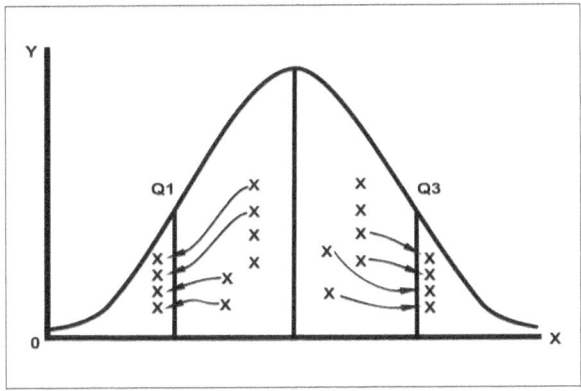

Measuring Kurtosis: the fourth moment (m_4) and the coefficient of kurtosis (β_2)

We define the basic measure of kurtosis as the fourth moment m_4 or $\dfrac{\Sigma D^4}{N}$. However, the most often used measure is β_2, a measure relative to the variance of the variable. The equation for measuring relative kurtosis, β_2 uses the fourth and second moments, m_4 and m_2 as given below.

$$\beta_2 = \frac{m_4}{m_2^2}$$ (Equation 3.12)[28]

The equation instructs us to find the fourth moment $\dfrac{\Sigma D^4}{N}$ and divide it by the second moment squared $(\dfrac{\Sigma D^2}{N})^2$. Putting our equation to work, we use the data in table 3.9.

Table 3.9 Calculation of m_4

Salaries in Thousands of Dollars	Algebraic Deviations from the Arithmetic Mean (D)	Deviations Squared (D²)	Deviations Cubed (D³)	Deviations to the Fourth Power (D⁴)
3	37	1369	50653	1874161
4	36	1296	46656	1679616
5	35	1225	42875	1500625
10	30	900	27000	810000
20	20	400	8000	160000
30	10	100	1000	10000
40	0	0	0	0
50	-10	100	-1000	10000
60	-20	400	-8000	160000
70	-30	900	--27000	810000
75	-35	1225	-42875	1500625
76	-36	1296	-46656	1679616
77	-37	1369	-50653	1874161
928, 369.5		10580	0	12068804
		662345.6		
				928369.5

28.   CC pp. 232-235

Notice that the sum of deviations cubed ($\sum D^3$) equals zero. This distribution is symmetric. Even though the variable is symmetric, this information is not sufficient to allow us to say that the variable approximates normality. Before we can make that claim, we must examine the variable's Kurtosis. To calculate the coefficient of kurtosis, β_2, we divide m_4 by $(m_2)^2$, absolute kurtosis by variation squared. Substituting for m_4, 928,369.5 and dividing by $(m_2)^2$, 662,345.6 equals about 1.4. What does this mean? A normal distribution, symmetrical and mesokurtic, has a β_2 value of three (3), a platykurtic distribution less than three (<3), and a leptokurtic distribution greater than three (>3). Our example is platykurtic, heavy in its middle, light in its tails. Kurtosis, tests the "middleness/tailness" of a variable. Its value is in detecting non-normality in a *symmetric distribution*; using it makes sense only if the distribution is symmetric or nearly so, because no skewed distribution is normal.

All the statistical programs we recommend calculate kurtosis using β_2 or some variation of that metric. Many statistical programs on our list also use combined measures of $\sqrt{\beta_1}$ and β_2 to test for normality.[29] We will show you some examples later.

SPECIAL GRAPHIC DESCRIPTIONS

BOX-WHISKER DIAGRAMS

Box plots show variables from an overhead perspective as compared, say, to the histogram's side view. The Box's construction gives a sense of variance, skewness, and outliers showing Q_1 (25th percentile), Q_3 (75th percentile), the median, the InterQuartile range, the crude range excluding outliers, and possible unusual observations (outliers).

Constructing a box plot follows a series of simple steps:

1. Locate Q_1, Q_3, and the median (Q_2, 50th percentile), and the largest and smallest observations. Build a scale with a vertical line showing a range slightly larger than the crude range. We do not scale the horizontal line, but simply label it with the name of the variable.

---

[29]. NCSS uses combinations of coefficients of skewness and kurtosis as an omnibus test of normality.

2. Calculate the InterQuartile Range ($IQR = Q_3 - Q_1$)

3. Vertically build a box. We constrain the width of the box by proportional aesthetics. The InterQuartile Range (IRQ) defines the box's length. We specify the lower bound of the box as Q1; its upper bound as Q_3. We place the median in the box as a horizontal line. The box can show the mean as a large star or some other clear symbol.

4. Identify outliers by defining a modest (suspect) outlier as any observation greater than $1.5 \times IRQ$. We label any observation greater than $3 \times IRQ$ as a more distant (confirmed) outlier.

5. Locate the largest observation *not* an outlier and construct a line from the upper bound of the box to this value and cap the line. This identifies the right whisker.

6. Locate the smallest observation *not* an outlier and construct a line from the lower bound of the box to this value and cap the line. This cap determines the left whisker.

7. We show outliers by using some symbol at the bottom and top of the whiskers, e.g., a dot (.) for modest outliers and a larger circle (°) for distant outliers.

Figure 3.6 shows a vertical box plot of a variable constructed to illustrate this notion.

Figure 3.6 Box-Whisker Plots

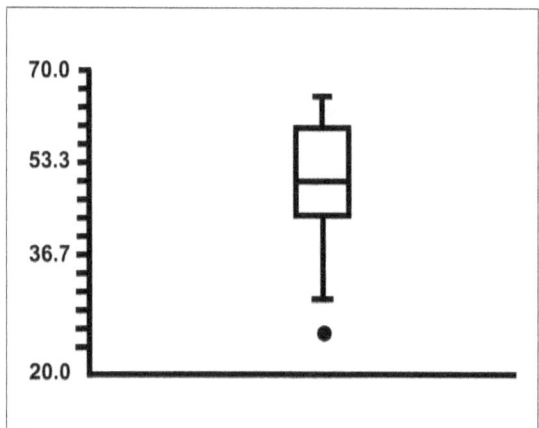

This single picture is a descriptive bonanza. Looking at the data from the top, we see the left skew, outliers, InterQuartile Range (50% of the data), and the highest and lowest observations not outliers. The vertical lines, capped by small horizontals, extending from the box represent the range of highest and lowest that are not outliers. Small circles are potential outliers, defined as 1.5 times the *IRQ*; larger solid circles identify probable outliers, defined as 3.0 times the *IRQ*.

Stem-Leaf Plots

Frequency distribution tables, histograms, and box plots lose their original data points. The histogram shows, for instance, the frequency of each class, but does not permit observation of the actual values. Stem-leaf plots show the general picture of a data set while preserving observation detail. Table 4.2 shows a stem-leaf plot of a variable constructed for illustration.

Table 3.10 Stem and Leaf Plot of Math Scores

Leaf Digit Unit = one   Minimum 53.

5 3 represents 53   Median 79.

Maximum 94

CUMULATIVE	STEM	LEAVES
2	5	34
5	5	778
7	6	34
13	6	588888
28	7	011122233333344
(24)	7	5555555566666688899999999
30	8	0000000000111111222233344444
15	8	55567777789999
1	9	4

This looks complicated, but some interpretation helps. The first part of the table gives the high (94), low (53) observations, and the median (79). We rounded all the observations to two digits. Each observation has a stem, its first digit, and a leaf, its second digit. Look at the first column in the table "CUMULATIVE." That number 2 gives the number of observations in this class. The second column "STEM," the number 5 stands for 50. The third column "LEAVES" shows the last two digits, 3 and 4, telling us the value of the two numbers in this class, 53 and 5 4. Thus, we see that there are two observations in the class, each in the lower 50s, with values 53 and 54.

Now look at the second row. The first number 5 is "CUMULATIVE," telling that there are five observations in the first two rows, two in the first, three in the second. The "STEM" in the second column, 5, and the leaves in the third column, 7, 7, and 8, telling us the values of the three observations in this row 57, 57, and 58. In the third row, the "CUMULATIVE" observations 7 show two observations in this row 63 and 64. We interpret each subsequent row in the same way until we reach the row with a parenthesis (24) which defines the median row and tells us that there are 24 observations in this row starting with 75 and going up to 79. Adding the cumulative 28 to the median row, the row identified with parentheses (24) tells that there are 52 observations including the median row, equal to or less than 79.

The row following the median row 30 is down "CUMULATIVE" telling us that from this row down there are 46 observations equal to or greater than 80. The next row shows 15 observations over 84 in the data set. Finally, the last row shows one observation 94 larger than 89. Looking at the pattern formed by the leaves, we can see the left skew. The $\sqrt{\beta_1}$ coefficient of skewness is about - 0.8.

Summary

The critical ideas discussed in this chapter are the following:

1. The concept of variation

    Variation refers to the degree to which a variable's observations vary, fall around, or spread about a central value. Variation tells how much error (noise) is associated with the mean. It is also the essence of statistical problems: explaining and reducing noise.

2. The standard deviation and variance (the second moment)

    The standard deviation (S) of a series measures the spread in data based on squared deviations from the mean. The equation for S subtracts the observations from the mean, squares those differences (removing negative signs), sums the squares, divides the results by the number of observations, and takes the square root of that quotient. The square of S equals the variance S^2 (or second moment, m_2 about the mean) of a variable. We use this variance as a fundamental building block in statistical inference.

3. The normal distribution

Commonly referred to as the bell curve, the normal distribution is the grounding notion of many (perhaps a majority) of analytic techniques. In a normal distribution or bell curve, there is perfect symmetry (balance) between the standard deviation and mean in the sides of the "bell," i.e., exactly symmetric on both the positive and negative sides of the curve / distribution. In a normal distribution the percentage of observations falling between ± 1 standard deviation from the mean is about 68%, ± 2 standard deviations from the mean is about 95%, and ± 3 standard deviations from the mean 99+%,

4. Camp-Meidell inequality

When a distribution is not normal, we use other methods to assess its spread. The Camp-Meidell Inequality $(1-(\frac{1}{2.25K^2}))$ can be applied to a set of observations that exhibit only one mode and the difference between the mean and mode does not exceed the standard deviation. Using Camp-Meidell we can say that the percentage of observations falling between ± 1 standard deviation from the mean is about 56%, ± 2 standard deviations from the mean about 89%, and ± 3 standard deviations from the mean about 95%,

5. Chebycheff's inequality

If a distribution does not follow the Camp-Meidell assumptions, we can use Chebycheff's Inequality $(1-(\frac{1}{K^2}))$ to assess a variable's variance. Chebycheff's formula, $(1 - (1/K^2))$ where K equals the number of standard deviations, generates a view of the distribution such that about 75% of the observations in a variable no matter how it's distributed, will fall within ± 2 standard deviations from the mean, while 89% fall within ± 3 standard deviations.

6. The coefficient of variation

We calculate the coefficient of variation (CV), a relative measure of dispersion, by dividing the standard deviation by the mean. This allows comparisons of variation across variables using different units of measure.

7. The Standard deviation and variance of nominal data

The notion of the standard deviation and variance are straight forwardly applicable to nominal data coded zero and one.

8. Skewness

Skewness means asymmetry, a departure from normality. If a variable is skewed (left or right), it is by definition not normal. The base equation for measuring skewness is the third moment (m_3). We measure absolute skewness by m_3 and relative skewness by some variation of m_3 divided by m_2.

9. Kurtosis

Kurtosis is another measure of normality. While skewness measures asymmetry, kurtosis measures middleness/tailness. When a distribution is platykurtic, it is heavy in the middle (e.g., 2nd and 4th quartile of data). A variable is leptokurtic when it is heavy in its tails (phallic).

We measure absolute kurtosis using the fourth moment and relative kurtosis using some variation of m_4 divided by m_2. β_2 is one measure of relative kurtosis. We calculate β_2 by dividing the fourth moment (m_4) by the square of the second moment (m_2)^2. A value of three identifies a mesokurtic variable. A value less than 3 identifies a platykurtic distribution. A value greater than 3 identifies a leptokurtic distribution.

10. Box-whisker Plots

The box plot is a graphic description of a variable that provides an overhead view with different information than provided by the side views of histograms and polygons

11. Stem-leaf plots

The stem-leaf plot is another graphic side-view description that captures the general shape of a variable while retaining the variable's detail

There are transformations that will make variables suitable for statistical methods based on normality assumptions. We will discuss many of these as we progress through subsequent chapters.

It is important to note that all the statistical programs on our list will do all of the calculations discussed in this and the previous chapter. The central idea is that you know what these calculations mean and how they are useful.

Appendix

Chapter 3

Your tasks: (1) Using the data set in chapter 2 and a spreadsheet, calculate S, S^2, TV, CV, $\sqrt{\beta_1}$, and β_2 for at least one of the variables and interpret your findings; (2) run the same analyses using your computer program to check your work; and (3) construct box-whisker and stem-leaf plots for one of the variables, interpret your findings, and explain how these two graphics provide information not provided by the histogram.

Chapter 4

OTHER USEFUL DESCRIPTORS

In addition to frequency distributions and the quantitative descriptors discussed in chapters 2 and 3—central tendency, dispersion, normal curves, skewness, and kurtosis-this chapter develops several additional useful quantitative and graphic descriptive devices.

Additional Measures of Central Tendency

Three specialized and useful measures of central tendency should have a place in your statistical toolbox: The weighted mean, the geometric mean, and the harmonic mean.

The Weighted Mean

Sometimes all observations on a variable are not equally important. Simply averaging the price of an identical auto across a three-city region will yield a biased value unless you take into account the different quantities of the cars sold in each city. If the price in Detroit was $20,000, in Hamtramck $22,000, and in Ann Arbor $24,000, a simple arithmetic averaging yields a mean of $22,000. But this does not reflect the number of autos sold at each of those prices, in each location. To make sure that your mean accurately reflects that additional data, you need to weight it.

So 400 cars were sold in Detroit, 300 in Hamtramck, and 200 in Ann Arbor. These are the weights on the sale prices. Table 4.1 shows how to calculate the weighted mean.

Table 4.1 Calculation of the Weighted Mean

City	Price	Number of Autos Sold (Weights)	Weights × Price
Detroit	$20,000	400	$8,000,000
Hamtramck	$22,000	300	$6,600,000
Ann Arbor	$24,000	200	$4,800,000
Sums	$66,000	900	$19,400,000
Means	$22,000		$21,556

To obtain the weighted mean, multiply the sale prices by the quantities sold in each city. Sum the three products and divide the total dollar value ($19.4 million) of the autos sold by the total number sold (900). These calculations yield the weighted mean, $21,556. Compare that to the unweighted mean of the three prices or $22,000.

The Geometric Mean (M_G)

We symbolize the geometric mean using M_G and calculate it with the following equation.

$$M_G = \sqrt[Nth]{X_1 * X_2 * ... X_N}$$ (Equation 4.1)

Where X_1, X_2, etc. represents individual observations, and N the total number of observations used in the calculation.

To obtain M_G, we multiply the X_i values together and extract the Nth root. For example, we obtain the geometric mean of the following variable, (2, 3, 5, 8, 10), by multiplying $2 \times 3 \times 5 \times 8 \times 10 = 2400$ and extracting the 5th root (or 2400 to the 0.20 power = $2400^{0.2}$) of the product which equals 4.74288.[30] The arithmetic mean is 5.60.

---

30. We note that we can extract any root by raising the value to a fractional power. For example, the square root of any number is that number, X raised to $X^{0.5}$ the 0.5 power. The fifth root of any number is that number raised to $X_2^{0.20}$ 0.20 power.

While it may not be obvious, we can calculate the geometric mean by transforming each observation in the variable to a logarithm, averaging the logarithms, and transforming this average back to its natural number. Thus, we can restate the equation for the geometric mean as

$$M_G = \text{Antilog} \frac{\sum LogX_i}{N}$$ (Equation 4.2)

Equation 4.2 does the same thing[31] as equation 4.1. Therefore, you should see that in a right skewed variable M_G will be smaller than the arithmetic mean because the log transformation pulls the larger values of the variable down.

One use, then, of the Geometric mean is in diminishing the effect of outliers at the larger end of the variable. Taking the same variable in the example given above, we change observation with the value 10 to 50 (2, 3, 5, 8, 50) and now the product of the observations is 12,000, heavily influenced by the outlier, 50. By extracting the fifth root of 12,000, $\sqrt[5]{2\times3\times5\times8\times50} = \sqrt[5]{1200}$ we get the geometric mean 6.54. A comparison of the geometric mean 6.54 to the arithmetic mean 13.60 shows how the geometric mean reduces the effect of outliers.

When we calculate percentages, ratios, or rates of change over time and we do not know the base on which these values were generated, geometric means provide a useful device for averaging. For example, suppose an investment portfolio yielded 8%, 5%, 3%, and 36% over the last four years. To calculate the average return, our first inclination takes the arithmetic mean $\frac{52\%}{4}$ = 13%. But if we applied the arithmetic mean to the base from which we calculated the rates, because of the outlier, the mean would exaggerate the actual returns.

Instead, we use the M_G. To calculate the M_G, we express the percents decimally, 1.08, 1.05, 1.03, and 1.36, multiply these together and extract the fourth root of the product. This calculation yields the geometric mean of 1.1227 or an average of 12.3% return on investment, an average that

---

31.   Adding the logs of the X observations is the same thing as multiplying untransformed observations together. Dividing that sum by N is the same thing as taking the Nth root.

reflects the actual returns when applied to the base. Using M_G instead of M has reduced the effect of the mean return by about 0.7%.

If, in one of the periods, the rate of return equaled zero, we would record the zero as one. If the series experiences a negative rate of return, say - 5%, we record the decline as 0.95. For example, if the rates of return of our small business over a three-year period were -3.43%, (.9647), 0% (1.0000), and -7.32% (.9268), what is our average loss rate? The geometric mean equals about 0.963 or a rate of decline as 1 - 0.963 or about -3.63%. The mean would show a decline of about 3.62%. The M_G has increased the loss rate by a small fraction, about 0.0001 because the log transformation has slightly increased the effect of the period with the largest loss.

Geometric means are also useful in calculating average rates of return if we know the period length and starting and end values. Here, we take the difference between starting and end observations and extract the period root. For example, if you wanted to know the average rate of return over a five-year period with a beginning value of $42,000 and the end value of $60,000 we take the difference between the two values, $18,000 and extract the fifth root. The average rate of return as measured by $M_{G'}$ $\sqrt[5]{18000} = 18000^{0.20} = 7.1\%$.

The Harmonic Mean

We calculate the harmonic mean by summing the reciprocals of each observation in the variable and dividing that sum into the number of observations. We give the equation below.

$$M_H = \frac{N}{\frac{1}{\Sigma X_i}} \qquad\qquad \text{Equation 4.3}$$

The equation tells us to sum the [reciprocals one divided by the observation $(\frac{1}{X_i})$] of each of the observations and divide that sum into the number of observations in the variable. How is such a measure useful?

Like the geometric mean, the harmonic mean diminishes the effect of outliers. Using the same example as stated in the discussion of the $M_{G'}$ rates of return on an investment portfolio over four years, expressed decimally as 1.08, 1.05, 1.03, and 1.36. The harmonic mean of that series equals 1.116 reflecting an average return of about 11.6%. If we compare that to the arithmetic and geometric means, it is smaller than both. Thus, we see that harmonic mean shrinks the outlier effect by even more than the geometric mean.

Speed provides another example. If, for half the distance of a trip, you travel at 50 MPH and for the other half 70 MPH, what is your average speed? The previous sentence gives distance, not time, at each speed, yet it asks for the average speed. When this is the case, the M_H is the preferred measure. Thus, the harmonic mean, 58.33 MPH. takes precedence. If the sentence had said, "you travelled half the time at 50 MPH and half the time at 70 MPH," the arithmetic mean, 60 MPH, would be preferred.

Reexpressions

Standardizing Variables: Z Scores

Standardizing any variable eliminates the unit of measure. We calculate the standard score (Z scores) of any variable by algebraically subtracting the mean from all observations and dividing these deviations by the standard deviation. We give the equation for Z scores below:

$$Z_i = \frac{X_i - M_X}{S} = \frac{D_i}{S} \qquad \text{Equation 4.4}$$

The equation instructs us to subtract the mean from each of the observations to realize D_i and divide D_i by the standard deviation of the variable; the mean of the Z score equals zero (0) and the standard deviation equals one (1). This reexpression does not change a variables mathematical structure.

We defer a discussion of the usefulness of a Z reexpression of a variable to another chapter, but you should see that if a variable approximates normality, its Z reexpression will immediately enable us to calculate quickly the proportion of observations falling above, below, or between Z scores.

Adding, Subtracting, Multiplying, or Dividing a Constant to a Variable

Adding, subtracting, multiplying, or dividing any constant, e.g., five (5) or - five (-5), to any variable, a reexpression, will not change its structure. For example, adding the constant five or minus five to a Z score will change its mean to five or minus five, but will not change the standard deviation or its basic structure. Multiplying by a constant of five (5) will change the standard deviation to five (a multiple of 5), but will not change the mean. Dividing by five will not change the mean but will reduce the standard deviation to 0.20 (a divisor of 5).

While all these operations on Z scores (adding, subtracting, multiplying, or dividing) will change either the mean or the standard deviation, the operations will not change the variable's mathematical structure. The coefficients of skewness and kurtosis will remain the same. Histograms portraying the effect of the operations will look the same. In addition, while we have not discussed scatter plots, any set of Z variables created by these various operations becomes a perfect surrogate for the original. Figure 4.1, a picture of the relationship between a Z score and a Z + five score, illustrates this idea. The Z score, on the horizontal axis and Z + five on the vertical axis, show how the two variables form perfect surrogates.

Figure 4.1 Plot of Z against Z + 5

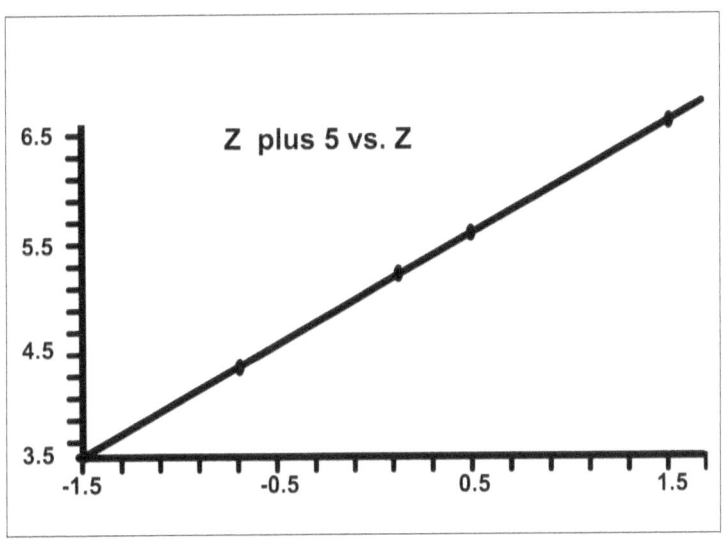

From this figure, you can see the identity of Z and Z+five. That is, any two variables plotted as X and Y in which the plot yields a perfect straight line are structurally identical variables.

One useful application of adding a constant to a variable is that it eliminates negative observations. Why is that useful? In the next section, we discuss logarithmic transformations of variables. If the variable contains negative observations, we cannot make log transformations.

Univariate Transformations to Approximate Symmetry

Previously, we introduced the idea of the Z score, emphasizing that changing data from its original metric to Z is a reexpression because it does not change the mathematical structure of the variable. By that, we mean that the reexpressed variable preserves the same equal scale differences and ratios as the original variable.

By comparison, a transformation changes the mathematical structure of a variable. In this section, we ask you to remember some high school algebra as we introduce the notion of transformations useful in changing skewed variables to approximate normality.

Square Root Transformations

Suppose we have a right skewed variable that we want to transform to approximate symmetry. The task is to *pull* the right tail to the left or back toward the mean. We can do this in several ways. The first method, the transformation with least effect, transforms the variable by taking a root $\sqrt[2]{X_i}$ of each observation of the variable. Taking a square root of each observation effects the larger observations more than the smaller, thus pulling in the right tail. For example, taking the square root of a variable (n = 5) consisting of observations 4, 6, 12, 36, and 180 produces values of 2, 2.4, 3.4, 6, and 13.42 pulling in the right tail and making a skewed variable more symmetrical. The arithmetic mean of the untransformed variable is 47.60. The arithmetic mean of the square root transformed variable is 5.47. To restore the transformed variable back to its original metric we square 5.47^2 to get 29.92. The difference between the two means (47.60 and 29.92) represents the reduction of the effect of the outlier, 180. The higher the root transformation, the greater is the effect on a variable.

Logarithmic Transformations

The second method for pulling in the tail of a right skewed distribution, a logarithmic transformation, has a more powerful effect than taking the square root. We use natural logs (LnX_i) in our example. Again, from your high school algebra, the natural *log* Ln of 4 is the power to which we must raise the base $e \approx 2.71828$ to give 4. Thus, the Ln transformed values of the series 4, 6, 12, 36, and 180 yields 1.39 1.79 2.48 3.58 and 5.19. Taking the sum of the transformed variable (14.44) and dividing by 5, gives the mean of the logs (2.89). To restore the natural log transformation back to its original matrix, we take the anti-log of 2.89, 17.96, the mean of the LN transformed variable. Notice that the LN transformation has a greater effect than the square root transformation in pulling in the right tail of a variable.

Reciprocal Transformations

The third method (the strongest) uses the reciprocal $\dfrac{1}{X_i}$ of each observation in the variable. Reciprocally transforming the series 4 6 12 36 and 180 produces 0.25, 0.167, 0.0833, 0.0278, and 0.00556. This transformation pulls in the tail even more, but it also reverses the order of magnitude, an issue that we must take into account if we use this variable in other calculations. The mean of the reciprocally transformed series is 0.107. To restore the reciprocal transformation back to its original matrix, we divide 1 by 0.107 to get 9.34, the mean of the reciprocally transformed variable. Notice that the reciprocal transformation is the most radical of the three.

To further illustrate, we created a right skewed variable ($N = 100$) called dv2 with the following characteristics. Table 4.3 shows the descriptives for the untransformed variable.

Table 4.3 Descriptives of the Untransformed Variable

dv2 ($N = 100$) Sum = 999.220
Mean = 9.992 Variance = 1.085 Std Dev. = 1.042
Range = 4.310 Minimum = 8.410 Maximum = 12.720
Skewness = 0.625 Kurtosis = 2.587

Note the value of the coefficient of skewness, 0.625.

For this variable, we tried a natural log transformation that produced a new variable, Ln dv2. Table 4.4 shows the descriptives for the L_N transformed variable. .

Table 4.4 Descriptives of Ln Transformed Variable

Ln DV2 (N = 100) Sum = 99.740
Mean = 2.30 Variance = 0.0103 Std Dev. = 0.102
Range = 0.413 Minimum = 2.13 Maximum = 2.54
Skewness = 0.442 Kurtosis = 2.347

Note that we have reduced skewness to 0.442, within the range of normality. Also, note that the transformation made the variable slightly more platykurtic, but still within an acceptable rang

Power Transformations Greater than One[32]

For a left skewed variable, we want to push the right side out. Here the calculation raises the series to some power greater than one, for example, squares the series. To illustrate we created a left skewed variable call "left." Table 4.5 shows the descriptives for the untransformed variable.

Table 4.5 Descriptives of the Untransformed Variable

Left (N = 100) Sum = 5177.590
Mean = 51.776 Variance = 81.042 Std Dev. = 9.002

---

32.   JCC Jacob Cohen pointed out that all of these calculations are power transformations and can be represented by $X \rightarrow X^\lambda$ where λ is the power to which X must be raised to approximate normality. We represent reciprocal transformations $\dfrac{1}{X_i}$ by a negative power X^{-n}, a root transformation, e.g., the square root, by a fractional power $X^{0.5}$, a log transformation by X^0, and raising a variable to some value greater than one, e.g., its square, X^2.

     Cohen also gave a simple explanation for X to the zero power $(X0)$ power as a log transformation. Cohen's explanation goes something like this: X to the zero power $(X0)$ transforms all values of X to 1.0. As the power, λ, approaches zero, the expression $\dfrac{X^\lambda - 1}{\lambda \rightarrow LnX}$, leads to the use of the natural logarithm as the transformation when λ = zero.

Range = 40.270 Minimum = 25.640 Maximum = 65.910
Skewness = -0.530 Kurtosis = 2.404

Note that the coefficient of skewness is -0.530
To transform "left" so that it approximates a normal distribution we squared each of the observations in the "left" variable and called the new variable "powtran." Table 4.6 shows the descriptives for the power transformed variable. .

Table 4.6 Descriptives of the Power (Squared) Transformed Variable

Powtran (N = 100) Sum = 276097.565
Mean = 2760.976 Variance = 801835.958 Std Dev. = 895.453
Range = 3686.718 Minimum = 657.410 Maximum = 4344.128
Skewness = -0.150 Kurtosis = 2.245

We can see that our transformation was successful (coefficient of skewness down to -0.15). Even though platykurtosis increased, declining from 2.404 to 2.245, it still lies within an acceptable range.

The Geometric and Harmonic Means as Power Transformations

At this point, we ask you to look back at the equations for the geometric and harmonic means. First, we look at the geometric mean.

$$M_G = \sqrt[Nth]{X_1 * X_2 * ... X_N} \qquad \text{Equation 4.5}$$

If we transform the equation to logs, it looks like this:

$M_G = \text{Antilog} \dfrac{LogX_1 + LogX_2 + ... LogX_N}{N}$. Summing the logs of each

observation is the same as multiplying them together. Dividing by N is the same thing as taking the N_{th} root. Thus, the geometric mean is simply the antilog of the mean of a log-transformed variable.

Now we look again at the equation for the harmonic mean.

$$M_H = \dfrac{N}{\Sigma \dfrac{1}{X_i}}$$

Rearranging this equation to

$$M_H = \dfrac{\Sigma \dfrac{1}{X_i}}{\dfrac{1}{N}}$$ makes it obvious that the harmonic mean is the mean of

reciprocals or simply a reciprocal transformation. Then we can take the equation back to the original by:

$$\Sigma \dfrac{1}{X_i} * \dfrac{N}{1} = \dfrac{N}{\Sigma \dfrac{1}{X_i}}.$$

Summary

The critical ideas discussed in this chapter are the following:

1. The Weighted mean

2. The geometric mean

   Geometric means, the means of variables transformed to logarithms, find their usefulness in calculating averages of percentage variables in which we do not know the base.

3. The harmonic mean

   The Harmonic mean reciprocally transforms variables transformed finding its usefulness in averaging variables given in one unit of measure in which interpretative need requires another unit of measure.

4. Reexpressions

A reexpression does not change a variables mathematical structure.

a. Z scores

Z scores represent a reexpression of variables by removing the unit of measure without altering the variables structure. Z scores have means of zero and standard deviations of one.

b. Adding, subtracting, multiplying, or dividing constants to a variable

Adding, subtracting, multiplying, or dividing any constant, e.g., five (5) or - five (-5), to a variable, another reexpression, will not change its structure. For example, adding the constant five or minus five to a Z score will change its mean to five or minus five, but will not change the standard deviation or its basic structure.

At the risk of boring reiteration, if you (our readers) are serious about doing sober statistical work, you must acquire of at least one good statistical package, e.g., SPSS, NCSS, Statistix, Systat, Limdep, or OpenStat (and a good spreadsheet, e.g., Excel or QuatroPro). All these statistical packages can, with a few keystrokes, do all the calculations, and make all the tables and pictures presented in these chapters. If financial resources are an issue, the two free packages, MyStat and OpenStat plus free spreadsheets available in the Net are more than adequate.

Appendix

Chapter Four

The data set has two variables. At least one of the variables is skewed. Your tasks are the following:

1. Using your computer program, construct box and stem-leaf plots for all the variables. Explain how these plots yield information not provided by the frequency distribution.

2. On any one of the variables, calculate the geometric and harmonic means and explain why they differ from the arithmetic mean.

3. Make transformations designed to cause the variables to be more symmetrical and explain why these transformations work are how they are useful.

ID	X1	X2
1	54.90	126.53
2	45.80	125.41
3	40.45	78.67
4	44.05	110.62
5	55.75	70.33
6	67.22	127.90
7	40.58	80.00
8	43.61	62.10
9	45.68	112.55
10	46.17	86.27
11	53.86	103.49
12	56.88	95.28
13	37.34	96.58
14	41.21	123.31
15	55.30	119.69
16	45.45	125.44
17	42.38	77.66
18	42.01	80.16
19	54.03	100.60
20	64.65	100.27
21	55.47	118.53
22	43.05	72.67
23	58.79	117.98
24	42.63	101.93
25	65.59	121.01
26	56.89	104.09
27	42.95	94.48
28	53.37	103.64
29	44.62	75.41
30	46.51	106.40

Inferential Statistics

Introduction

In the next four chapters, we take you into the realm of sampling and statistical inference. When people use the term *inference*, they usually refer to one of two meanings: deduction or induction. Deductive inference, using a thinking machine called a syllogism, goes from the general to the specific. Here is an example.

All Democrats are liberals.	(major premise)
Johnny is a Democrat.	(minor premise)
Therefore, Johnny is a liberal.	(conclusion)

If the major and minor premises are true, then the conclusion must be true. Deduction, then, carries with it inferential certainty.

Induction, on the other hand, goes from the specific to the general. Here is an example of induction. If we take a sample of swans and all the swans are white, we can say with a high degree of certainty (or with a high probability) that all swans are white. However, note the language, "with a high degree of certainty," not the same thing as certain. This means that inductive inference always carries with it inferential risk or worse inferential uncertainty. Its conclusions are always subject to error. The next swan observed may indeed be black or blue or red.

Thus, the fundamental issue is as follows: When we take a sample from a population whose variance is greater than zero and calculate some descriptive measure, say the arithmetic mean, our sample mean (unless we are incredibly lucky) will almost certainly differ from the population mean. This difference is sample error or error due to chance. The task of statistical inference is to make a probabilistic estimate of that difference. In other words, the statistical method, reliant on sampling, is an inductive method. Moreover, because it is inductive, we must have technologies that estimate the risk of being wrong. These technologies must provide a known shield from claiming that we have found a signal when all we have is random error. Or, conversely, concluding that that all we have is random error when we probably have a signal.

Chapter 5 introduces the concepts of probability and sampling. The chapters that follow build on these notions showing how the statistical method's technologies that cope with various problems arising from inductive inference. In chapters 6, 7, and 8, we expand our explanation of the necessity of using sampling over a census and develop the tools for statistical inference: estimation and hypothesis testing. The practical issues are clear. Taking a census to calculate the national monthly unemployment rate, for example, is physically impossible. Putting aside cost issues, it would take a year to calculate a monthly rate that, by that time, would be hopelessly out of date. In the same vein, if we wanted to test the hypothesis that unemployment had changed from one month to the next, for policy purposes, a census would be useless.

In our study of descriptive statistics, our examples assumed that the variables were measures of populations, for example, all the people who attended all the college football games during a recent season. But if we were to do a statistical study of college football fans who physically attend games rather than watching on TV, time and resource limitations would force drawing a sample from the population of game go-errs, and using the data in that sample to estimate the attitudes and characteristics of the entire population.

In addition, for the same reasons, we would probably not take on the monumental task of hunting down every one of those dedicated football game attendees to ask them—probably among many other things—whether they prefer bratwurst to barbecue as the main course for their tailgate parties.

Therefore, we take a sample, count the number who prefer bratwurst to barbeque, calculate the percentage, and inductively generalize the sample percentage to the value of the entire population. Our sampled value, say, 55% of college football fans prefer bratwurst over barbeque, will deviate by some amount from the actual or "true" percentage of college football bratwurst lovers, if by some strange voodoo we were granted the almost superhuman ability to interview every last one of them.

Put more simply, the estimates of population parameters derived from samples will always contain error attributable to the act of sampling because they are inductive. Expressing errors as probabilities enables us to avoid claiming that we have found a signal when all we have are findings confined to the sample. It also enables us to infer population signals instead of confining our findings to the sample. Therefore, we study probability theory as the theoretical foundation of inferential statistics. It is important to note that we have introduced another concept of error. While the standard deviation and variance represent the degree to which the mean does not represent a variable, for clarity of purpose, let us call that error one. Let us call sample error, error attributable to random sampling or chance, error two.

CHAPTER 5

PROBABILITY AND SAMPLING

This chapter introduces the basic concepts of probability theory underlying inferential statistics and presents the basic concepts of sampling.

PROBABILITY—THE FOUNDATION OF INFERENTIAL STATISTICS

PROBABILITY DEFINED

The probability of any outcome A, call it $P(A)$, is the proportion of times A will occur in a random process. The usual example is coin flipping. We express $P(A)$ in a formula as

$$\frac{NumberofTimesEventAOccurs}{InfiniteNumberof\ \mathbf{Re}\ peatedTrialoftheSameKind} \qquad \text{(Equation 5.1)}$$

Therefore, when we state the probability of an outcome or event, we mean the *frequency* of its occurrence relative to a certain number of trials. For example, what is the probability that a flipped coin will land heads-up?

Estimating Probability

The answer to the question about the flipped coin is deceptively simple. If the coin is unbiased ("fair and balanced, unrigged"), and if you flipped it an infinite number of times under the same conditions, 50% of the time the outcome would be heads. Thus, the probability of flipping a head on one flip of an unbiased coin is 0.50.

We said "deceptively simple" because we cannot flip a coin an infinite number of times, nor flip it exactly the same way each time. We also cannot be sure that the coin itself does not change. Because we cannot meet these conditions, which are even more difficult than carrying out a "true" census of a population, you cannot know the "true" probability of any event. All you can do is estimate it.

There are two methods for estimating probability: empirically and using the assumption of equiprobability.

Empirical Estimation of Probability

To estimate a probability empirically, as we saw with bratwurst, barbeque, and tailgating football fans, we take a sample, count the number of events or outcome A's in the sample, and divide that number by the size of the sample. You can estimate a probability with an experiment (of a sort). With a six sided die, you estimate the probability of rolling a six (6) on any one roll by rolling the die one hundred times, observing the number of times a six comes up, and dividing by one hundred. If a six comes up 17 times, the estimated probability of rolling a six is $\frac{17}{100}$ or 0.17.

However, if we manufacture bricks and are more concerned with brick quality than bratwurst, we want to estimate the probability that any one bricks fails—cracks in two—under certain conditions of pressure and stress. Over the course of a couple of weeks, we test 1,000 bricks and find fifty that break. The estimated probability of making a defective brick—P (defective) = $\frac{50}{1000}$ = 0.05 = 5%.

That may or may not be good enough for your customers, but at least you have a baseline: if the production process is not changed, the probability of making a defective brick is 5%.

Now, we are back to the original problem of violating the conditions we used to define probability. To estimate the probability of a bad brick, your inspection process destroys the bricks. While this experimental approach uses pretty much the same *process* for *testing* the bricks—within the tolerances established by modern engineering standards—it rules out repeating the process with the *same* brick. This highlights again that the best we can do in empirically estimating probability is to use cross-sections, or experimental trials of approximately the same kind. We have to test many bricks and assume that they are approximately the same. In addition, we assume (more confidently, we suppose, but with our fingers crossed behind our backs)—that the manner of breaking the bricks is approximately the same.

Equiprobable Estimation of Probability

To use the second method for estimating probability, we have to make another assumption, that of equiprobability, based on the nature of the estimating process itself. Let us assume that a single die has only six sides and is unbiased. Equiprobability tells us that in a large number of rolls of the die, the number six will come up as often as any other number. Our equiprobable estimate of the probability of rolling a six on one roll, then, is 1/6 (or 0.1667, or 16.67%).

Using equiprobability is convenient and simple; but again, as it is with any estimation method, there is a catch: it requires you to be mindful of the number of mutually exclusive outcomes you are dealing with. That is, with a balanced die, there are only six possible outcomes, and they are mutually exclusive. On one roll, you can roll a six or a one, or a three, or a four (or a five or a two), but you cannot roll a one *and* a six. Again, the method also requires us to assume that the die is unbiased (there is that "fair and balanced, unfixed" thing again), i.e., that it is not weighted in favor of a particular outcome.

Characteristics of Measures of Probability

Numerical Limits

A probability measure has numerical limits. It is always expressed as a number bounded by zero and one; the measure cannot be negative or greater than one.

If an outcome never occurs in a large number of trials (since, as we said earlier, probabilities are expressions of relative frequencies), its relative frequency is zero and its probability is zero. If an outcome always results, its relative frequency is equal to the number of trials and probability is one.

You are a supervisor of OB-GYN nurses in a hospital district. You witness a large number of births over your 30 years of service, and not one of those babies is born with flaming green hair. You estimate a zero probability of a flaming green-haired baby being born. Now, you are not completely certain that no baby will ever be born with flaming green hair, because such a thing might occur, but at least you know that the probability of this happening is very close to zero. In the same spirit, you might also say that the probability of a baby being born with an umbilical cord is one.

The Sum of the Probabilities of all Possible Outcomes Equals One

A second characteristic of a probability measure is that the sum of the probabilities of all possible outcomes is equal to one. If the probability of $A = P(A)$, and the probability of non-$A = P(Q)$,

Then $P(A) + P(Q) = 1$.

Conversely $1 - P(Q) = P(A)$.

If you are a bookkeeper, you might want to experiment a little bit and randomly select an entry from an accounts payable ledger to check its accuracy. The two possible outcomes are the selection of an incorrect entry, and the selection of a correct entry. Because the sum of the probabilities of all mutually exclusive outcomes equals one, if the probability of an incorrect entry is 0.05 and the probability of a correct entry is 0.95, then the sum of the probabilities must equal one, if correct and incorrect are the only outcomes. Similarly, if the probability of selecting a correct entry is 0.8, the probability of selecting an incorrect entry must be 0.2, since they are the only possible outcomes.

Mutually Exclusive Outcomes are Additive (The Addition Rule)

The Addition Rule constitutes the third characteristic of a probability measure. In calculating the probability of mutually exclusive events—i.e., those that cannot occur simultaneously—we simply add the two probabilities together. Assuming equiprobability, the probability of rolling a one on a single roll of an unbiased six-sided die is 1/6. The probability of rolling a six is 1/6. Hence the probability of rolling a one or a six = 1/6 + 1/6 = 2/6 = 1/3, or about 33.33%.

Repeated Trials

So far, the probabilities we estimated were outcomes from a single trial. Now, we show how to estimate the probability of a *sequence of outcomes in repeated trials*. To do this, we will distinguish between independent and dependent repeated trials (or series), and in the process, calculate the probabilities of single-event outcomes, something you already know how to do.

Repeated Independent Trials (The Multiplication Rule)

If each roll of an unbiased die or flip of a coin does not *influence the probability of a subsequent* flip or roll, we describe the repeated rolls or flips as independent. More formally,

> two events, A and B, are independent if the occurrence of A does not affect the probability of the next occurrence, B.

So to find the probability of two or more independent events occurring one after the other, find the probability of each event occurring separately, and multiply the two probabilities together. This is the *multiplication* rule. The probability of rolling three straight ones with an unbiased six-sided die, then, is

$$1/6 \times 1/6 \times 1/6 = 1/216 = 0.00463,$$

or about 5 in a thousand.

So again, more formally,

> When two events, A and B, are independent, the probability of the two events occurring in sequence, P(A and B) equals P(A) × P(B).

Now, we estimate the probability of casting a one on two throws of an unbiased die. Since the two throws are independent, the probability of a one on the first throw is 1/6 and 1/6 on the second. What is the probability of a miss (a non-one) and a hit (a one), in two consecutive throws? Using the multiplication rule:

$$= 5/6 \times 1/6 = 5/36$$

Because hitting on the first, and missing on the first and hitting the second, are mutually exclusive, so the probability of rolling a one on two rolls is as follows:

$$= 5/36 \text{ (a miss on the first roll)} \times 6/36 \text{ (a hit on the second roll)}$$
$$+ 6/36 \text{ (a hit on the first roll)} = 11/36.$$

This is approximately 0.139, or about 14%. If you were a betting woman, your probability of rolling a one on two rolls equals about 0.14, or about 14 times out of every 100 times you repeated the rolls. In this example, we used both the addition and multiplication rule.

Repeated "Conditional" or Dependent trials

Draw two cards from a deck of 52 cards *without* replacing the first card before drawing the second. The outcome of the first draw influences the outcome of the second because the value of the first card—let us say the jack of hearts—changes (reduces) the probability of drawing a heart or another jack on the next draw. We call these draws (trials) conditional or dependent.

Summarizing the rule, then, when two events, A and B, are dependent, we calculate the probability drawing A and then B in that sequence as:

$$P(A \text{ and } B) = P(A) \cdot P(B|A)$$

Notice the element $P(B|A)$. The vertical line separating A and B means "given." The equation says the probability of (A and B) is the probability of A $P(A)$ times the probability of B, given A, $P(B|A)$.

Now, we apply the rule again. Calculate the probability of missing the king of spades on the first draw and drawing it on the second without replacing the card drawn first. The probability of missing the king of spades on the first draw is 51/52. The probability of drawing the king of spades on the second draw, if you miss on the first = 51/52 × 1/51 = 0.981 × 0.0196 = 0.0192 or about 1.9%.

Now we ask, "What is the probability of drawing the king of spades on two straight draws without replacement." The probability of drawing the king of spades on the first draw equals 1/52. The probability of missing on the first draw and drawing the king of spades on the second equals 51/52 × 1/51 because not replacing the card drawn on the first draw influences the probability of drawing the desired king on the second. Because drawing the king on the first draw, missing on the first, and drawing the king on the second are mutually exclusive, makes the two probabilities additive. Therefore, the probability of drawing the desired king on two draws equals 1/52 (0.0192) + 51/52 × 1/51 (0.981 × 0.0196 = 0.01923) and 0.0192 + 0.1923 = 0.0384. Therefore, the probability of drawing the king of spades without replacement on two draws equals about 3.8%

For all Texas Hold 'Em fans out there, the decimal fraction, expressed as about 3.8%, is a more readily understandable number. But then again, other players, all drawing cards without replacement, complicate things, so consult a poker king.

The Odds Ratio

Often the probability of winning anything is given as an odds ratio. What are the odds of Sea Biscuit winning the Kentucky Derby? Based on recent previous performances, the bookies, railbirds, and other would-be semicriminal elements are giving him a one-in-four shot, or about a 25% chance of winning this or any other race. However, these are not the odds on which you base your betting decision. Instead, you calculate the odds in the following way:

$$\text{Odds} = \frac{P_{Winning}}{1 - P_{Winning}} \qquad \text{(Equation 5.2)}$$

The equation says to divide the probability of winning by 1 minus the probability of winning. So the odds

$$= \frac{.25}{1 - 0.25} = \frac{0.25}{0.75} = 0.333$$

or approximately 1 in 3. The odds, then, are three-to-one, and you bet $1 to collect $3 if he wins. Importantly though, note well, that there are four outcomes rather than three. The probability of Sea Biscuit wining any given race is one in four. Sea Biscuit is likely to win one time in four and lose three times in four races.

Importance of Probability to Inferential Statistics

This has been a brief and incomplete overview of probability theory. We did not intend to make you an expert, and do not run to the nearest craps table based on what we said. We simply wanted to convey the meaning of probability, and to set the stage for your use of basic probability theory to make logically valid and statistically reliable inferences.

We ask you to consider the following example. For marketing purposes, you want to estimate the annual mean income of some population in a particular region. You take a sample, and calculate the mean income of those sampled as $50,000/year. You know that it is highly probable that this estimate will contain error. So the question is, how much error?

The power of probability theory underlying inferential statistics comes into play here. In the chapters that follow, we show you how to make an interval estimate of the mean, e.g., $45,000 to $55,000, between which we can state how confident we are that the population mean will be in that range. We can create limits between which we are 90%, 95 percent, more than 99% confident, or any degree of confidence between which the population mean will fall. The confidence interval shows the amount of sample error in our estimate, and is simultaneously a probability statement about our confidence in that estimate.

Sampling

Definitions

Population: the totality of observations in which we have a research interest.

Sample: A sample is a smaller part or subset of a population.

Parameter: A parameter is a descriptive or relational measure of a population, for example, an average, proportion, skewness coefficient, or kurtosis coefficient.

statistic: In this context, we make another use of the word Statistic: A statistic is a descriptive measure of a sample. We use this word to distinguish it from population measures (parameters). A reminder: In our use of formulae, we make no distinction between a statistic and a parameter assuming that all the variables in our examples are sample data.

Population frame: A population frame is an operational definition of a population. By operational definition, we mean some device from which we can choose a sample, e.g., a list or map. Obviously, we can take neither a census (complete count) nor a sample without some way to identify the individual observations in a population. Except for convenience samples, we cannot take a sample without a frame.

Bias: Bias produces a difference between statistic and parameter. When a statistic is consistently larger or smaller than its comparable parameter, we call the error bias. An invalid measurement (does not measure what it is supposed to measure) will produce a biased statistic. For example, if our interest was political preference and we measured racial or ethnic prejudice, the wrong measurement would bias our statistic.

Population frame misspecification: Frame misspecification is another cause of bias. For instance, if our research interest was predicting an election outcome and we use a telephone directory as a frame, we will miss those who do not have a telephone, use a cell phone, or have unlisted numbers. These people might have no-random voting preferences different from those listed in the directory, thus biasing our statistic.

Mistake: Another source of difference between statistic and parameter is mistake. People make mistakes in recording and calculating, and such mistakes can produce large differences between a statistic and parameter.

Bias and mistake are nonsampling errors. They can and do exist even when taking a census. Once bias and mistake become part of a statistic, we can do nothing to compensate or correct the error because there is no basis on which to estimate their direction and/or magnitude. It is, therefore, important to eliminate or reduce bias and mistake through sample design and quality control.

Sample error: Chance or sample error, occurring through random selection, is significantly different from mistake and bias. Because random errors are the product of chance, they are unbiased (they tend to cancel). Equally important, using inferential statistical methods, we can estimate and control their magnitude with precision (given resource constraints). In other words, if we eliminate bias and mistake, the only other source of error is predictable random error.

The Purpose of Sampling

The purpose of sampling is to make inductive estimates of parameters or test hypotheses about parameters.

The Case for Sampling—If samples introduce random error (a difference between the statistic and the parameter), why do we take samples? Why not eliminate sample error by taking a complete count (census)?

First, sampling might be the only possible alternative. If a population is theoretically infinite, i.e., a continuous process like an assembly line, a complete count is impossible. Until it is permanently shut down, an assembly line is theoretically an infinite population.

Second, sampling may be the only practical alternative. For large populations, e.g., one million, or limited resources are limited (usually the case) or critical timelines (the need for current data), sampling represents the only practical alternative. If the act of measurement destroys the sample unit, as in some statistical quality-control procedures, sampling represents the only practical alternative.

Third, a sample can be more efficient. If, in a properly designed sample, we can reduce the error (the difference between the statistic and the parameter) to satisfactory limits, based on some cost-benefit criterion, the sample can be more efficient. For example, if a project's error tolerance is manageable, the amount of sample error may be reduced sufficiently to allow use of a sample instead of a complete count.

Finally, sampling may be more accurate than a census. If measurements requiring close tolerances are tedious and tiring, a complete count may result in so many mistakes that the random error in the sample is smaller than the error introduced by mistake.

Methods of Choosing Samples

There are three ways to choose samples: Random, purposive, and convenience.

Random Samples

The individual observations in a random sample are chosen by random methods. An example is numbering all the items in a population, putting the numbers in a capsule, putting the capsule in a barrel or any large container, shaking the container thoroughly, and drawing the sample units from the container. Modern computer technology through simulation has replaced the barrel and the capsules, but the principle remains the same. There are two types of random samples: unrestricted and restricted.

Unrestricted Random Samples

An unrestricted random sample allows all combinations of n items (n equals sample size) an equal probability of selection. For example, taking a sample of three from a population of ten, we can draw 120 different samples. The calculation is as follows:

$$\frac{N!}{n! \times (N=n)!} = \frac{10 \times 9 \times 8 \times 7 \times 6 \times 5 \times 4 \times 3 \times 2 \times 1}{(3 \times 2 \times 1) \times (7 \times 6 \times 5 \times 4 \times 3 \times 2 \times 1)} = \frac{10 \times 9 \times 8}{3 \times 2 \times 1} = \frac{720}{6} = 120.$$

In other words, there are 120 different combinations of size three samples that can be taken from a population of ten.

To take an unrestricted random sample of three from a population of ten, we could put ten tokens labeled A through J in a jar, shake the jar vigorously, and draw one token, say G. Then replace the drawn token, shake the jar, and draw again. If, by chance, we again draw G, we put it back in the jar and draw another token until we have a sample of three different tokens. This procedure assures that each of the possible 120 combinations of three have an equal probability of selection. Any random sample that does not meet this criterion is a restricted random sample.

Another example of an unrestricted random sample involves 2,000 firms in a particular community. To select an unrestricted random sample of 50 firms, we could use a computer program that numbers the firms 0001 through 2000 and randomly select 50. The first number might be 1823, the second 0602, and so on until it has a sample of 50. It is possible that the same number will appear more than once. Computer programs will exclude duplicates and continue the process until it selects 50 nonduplicates.

Restricted Random Samples

A restricted random sample does not give each item in the population an equal chance of selection. Here are some examples of restricted random samples.

Stratified Random Samples

Deviating from the unrestricted random sample makes sense in some conditions. The first of these is a stratified random sample, a procedure in which we create strata (groupings) within a population from which we can take subsamples at random. Stratified sampling is rational and effective when relevant groupings with different variances exist in a population. The size of samples within strata should be proportional, not to the size of the groupings, but to the size of the variance in the stratum. Allocating subsamples of group variance reduces sample error. It also ensures that all groups in the population will be represented in the sample. The obvious advantage of stratified sampling, if the groupings are relevantly homogeneous, is that it assures the reflection of all groups in the sample

Systematic Random Sampling

If we wanted to ensure that a random sample provided representative geographic distribution (or some other criterion), a systematic sample (sometimes called a kth item sample) is appropriate. Systematic random sampling begins with a random start. For example if we wanted a sample of 100 from a population of 5,000, we divide the population size by the sample size, 5,000/100 = 50. After a random start between one and fifty, we would take every 50th (kth item) in the list. If the first item selected from a random start were 35, the next items would be 85, 135, 185, 235, and so on. This method works when there are no *unique periodic* variations in the data.

Multistage Area Random Samples

Another example of a restricted random sample is a multistage area sample. When confronted by the absence of a traditional frame, e.g., a list, an effective alternative is to use a map as a frame. If, for example, we wanted to estimate the unemployment rate in a particular community, we could use census tracts as our frame. Then randomly select blocks from census tracts and randomly select houses from blocks.

Purposive or Judgment Samples

We take a purposive sample by expert opinion. For example, a person regarded as the world's foremost expert in poultry diseases might choose a sample of fowl, not at random, but on expertise and experience. Of course, it is difficult to quarrel with genuinely expert judgment, but the error that results from this method is not random; it is *biased* and, therefore, not predictable, controllable, or reducible.

Convenience Samples

We select some samples because of their accessibility. We call these convenience samples, because they are handy. For instance, a research project concerning methods of teaching reading might take a classroom in a nearby school because it is easily accessible and because the school authorities are sympathetic to the project. Convenience samples are inevitably subject to bias.

This said, much (perhaps most) academic research in the social and behavioral sciences uses convenience samples. Lack of money, time, and other resources are the reasons usually given for using convenience samples. Even with limitations, there are good reasons for treating convenience samples as if they were random. That is, we often treat convenience samples as if they were random samples from a hypothetical population with the same characteristics as the sample. This makes sense if we limit our generalizations to the hypothetical population.

Population Size, Population Variance, Sample Error, and Sample Size

The size of the population is not the determinant of sample error. In a population of 1,000,000, if the variable has zero variance, a sample size of one (1) yields perfect results. Perfect results would still obtain even if we doubled or tripled the population size.

The determinants of sample error are population variance and sample size. Assume a population, (A) (N = 50,000), has an arithmetic mean family income of $30,000 per year with a standard deviation of $4,000. Assume a population, (B) (N = 10,000), has a mean income of $40,000 and a standard deviation of $7,000. Sample error in population B (N = 10,000) is greater than population A (N=50,000). We see the rationale for this assertion more clearly by observing polygons of the two populations, figure 5.1

Figure 5.1 Sample Error Determined by Population Variance and Sample Size

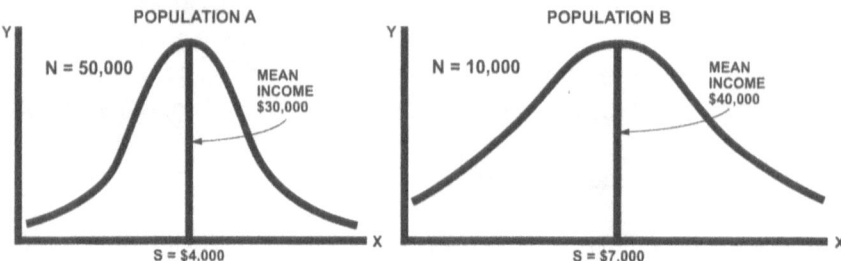

Intuitively, we can see that a random sample taken from the smaller population B is more likely to have a larger sample error than a sample taken from the larger population A because the size of the population does not determine sample error. The smaller population

has a coefficient of variation equal to $\dfrac{S}{M} = \dfrac{7000}{40000} = 0.175$, the standard

deviation divided by the mean. The larger population has a coefficient

of variation equal to $\dfrac{S}{M} = \dfrac{4000}{30000} = 0.133$. We can also see, intuitively,

that sampling from the smaller population B will require a larger sample than from population A because population B has greater relative variance. Thus, population variance and sample size determines sample error.

Summary

The critical concepts that we want you to take away from this chapter are the following:

1. Calculation and meaning of probability

The probability of any outcome A, call it P(A), is the proportion of times A will occur in a random process. The usual example is coin flipping. We express P(A) in a formula as

$$\frac{NumberofTimesEventAOccurs}{InfiniteNumberof\ Re\ peatedTrialsoftheSameKind}$$ (Equation 5.1)

2. Types of samples

There are three types of samples: random, purposive, and convenience. Purposive sampling introduces bias that cannot be estimated or reduced. We can treat convenience samples as random from a hypothetical population with the same characteristics as the sample.

3. Importance of random sampling

Random samples, unrestricted or restricted, produce random error. If the design and execution of the sample eliminates bias and mistake, the only errors remaining in our sample are random, errors that can be estimated and reduced (not eliminated) to an acceptable range of tolerance.

4. Sample (random) error

Random samples produce random error, error attributed to taking the sample. Unlike bias and mistake, we can estimate and reduce random error.

5. Bias and mistake

Bias and mistake are nonsampling errors. Once we introduce bias and mistake, we can neither estimate nor control it.

6. Determinants of random (sampling) error

Random error is determined by population variance and sample size.

Chapter 6

SAMPLING DISTRIBUTIONS, STANDARD ERRORS, AND ESTIMATION

Our discussion of sampling in chapter 5 made the important distinction between bias and mistake (as nonsampling errors) and random error, emphasizing random-selection methods. If we eliminate bias and mistake, making chance the only source of error, we are able to estimate and reduce that error. This chapter focuses on the basic tools of inferential statistics, sampling distributions, and standard errors and the application of these tools to estimate population parameters from sample statistics.

Basic Tools of Inferential Statistics

The Empirical Sampling Distribution of the Mean

Assume that we have a population (N = 10,000) of aptitude test scores measuring capability for a job requiring abstract reasoning. We took these scores, extending over four years, from the personnel files of a hypothetical employment agency.

A census (complete count)[33] of the population yielded a population mean of 100.00 and standard deviation of 17.00. Now suppose that we take 32 unrestricted random samples[34] of 100 from this population

---

[33.] A census in this example was not difficult because we recorded all the scores in a computer readable file.

[34.] Throughout this discussion, we will assume unrestricted random samples. As we discussed in chapter 5, it is sometimes useful to use stratified sampling because costs, information, research interests, and variances differ across strata. In these situations, a stratified sample will be more efficient, i.e., the sample error will be smaller.

and calculated the sample means and standard deviations of each. Table 6.1 displays these 32 means and standard deviations.

Table 6.1 32 Sample Means and Standard Deviations (N = 100)

Means	SDs	Means	SDs	Means	SDs	Means	SDs
99.24	18.97	101.37	18.08	99.35	18.36	101.9	16.25
99.49	17.3	102.35	17.3	99.18	18.23	101.99	16.03
101.17	17.93	100.4	16.44	102.06	17.76	101.67	18.73
99.92	16.45	100.48	18.14	100.16	16.16	99.76	15.85
101.38	16.05	100.11	19.1	101.05	17.66	102.78	17.41
100.14	16.37	98.22	15.98	101.23	16.28	101.58	15.15
101.3	14.48	100.16	17.17	103.1	16.4	100	17.13
100.34	17.19	99.34	17.52	98.9	16.51	103.29	17.97

Using these sample means, we construct a frequency distribution and histogram called the empirical sampling distribution of the means reproduced as figure 6.1.

Figure 6.1 Histogram and Polygon of the Empirical Sampling Distribution of Means

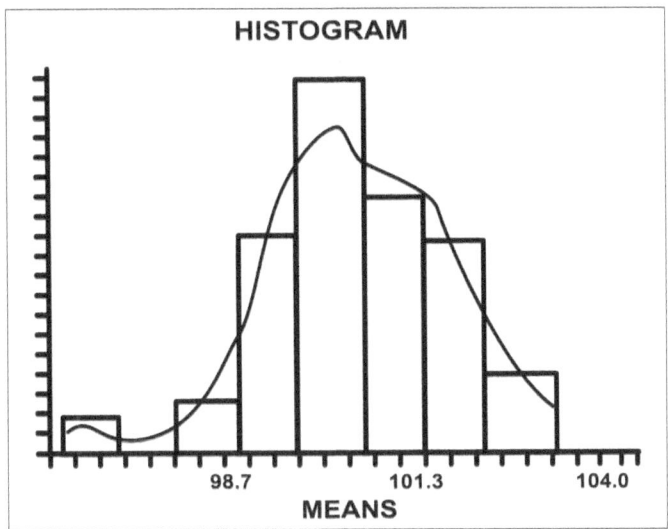

Then we calculated the mean and the standard deviations of these 32 sample means and 32 standard deviations and compared them to the population mean of 100 and population standard deviation of 17.00.

The differences between these two were 0.732 and 0.07 respectively. Assuming no bias or mistake, the differences between the two, the population mean and the mean of the 32 sample means, 0.732, and the population standard deviation and the mean of the standard deviations of the 32 sample means, 0.07, equal chance error, error attributable to the act of sampling.

The standard deviation of our empirical sampling distribution of means, called the empirical standard error of the mean, SE_{EM} , 1.277, is an empirical estimate of random error in the means. Please note: This value does not equal the standard deviation of the variable; it equals the standard deviation of the sample means telling us how much random error is present in the sample means. The E in the sub script of the symbol signifies empirical. Our skewness and kurtosis measures of the 32 means ($Sk = 0.181$; $Ku = - 2.280$) and 32 standard deviations ($Sk = -0.132$; $Ku = -2. 53$ indicate that the distributions approximate normality.

This exercise shows how close the mean of a relatively small number of random sample means and random sample standard deviations (32) come to the actual mean and standard deviation of the population. In addition, it shows the approximate normality of the sampling distributions of means and standard deviations. From this demonstration, we can draw a remarkable conclusion: as we increase the number of samples, say from 32 to 60, the means of the empirical sampling distribution of means and standard deviations will come closer and closer to the population mean, the standard error of the mean will decline, and the distribution will come closer to normality.

The Theoretical Sampling Distribution of the Mean

Now suppose we take all the possible random samples of size $n = 100$ from the specified population and calculated the mean and standard deviation of this enormous set of sample means.[35] Obviously, such a large calculation would take a colossal amount of time and effort, but let us assume we did it. We call this distribution the theoretical (vs.

[35.] A combination is an unordered grouping of a set in which *duplicates are not allowed* and *order does not matter*. The formula for calculating the number of combinations is:

$$C = \frac{N!}{(n!) \times (N-n)}$$

(equation 6.1)

empirical) sampling distribution of the mean and it has the following characteristics:[36]

1. For large samples (sample sizes equal to or greater than thirty), the theoretical sampling distribution of the mean[37] is normal. This holds without reference to the distribution of the population, skewed, bimodal, leptokurtic, and platykurtic. We note that in future discussions of normal distributions, we will refer to them as T or Z distributions.

2. For small samples, sample size less than thirty, the sampling distribution of the mean no longer approximates a normal T (Z) distribution. It is still symmetrical, but with greater variance. We refer to these distributions as t distributions. Figure 6.1 shows examples of these two distributions.

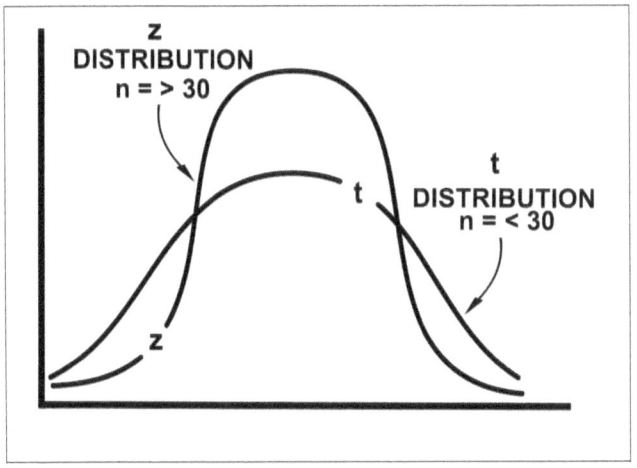

The symbol N is the population total. The symbol n is the sample size. Example: What are the number of combinations of a sample size of two can we take from a population of five:

$$C = \frac{5 \times 4 \times 3 \times 2 \times 1}{(2 \times 1) \times (3 \times 2 \times 1)} = \frac{120}{12} = 10$$

Referring to equation 6.1, you can see how large C would be if the population equaled 10,000.

[36.] We make these assertions without mathematical proof.

[37.] This is also true of many of the other descriptors discussed in chapters 2, 3, and 4, e.g., the proportion, with qualifications and standard deviation. It is not true of the crude range.

3. For sample sizes less than thirty, the sampling distribution becomes more widely dispersed, increasing sample error.
4. The horizontal axis (X) of these distributions shows every possible sample mean of sample size N. This means that any sample mean must fall somewhere on the X axis of the theoretical sampling distribution. Figure 6.2 depicts the sampling distribution of the mean.

Figure 6.2 The Sampling Distribution of the Mean

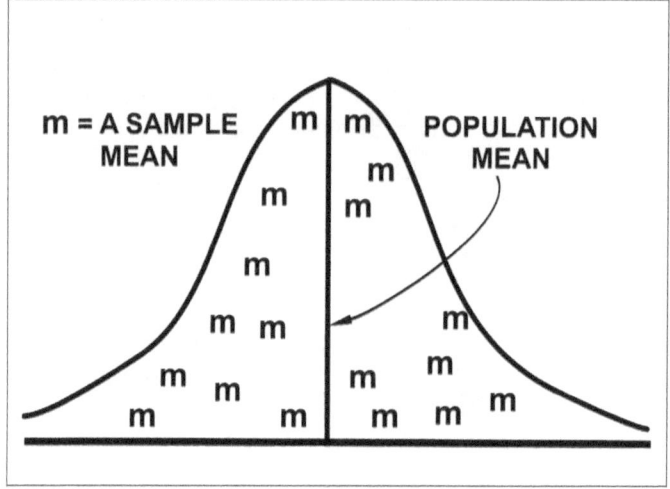

5. The mean of the sampling distribution of the means equals the population mean.
6. The standard deviation of the theoretical sampling distribution of means, called the standard error of the mean, is equal to sample error. We calculate the standard error of the mean as follows:

$$SE_{TM} = \frac{S}{\sqrt{N}} = \frac{17}{\sqrt{100}} = 1.70 \qquad \text{(Equation 6.2)}$$

The equation instructs us to calculate the theoretical standard error of the mean by dividing the population standard deviation by the square root of the sample size. Remember, our calculation produces the standard error, the standard deviation of the sampling distribution, not the population standard deviation. The standard deviation of the population, S_p, is 17. Dividing 17 by 10, we find the theoretical standard error of the mean, 1.70.

7. The standard deviation of a sample provides a good[38] estimator of the standard deviation of the population.

The practical significance of these assertions is that in everyday world, we usually do not know the mean and standard deviation of a population, so what difference does all this make? To answer this question, let us assume that we do not know the population mean, M_p. Nor do we know the population standard deviation, S_p. We draw an unrestricted random sample of 100 from a population and calculate its mean and standard deviation. The sample mean equals 97.68; the sample standard deviation equals 19.07. At this moment, what do we know? We know from assertion 1 that with a sample size > 30, the sampling distribution of the mean approximates a normal distribution. We know from assertion 5 that the mean of the sampling distribution equals the population mean. We know from assertion 4 that our sample mean falls somewhere on the x-axis of this normal distribution. In addition, we know from assertion 7 that the standard deviation of the sample is a good estimator of the population standard deviation.

Estimation

The conclusions in the preceding paragraph represent a substantial accumulation of knowledge. We can estimate the standard deviation of the population by using a sample standard deviation, (assertion 7). Here, we encounter one of those situations in which equation for the calculation of a sample statistic differs from the calculation of a population parameter. We calculate a sample standard deviation using the equation shown below.

$$S_s = \sqrt{\frac{\Sigma d^2}{N-1}}$$ (Equation 6.2)

Note that the difference between the calculation of the population standard deviation, S_p and the sample standard deviation, S_s, is in the use of N - 1 in the denominator.

---

38. There are many estimators for any parameter. Therefore, the choice of an estimator requires a set of criteria. The term *good*, as used here, means that the mean of the sampling distribution of the means is the population mean. The term *good* means that the estimator is unbiased. *Good* also means that our estimator is efficient in the sense that the estimator will minimize the standard error. Finally, *good* means consistent, i.e., the statistic will equal the parameter when the sample size equals the population size or $N_s = N_p$.

The reason for using N -1 lies in the notion of degrees of freedom, df. We can understand this concept in the following way: If the sum of any four observations equals twenty-five, we can specify any three values of the four freely or randomly, but the fourth must satisfy the restriction of the total to 25. Specifying the sum, 25, and randomly choosing values of 6, 10, and 2, removes one degree of freedom for values to enter the sample, i.e., the fourth observation must equal seven. We have lost one degree of freedom in randomly choosing observations. In computing the sample standard deviation, S_s we need to know the sample mean. We obtain the sample mean from the total sample size N. This uses up one degree of freedom. Hence, when we compute the sample standard deviation and variance, we have only N - 1 degrees of freedom remaining. The use of N - 1 has proven itself in providing the best estimate of Ss.

Therefore, we can estimate the standard error of the mean, Se_M, based on a random sample of 100 with S_s = 19.07 and M_s = 97.68. The estimate of the sample standard error of the mean is

$$Se_M = \frac{Ss}{\sqrt{N}} \qquad \text{(Equation 6.3)}$$

$$= \frac{19.07}{\sqrt{100}} \text{ or about } 1.91.[39]$$

By definition, because our sample mean lies somewhere on the x-axis of the sampling distribution of the mean and the sampling distribution of the mean approximates normality, and, because we have a good estimate of the standard error of the mean, we can make a probabilistic estimate of the population mean.[40] We call this estimate a confidence interval estimate. If we add ± two standard errors (2 × 1.91), 3.82 to the sample mean, 97.68, this interval, 101.5 - 93.86, provides a 95.% confidence interval estimate of the population mean.

This interval represents sample error. If we add ± 3 standard errors to the sample mean, this interval provides us with 99+% confidence that the

---

[39] For small samples (N < 30), the sampling distribution of the mean is a t distribution. That is, as the sample size gets smaller random error gets larger.

[40] We could simply make a "point" estimate of the parameter using the sample mean. The problem with the point estimate is that it does not tell us how much chance error is in the estimate.

population's mean falls between 103.41 and 91.95. Why? Because +/- 2 standard deviations in a normal curve encompasses about 95% of the observations and +/-3 standard deviations includes 99+% of the observations in the distribution. We can decrease the sample error by increasing the sample size because such an increase reduces sample error.[41]

Figure 6.3 illustrates this idea by showing a range of two and three standard errors from the mean. Please note: while these confidence intervals give us 95% (two standard errors) and 99+% (three standard errors) assurance that our interval estimate encompass the population mean, we still do not have certainty because 4.55% and <1% of the time our estimates will miss the population mean.

Figure 6.3 Estimation with 95 and 99% Confidence Intervals

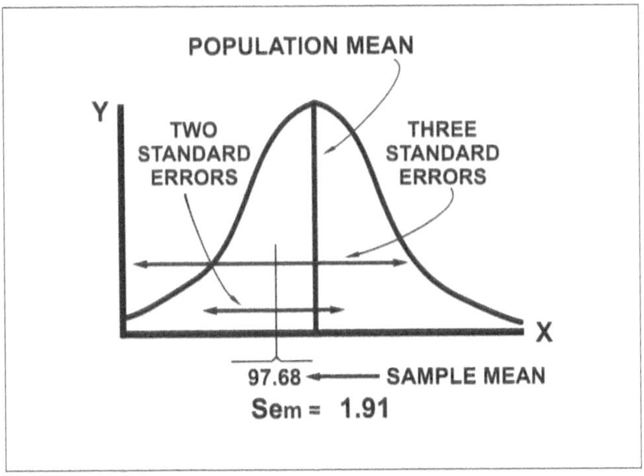

Other Sampling Distributions

The Theoretical Sampling Distribution of Proportions

Earlier we said that the notion of sampling distributions applies to other descriptive parameters including proportions, medians, and standard deviations. In this section, we address the most commonly used parameters aside from the mean, proportions and medians. We use the symbol P for proportions.

---

[41] Or, to put it more precisely, the standard error declines as a function of the square root of the sample size. Alternatively, to reduce error by half, we would have to quadruple the sample size.

Some of the characteristics of the sampling distribution of means apply with equal force to the sampling distributions of proportions. There are, however, critical exceptions. We outline the characteristics of the sampling distribution of proportions as follows:[42]

1b. When the population proportion equals 0.50, a sample size equal to or greater than 30 permits us to assume that the sampling distribution of proportions follows a normal distribution.

2b. If a population proportion lays between 0.40 and 0.60, a sample size equal to or greater than 50 permits us to assume that the sampling distribution of proportions follows a normal distribution.

3b. When the population proportion lies between 0.30 and 0.70, a sample size equal to or greater than 80 permits us to assume that the sampling distribution of proportions follows a normal distribution.

4b. When the population proportion lies between 0.20 and 0.80, a sample size equal to or greater than 200 permits us to assume that the sampling distribution of proportions follows a normal distribution.

5b. When the population proportion lies between 0.10 and 0.90, a sample size equal to or greater than 600 permits us to assume that the sampling distribution of proportions follows a normal distribution.

6b. When the population proportion lies between 0.05 and 0.95, a sample size equal to or greater than 1600 permits us to assume that the sampling distribution of proportions follows a normal distribution.[43]

7b. The sample proportion, P_s, represents a "good" estimator of the population proportion P.

---

42. From your high school algebra, this is an application of the binomial expansion. If we raise (0.5 + 0.5) to the 30th power the resulting expansion approximates normality. The binomial $(0.4 + 0.6)^{50}$ must be raised to the 50th power to approximate a normal distribution.

43. The first six assertions are taken from Cochran, William G. Sampling Techniques. New York, John Wiley, and Sons, 1953, p. 41, cited in KGF p. 160.

8b.   the sample standard error of proportions, Se_p equals:

$$Se_p = \sqrt{\frac{P*Q}{N}}$$ (Equation 6.4)

To estimate the standard error of P, Se_p, the equation instructs us to multiply the sample P_s by (1 - P) or Q, divide that product by N, the sample size, and extract the square root of that quotient.

We note that understanding equation 6.4 requires returning to chapter 3 and our discussion of the variance and standard deviation of a nominal variable. We showed that the variance of a nominal variable coded one and zero (a percentage) is P × Q. Dividing P × Q by N gives the standard deviation of the nominal variable. Taking the square root of the standard deviation gives the standard error. Compare that to the calculation of standard error of the mean.

Interval estimation of a population proportion, P_p from a sample proportion, P_s, then, is straightforward. Suppose we take an unrestricted random sample of fifty and calculate the sample P = 0.42. Because the sample p falls between 0.40 and 0.60 we can make the reasonable inference (see assertion 2b) that a sample size of 50 allows us to assume that the sample sampling distribution of p is approximately normal.

Based on that assumption we calculate the standard error of P,

$$Se_p = \sqrt{\frac{0.42*0.58}{50}} = 0.07.$$ Adding and subtracting 2 standard errors, Se_p

(0.14) to the sample P, produces an interval estimate of 0.28 to 0.56 and a 95.45% confidence interval estimate of the population proportion, P. Alternatively, we can affirm, with a probability of 0.9545, that the size of the error does not exceed 0.28. If that range of error exceeds our requirements and if resources permit, we can reduce the error by increasing the sample size.[44]

---

[44]   Suppose that in this example, your needs could not tolerate an error greater than +/- 5 percent. This would mean an increase in sample size. How big would the sample have to be to give an error of +/- 5%? We could return to equation 6.4, specify our error tolerance, and solve for N.

$$N = \frac{p \times q}{(s_p)^2} = \frac{0.42 \times 0.58}{.05^2} = 98$$

If our sample P_s fell within the range of 0.30 to 0.70, in order to assume normality of the sampling distribution, we would have to increase our sample size to at least 80 (see assertion 3). Most statistical programs will make these interval estimates adjusting the standard errors to the sample size, even when we cannot assume normality of the sampling distribution.

The Theoretical Sampling Distribution of the Median

Fortunately, when we have a large sample, e.g., 50, the sampling distribution of the median follows the normal distribution. The equation for the standard error of the median, SE_{Md} equals:

$$Se_{Md} = 1.25 * \frac{S_S}{\sqrt{N}}$$
(Equation 6.5)

The equation instructs us to divide the sample standard deviation by the square root of N, the sample size, and multiply that quotient by 1.25. The standard error of the median is subject to greater random variation than the mean.

Summary

The important ideas discussed in this chapter are the following:

1. Sampling distributions

    Sampling distributions represent the bedrock of statistical inference. In developing the notion of an empirical sampling distribution of means we showed that taking 32 random samples of size 100 from a population of 10,000, calculating the means and standard deviations of each sample, and averaging these means and standard deviations, produces values close to the population means and standard deviations of the population. In addition, the distributions of the 32 averages and standard deviations approximate a normal distribution. In turn, we asserted that if we took all the possible samples of size 100 and calculated the means of those samples, this would produce the theoretical sampling distribution of means, a normal curve with a mean equal to the population mean. The standard deviation of the theoretical sampling distribution of means, called the standard error of the mean, equals the random error in the sample means.

2. Standard errors: Standard errors measure the error in the sample means, proportions, and medians attributable to sampling. We estimate standard errors by using the standard deviations of samples.

3. Probabilistic interval estimates: We can make interval estimates with varying degrees of confidence using our sample values—means, proportions, medians—and standard errors.

It is critical to note that the estimation procedures using sample data discussed in this chapter do not eliminate risk. With an interval estimate, we can say that within two standard errors we are 95.45% confident that our sample statistic lies between two values; we cannot say that we are 100% certain. There is still a 4.55% probability that we are wrong. Our technical apparatus has enabled us to make estimates that assign probabilities to our estimates and that is all we can ask (although it is quite a lot) from the inductive, inferential statistical method.

We base probabilistic estimation on two vital concepts: the sampling distribution and the standard error.

Appendix

Chapter 6

The data set represents three random samples of size 50: Variables X1, X2, and X3 The first column is an identification number. Your tasks: (1) Using your computer program, for samples 1 and 2, make interval estimates of the mean and median between which you are 95.45% and 99+% confident; (2) for variable 3, make an interval estimate of the proportion between, which you are also 95.45% and 99+% confident. Explain how the respective standard errors provide that degree of confidence.

ID	X1	X2	X3
1	87.9	77.03	0
2	86.59	85.9	1
3	128.88	99.09	1
4	66.71	73.14	1
5	108.45	101.79	0
6	109.17	113.59	1
7	76.57	77	1
8	89.7	112.2	0
9	136.06	115.91	1

10	98.73	105.8	0
11	105.22	98.58	1
12	96.6	117.84	1
13	98.71	85.63	1
14	132.81	76.24	1
15	85.23	105.92	0
16	96.39	108.42	0
17	81.26	82.54	0
18	109.13	96.84	0
19	110.87	114.26	1
20	108.66	66.95	1
21	87.39	70.64	1
22	111.12	106.19	0
23	92.64	84.07	0
24	106.13	93.14	1
25	101.08	107.71	1
26	131.18	122.15	1
27	98.07	112	1
28	91.76	110.37	1
29	102.35	101.9	0
30	85.05	89.55	1
31	79.36	81.7	1
32	101.17	96.89	0
33	128.02	108.39	0
34	129.02	102.67	1
35	123.66	117.44	0
36	112.12	105.89	0
37	108.19	113.82	1
38	125.46	91.91	1
39	131.94	110.75	1
40	74.69	99.89	1
41	84.29	100.26	1
42	100.44	110.7	0
43	113.53	96.79	0
44	121.85	105.3	1
45	65.49	86.36	0
46	141.43	80.42	1
47	116.36	118.42	0
48	122.56	112.39	1
49	139.32	120.85	1
50	89.06	103.09	0

CHAPTER 7

TWO ALTERNATIVES: HYPOTHESIS TESTING AND STATISTICAL SIGNIFICANCE

In this chapter, we introduce the problem of choosing between two alternatives in the face of uncertainty. As an example, organizations frequently must decide whether to advertise using television or direct mail or whether to adopt a new system for responding to customer inquiries or stick with the old. No matter what basis an organization uses to make these choices, the enterprise always faces the risk of making the wrong choice. When we use sampling to guide these decisions, the testing process is called hypothesis testing.

A hypothesis is a conjecture expressing an opinion based on incomplete evidence. We can base this conjecture on theory, empirical evidence, a confederacy of hunches, or a rough guess. Statistical hypothesis testing provides the tools for minimizing the probability of making the wrong choice. While this technology does not free us from risk of making the wrong choice, it enables us to specify and control the risk of making the right choice.

Type I and Type II Errors

Before we begin a discussion of the tools (the technical apparatus) used to minimize the risks of making wrong choices or maximizing the probability of making the right choices, we need to consider two important ideas. First, using sample data, whatever we choose, we run the risk of being in error because we base our conclusions on inductive probabilities, not certainties.

Second, when using samples to make choices, there are two kinds of errors and, by necessity, two kinds of risk. The first error is concluding that a difference between alternatives is *real* when chance actually caused the difference. For example, if we use a sample to test the effect of a new method against an old method and our sample shows that the new method is better than the old, how do we know that the difference between the two is not attributable to chance? If we conclude that the new method is better and it is not, we have committed a type I error, i.e., saying that the difference is real when it is actually attributable to chance.

Alternatively, the second error is concluding that chance caused the effect when it is actually *real*. This is a type II error. Thus, the type I error is saying there is difference when it actually is attributable to chance; a type II error is saying that an effect is attributable to chance when the effect is real. Although it is true that type II errors can be just as costly as type I errors, generally our position is conservative; we would rather miss an effect when it is real (type II) than claim an effect is real when it is not (type I). For example, if we were testing the effect of a new treatment of a disease, we would rather miss a real positive effect of the treatment than claim a positive effect when the effect is due to chance.

Alpha and Beta Risks

These two errors, types I and II, carry with them risks. The risk of making a type I error is called a (alpha); we call the risk of making a type II error β (Beta).

Testing the Difference between a Sample Statistic and an Assumed Parameter

Suppose that an enterprise wishes to improve its telephone customer service by reducing the time that customers must stay on the line before service representatives answer. Customer surveys consistently show that waiting time is an issue of major concern to inquiring or potentially new customers. Management literature tells the firm's managers that, on average, four minutes are the maximum acceptable waiting time. Waiting more than four minutes causes customers to lose patience and hang up.

If, on the average, customers are waiting longer than four minutes and the resulting revenue loss exceeds the cost of a new answering system, management must develop an alternative. So if management implements a new system when the waiting time is acceptable (actually four minutes or less), they have made a type I error. If they decide not to put in place a new system when customer-waiting time exceeds four minutes, they have made a type II error.

To test the hypothesis that that the mean waiting time is greater than 4.0 minutes, management employs a consultant who draws a random sample of 100 calls and calculates the sample mean (M_s) and standard deviation (S_s); M_s, is 4.60 minutes and S_s, is 2.7 minutes.

If the sample mean taken by the consultant is not attributable to chance, the firm's customers (or potential customers) are, on average, waiting more than 36 seconds (six tenths of a minute) for a representative to answer, a longer time than acceptable. The dilemma facing management centers on whether the difference between an acceptable mean of 4.0 (the maximal acceptable time that a customer is kept waiting for someone to answer) and the sample mean, 4.60 is real or probably attributable to chance?

We will use this example to illustrate how we minimize the probability of making the wrong choice.

Step 1: Generate the Hypothesis

Generating the hypothesis requires an evaluation of risk. Would we rather say the effect is real when it is due to chance (a type I error) or say the effect is due to chance when it is real (a type II error)? Which is most costly, type I or type II? That question is answerable only in terms of the consequences of changing or not changing the process. Let us say that we want to minimize the risk of changing to a new system when we do not need it.[45] There are several possible reasons for choosing this approach. The first is, in the absence of any other information, it is easier, less costly in resources, training, and employee morale, to stand pat even if change is better, than to change to something that does not work. A second reason for choosing to miss an effect when it is real rather than changing to a new system that is not better, is that management has valid and reliable data showing that the marginal loss of customers and revenue ascribable to waiting too long is less than installing a new system that does not work better.

---

45.   This is typical of how most people approach decisions requiring any kind of change

Given that choice, we formulate something called the null hypothesis. The null hypothesis is that any observed difference is due to sampling error. Here is an example of a null hypothesis.

> H_O: The difference between a safe mean (4.00) and our sample mean is probably zero or is probably due to chance. We can operationally translate the null as $H_O = 4.00$ or $H_O = D = 0$

The symbol H_O signifies the null hypothesis. This convention does not necessarily reflect what management actually believes is true. We use the null convention because it makes it easy to formulate the alpha risk.

Step 2: Devise an alternate hypothesis

We couch the alternative hypothesis so that the rejection of the null is equivalent to acceptance of the alternative. In this case, we state the alternative as: The observed sample mean is greater than the safe mean and this difference is not the result of chance.

H_A: The actual mean is greater than 4.00

The symbol H_A signifies the alternative hypothesis. In this example, we have expressed the alternative as greater than, called a one-tail test. In other words, if we reject the null, we are saying that we accept that the mean of the population is greater than the hypothesized mean.

Alternatively, we can specify the alternative that sample mean is not equal to ($\neq$) the safe mean, a two-tail test. We chose the one-tail test because a sample mean less than 4.00, leaves management home free. The reason for using a one-tailed test is that management's concern is with customers having to wait too long. If they are waiting less than four minutes, then all is well. As we shall see, the way we state the alternative is critical.

Step 3: Specify the alpha (a) risk

In this step, we establish the acceptable probability of committing a type I error, saying that an effect is real when the effect is due to chance. Again, the value assigned to the alpha risk depends on the consequences of making a type I error. The firm will make a major revision in the customer calling process based on this finding so it is vitally important not to make a type I error. To avoid making a type I error, we set the alpha risk at 0.025. We call this probability the test's level of significance.[46] With this level of significance, the firm's managers are saying that they want to be 97.5% confident that the waiting time is greater than 4.0 minutes.

The firm could set the risk at 0.01, but that would increase the beta risk, which depends on other possible values of the population mean, a point explained later.

Step 4: Make the Test

This step follows taking the sample. First, we compute the sample mean, standard deviation, and standard error. Our computed sample mean is 4.60 minutes and the sample standard deviation is 2.7 minutes. The standard error of the mean is

$$Se_M = \frac{S_S}{\sqrt{n}} = \frac{2.7}{\sqrt{100}} = \frac{2.7}{10} = 0.27 \qquad \text{(Equation 7.1)}$$

We then take the difference between the sample mean, 4.60 and hypothesized mean, 4.00 (4.60 - 4.00) = 0.60) and divide by the standard error, 0.27 to yield—2.22. The way we express this calculation is

$$Z\ (T) = \frac{D_S}{Se_M} = \frac{0.60}{0.27} = 2.22$$

Z equals how far the sample mean falls from the hypothesized mean in a normal distribution.

Remember a critical piece of information: If the sampling distribution contains *all* possible sample means of N = 100, the sample mean, 4.60, must fall some place on the x-axis of the sampling distribution. As a reminder, here is a picture of the Sampling Distribution and how far our sample mean falls from the hypothesized mean of 4.00.

[46.] Note: The term significance means that the difference between the hypothesized mean and the sample mean probably did not happen by chance. The term carries no message as to the degree to which the difference is important. A difference can be statistically significant but behaviorally trivial. For an elaboration of this point, see below.

Figure 7.1 Sample Mean Falls +2.22 Standard Errors form the hypothesized Mean

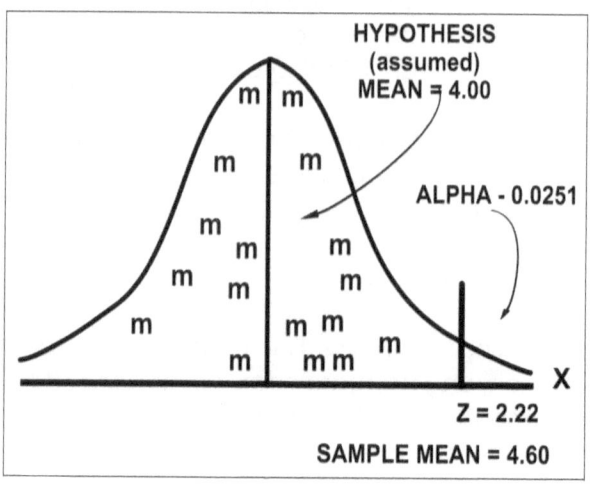

Because the sample size is 100, we use the symbol Z (T) to indicate that the sampling distribution of the mean is normally distributed. If the null hypothesis is actually true, this calculation produces a Z score that will yield the probability of our making a type I error. If the sampling distribution of the mean is normal and if the population's mean is really 4.00, then this sample mean, 4.6, lies 2.22 standard errors from the population mean. Since 2.22 is larger than T = 1.6449, we reject the null and accept the alternative. We use T = 1.6449 because this is a one-tail test, i.e., we are interested only if our sample mean is greater than 4.00. With an alpha risk of 0.025, 97.5% of the observations in a normal distribution fall below the Z value of 1.6449. Only 2.5 percent fall above 1.6449. A sample mean this large could happen purely by chance only 13 times in 1,000. To put the issue another way, what is the probability of getting a sample mean this large, 4.60, purely by chance, if the true mean is 4.00? The answer is 0.013209 or 132 times in 10,000. A betting person would say that it is highly unlikely that the true waiting time is, on average 4.00 minutes or less. It is more likely that the average waiting time is closer to 4.6 minutes. Therefore, we reject the null hypothesis and accept the alternative.

Again, the reason that we are interested in only one end of the sampling distribution is that management's concern is waiting times in excess of 4.00 minutes. If the sample mean had been 3.4 minutes, the process would have stopped there.

Testing the Difference between Two Sample Means

The result of the test confronts management with a dilemma. Should they stand pat on the waiting time or try to reduce it. Because management believes that a 4.6 waiting time will cost more in lost revenue than it will cost to fix it, they commission another consulting firm to devise a solution.

The new consulting firm develops a revised answering system, tests the new system by taking another random sample of 100 representatives, training the representatives in the new system, and measuring waiting times. The new sample's mean, M_s is 3.8 minutes with a sample standard deviation, S_s of 2.00 minutes. Now we have two sample means (4.6 and 3.8) raising the question as to whether this improved waiting time difference is *real* or attributable to chance.

In this situation, we are dealing with a *sampling distribution of mean differences* between two sample means. If we take a large number of sample means and calculate the differences between each pair of means, we can organize these differences as a frequency distribution, and estimate the standard error. We call this distribution the sampling distribution of mean differences. The arithmetic mean of the sampling distribution of differences is the true difference between the samples, and the standard deviation of the sampling distribution of differences is the standard error of the mean differences (Se_{Mdiff}). Figure 7.2 illustrates a null hypothesis of no difference.

Figure 7.2 Mean differences

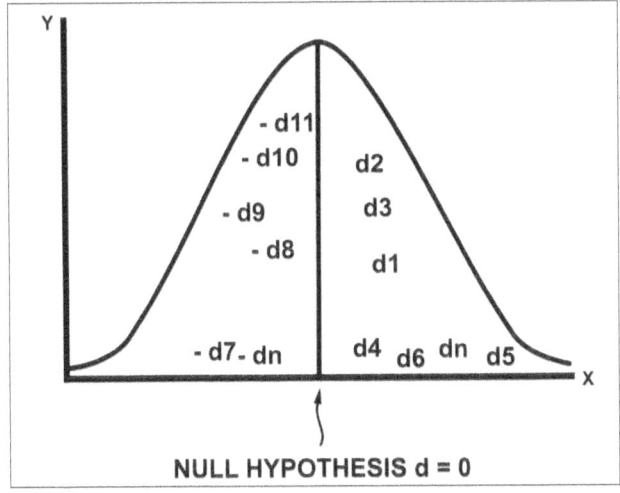

We estimate the standard error of mean difference (difference between two sample means) using the variances of the two samples, equal to the following:

$$Se_{Mdiff} = \sqrt{\frac{S_{S1}^2 + S_{S2}^2}{(N_1 - 1) + (N_2 - 1)}}$$ (Equation 7.2)

where Se_{Mdiff} is equal to the standard error of mean differences; S^2s_1 the variance of the first sample, N_1 equals the sample size of the first sample, S^2s_2 the variance of the second sample, and N_2 the sample size of the second sample.

Here is the test. The null hypothesis, H_o is $M_1 = M_2$. The alternative hypothesis H_A is $M_2 < M_1$, or the difference between the mean of sample 2, 3.8 is less than the mean of sample one 4.60. The alpha risk is 0.01.

Our two sample means are $m_1 = 4.60$ and $m_2 = 3.80$. Our sample standard deviations are $s_1 = 2.7$ and $s_2 = 2.0$. Using the equation given above, we calculate the standard error of mean difference as follows:

$$Se_{mdiff} = \sqrt{\frac{7.29 + 4.0}{100 - 1 + 100 - 1}} = \sqrt{\frac{11.29}{198}} = \sqrt{0.057} = 0.2388$$

The question: Is the difference between the two samples attributable to chance or is it statistically significant?

We answer the question by taking the difference between the two sample means, 3.8, and 4.6, 0.8 and dividing by the standard error of the mean difference.

$$Z = \frac{M_2 - M_1}{Se_{Mdiff}} = \frac{-0.80}{0.2388} \approx -3.35$$

Our interpretation is this: if the sample means do not differ, what is the probability of finding a difference this large purely by chance? If there is no difference, i.e., if the true difference is zero (0), our sample difference lies about -3.35 standard errors from zero. The probability of a difference this large occurring by chance in a one-tail test is less than 0.0004, certainly less than our alpha risk, 0.01. Therefore, we reject the null hypothesis and accept the alternative. Given this information, management would probably implement the new calling system.

Testing the difference between sample proportions

There are many other parameters that we can estimate and test with the same methods spelled out in this chapter. The most often used is the proportion. We give the equations for the standard errors below.

The Standard Error of Proportions

The standard error of a proportion is

$$Se_p = \sqrt{\frac{PQ}{n}}.$$
(Equation 7.3)

This equation permits the construction of confidence limits of p and tests of hypotheses involving a statistic and an assumed parameter. We discussed the Se_p in chapter 6.

The Standard Error of Proportion Difference

The standard error of a proportion difference is

$$Se_{pdiff} = \sqrt{\frac{s_{1p} + s_{2p}}{(n_1 - 1) + (n_2 - 1)}}.$$
(Equation 7.4)

We use this equation in testing the significance of difference between two sample proportions. In chapter 6 we gave the sample sizes for various levels of P necessary to assume normality of a proportion's sampling distributions. If sample sizes fall below that level, we cannot assume a normal distribution, and must use another method described in the next chapter.

More on the Notion of One and Two Tailed Test

The One-Tail test

Both of our examples used the one-tail test because we believe that the circumstances underlying the hypotheses test strongly suggest a directional alternative. A two-tailed test is a confession of ignorance, a condition in which an analyst is simply unable to hypothesize the direction of the difference or effect. Our position contradicts the usual conservative practice of using two-tailed tests. In other words, if the circumstance surrounding a test dictate or if you have good reasons to believe that you know the direction of a difference or effect, then the one tail test is correct and, in our opinion, preferable.

With the one tail test, our analytical focus is on only one end of the sampling distribution. In this context, the Z score that emerges from the analysis is a marker that enables us to calculate the percentage of observations in a normal distribution that fall outside the marker and, by definition the percentage of observations in the normal curve that fall inside the marker. For example, if we define the alpha risk as 0.05 in a one-tail test, the Z score that is the appropriate criterion for rejecting the null is 1.6449. For rejecting the null at alpha = 0.01 and 0.001, the critical Z values are 2.3269 and 3.0902 respectively. Negative values for Z represent the same probabilities at the other end of the normal curve. If, of course, your Z score is even larger than those specified say 4.3567, you can easily calculate the alpha risk (0.000007) using one of the applets available on the Net. Figure 7.3 illustrates the notion of the one-tail test.

Figure 7.3 The One-Tail Test

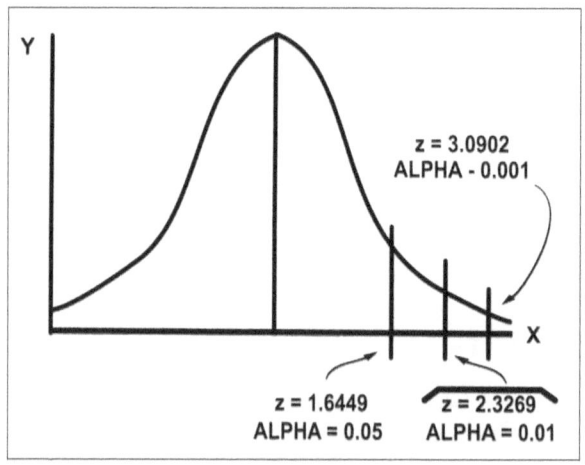

With a two tail-test, we are, by definition, interested in both ends of the sampling distribution. For example, given an alpha risk of 0.05, we are interested in the Z score outside of which five percent of the observations fall. The Z score marker that meets that requirement is 1.96. In a normal distribution 2 1/2% of the observations fall outside at $Z = -1.96$ and 2 1/2% at $Z = +1.96$. For alpha = 0.01 and 0.001, the Z score markers are -2.5758 and +2.7758 and -3.2905 and +3.2905 respectively. These illustrations do not preclude alpha risks at 0.1 or 0.2 if appropriate. Figure 7.4 helps differentiate the one tail and two tail tests.

Figure 7.3 Two-Tail Test

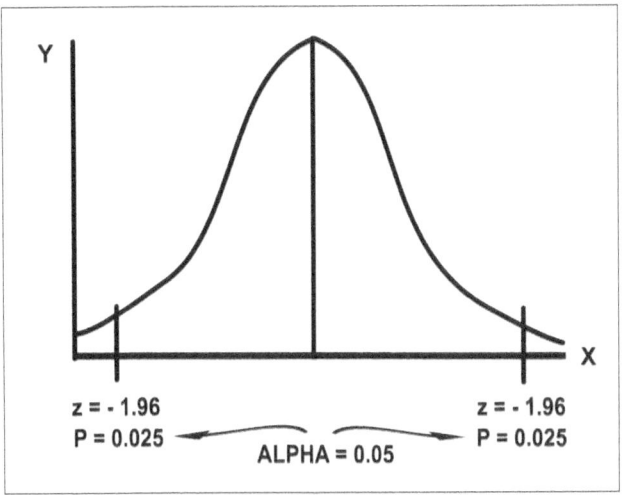

Again, there are many free applets available in the Net that will make the calculation of the proportion of observations in a normal distribution falling outside or inside any value of Z.[47]

More on the Idea of the Beta Risk

Earlier, we said that the beta risk is equal to one - alpha risk. Our interest in the beta risk, however, extends beyond that simple statement. We need to think about the type I and type II errors a little more carefully because they depend upon how we formulate the null hypothesis, H_o. In our example of telephone times we formulated the hypothesis that

---

[47.] We like David M. Lane's *HyperStat Online*. Lane's e-mail address is dmlane@davidmlane.com. You can download a large number of free statistical analytical tools from this Web site.

the telephone system was OK. If we had formulated the hypothesis that the system was not OK, this reformulation would reverse the type I and type II errors.

Consider again our example of telephone waiting times in which we tested the difference between two sample means, 3.80, and 4.60. In that example we saw that the probability of committing a type I error, switching to a new system, when, actually, the new system was no better than the old, was 0.000568, thus, causing the firm to make the switch.

The probability of committing a type II error is the probability of retaining the old system when the new is better. That probability depends on the actual value of the population mean waiting time. Let us suppose that management would have implemented the new call system even if the system had only reduced the mean to 4.00. What is the beta risk, i.e., the probability of accepting the null if the true mean is 4.00, high enough to cause management to implement a new call system.

So we need to calculate the beta risk if the true difference is - 0.20. The Z score is

$$Se_{Mdiff} = \sqrt{\frac{Ss_1^2 + Ss_2^2}{(N_1 - 1) + (N_2 - 1)}} = \sqrt{\frac{2.7^2 + 2^2}{99 + 99}} = \sqrt{\frac{11.29}{198}} = 0.238$$

$$Z = \frac{d}{S_{Mdiff}} = \frac{3.8 - 4.0}{0.238} = \frac{-0.2}{0.238} = -0.84$$

If the true mean was actually 4.00, the Z score for a difference of -0.20 = - 0.84, meaning that the alpha risk is 0.590. Thus, the beta risk if the true mean is 4.00 is one - alpha (one - a) or 0.41 because the beta risk is calling a true effect a random effect. We could go on calculating the beta risk for values of 4.10, 4.20, and 4.30 minutes if these values were also not acceptable to management.

More on the Notion of Statistical Significance

When we say that a difference between a statistic and a parameter is significant, we mean that the difference probably did not happen by chance. Please note that statistical significance is not the same thing as behavioral significance. Behavioral significance means that if we act on the information that a difference or an effect is probably true, our action will result in making a practical difference in whatever process we are testing. Or, as William James put it, that which makes no difference, makes no difference. Will a reduction of six tenths of a minute result in better customer relations, fewer switches, and greater profit?

To put the issue another way, with very large samples almost any difference, however small, will be statistically significant, probably not attributable to chance. A small difference, although probably not a chance effect, might have no behavioral or practical difference. The point is that some statistically significant findings are trivial, not practically important. Conversely, with small samples, e.g., in some medical research, large differences, not statistically significant, may demand further investigation with a larger sample. The point: some research can produce interesting, but insignificant behavioral results that we should not cavalierly discard without recommendations for continued study with larger samples. Do not confuse statistical significance with behavioral significance.

Hypothesis Testing and Theory

We need to insert another important caveat. Willy-nilly testing of a large number of differences, with the hope of finding something, is taboo. These kinds of fishing trips will result in statistically significant effects (the additive effect of independent probabilities), purely by chance. Therefore, we must accompany significance tests by, at minimum, good hunches. Hypotheses about differences or effects are not as that Mandrake the Magician pulled out of a hat.

Our position is that all inquiry emerges from confrontation with some blockage of a felt need. As our first rational response to goal-blocking problems, we formulate some explanation of the forces causing the problem. Formally, we call these explanations theories. Useful theories, theories that are not tautological, lead to testable hypotheses that yield solutions. Theories that do not logically produce hypotheses may be interesting, but from a scientific point of view are circular.

Summary

The important ideas in this chapter are the following:

1. Type I and type II errors

   The type I error is concluding that a difference between alternatives is *real* when chance caused the difference. The type II error is concluding that chance caused an effect when it is actually *real*.

2. Alpha and beta risks

   The alpha risk is the probability of making the type I error. The beta risk is the probability of making the type II error.

3. The null hypothesis

   The null hypothesis is that any observed difference is due to sampling error. This convention does not necessarily reflect what we actually believes is true. We use the null convention because it makes it easy to formulate the alpha risk.

4. The alternative hypothesis

   We couch the alternative hypothesis so that the rejection of the null is equivalent to acceptance of the alternative

5. standard errors of difference.

   The standard errors of difference are the standard deviations of sampling distributions of differences (means, proportions, medians).

6. One-tail tests

   We use one-tail tests when we have strong enough theory to hypothesize the direction of differences.

7. Two-tail tests

   We use two-tail tests when we cannot hypothesize the direction of differences.

8. Statistical significance

When we say that a difference is statistically significant, we mean that the difference is probably not due to chance. Statistical significance is not the same thing as behavioral significance. Behavioral significance means that if we act on the information that a difference or an effect is probably true, our action will result in making a practical difference in whatever process we are testing.

Finally, at the continued risk of nagging, all our listed statistical packages can do all these calculations. Your job is not memorizing each equation, but interpreting a program's output. Effective interpretation requires understanding of what the program is doing.

Chapter 7

Appendix

The data set represents five random samples with sample sizes of 30. The variable labeled Single Sample was taken this month. The four variables labeled sample1 and sample2 and sampleA and sampleB were also taken at approximately the same time this month. The quantities in the first three variables are Cardinal; in the last two the variables are nominal. Your tasks are the following:

1. Assume that the sample labeled Single Sample represents income in thousands of dollars taken from a population of $N = 10,000$. A census of this population taken four years ago showed the mean income of the population to be 125,000 with a standard deviation of 20,833. Your sample of 30 was designed to test the hypothesis that the mean income had changed.

2. Assume that the two unrestricted random samples labeled sample1 and sample2 represents income in thousands of dollars taken independently from two communities, 1 and 2. Test the hypothesis that the mean income of the two communities are not different.

3. Assume that the two nominal variables are unrestricted random samples of political preference from two communities, A and B of

about the same size where 1 = Democrat and 0 = Republican. Test
the hypothesis that there is no difference in political preference
between the two communities.

Single Sample	Sample1	Sample2	SampleA	SampleB
107.77	99.07	207.1	0	0
120.46	104.22	-18.61	0	0
115.14	134.94	96.36	1	1
96.04	104.13	221.44	0	1
78.2	122.78	-45.06	1	1
118.72	109.13	-161.81	0	0
103.31	134.9	78.93	0	1
109.51	131.26	164.41	0	0
131.16	139.7	-55.7	1	0
154.16	118.98	131.53	0	0
110.28	127.11	-65.62	1	0
86.7	89.7	134.47	0	1
126.51	145.59	270.29	0	0
139	143.06	-30.74	0	1
146.19	111.32	182.23	0	1
112.27	123.89	228.43	1	1
115.48	93.15	-142.24	1	0
134.18	118.98	7.32	0	1
108.88	121.4	236.22	0	0
122.84	125.85	332.35	0	0
110.65	106.43	145.98	0	1
63.53	100.67	15.02	0	1
129.79	129.33	299.65	1	0
133.77	153.88	-3.45	0	0
158.19	137.74	388.93	1	0
110.58	94.73	17.43	0	1
142.37	124.46	168.04	1	0
130.39	147.52	60.04	0	0
142.38	139.4	-24.21	0	0
117.42	95.04	-74.85	0	0

CHAPTER 8

HYPOTHESIS TESTING:
MORE THAN TWO ALTERNATIVES

In chapter 7, we focused on testing hypotheses concerning two alternatives: a sample statistic and an assumed parameter and two sample statistics. To generalize that discussion, we now consider problems of testing differences between more than two statistics. In the first part of this chapter, we concentrate on problems testing differences between more than two sample means. This procedure, called analysis of variance (AOV), uses a sampling distribution called F. In the second part, we address testing differences between proportions. This procedure, called chi-square, uses a sampling distribution called chi-square symbolized by X^2.

Analysis of Variance[48]

Test Among More Than Two Means

Suppose that we wanted to compare the gas mileage of three automobiles of the same type, A, B, and C at a constant speed of 60 miles per hour. To make the comparison we ran random tests for each car four times. Table 8.1 shows the results.

Table 8.1 Miles Per Gallon for Three Brands of Automobiles on Four Tests at 60 MPH

	Auto A	Auto B	Auto C	Means of All Three Autos Per Test
Test 1	22	22	25	23
Test 2	21	25	29	25

---

48. As we will see, AOV has many more uses than the tests described in this chapter.

Test 3	26	24	28	26
Test 4	23	25	30	26
Means of Four Tests per Auto Type	23	24	28	25

The values in the three columns show the miles per gallon for each car. The four rows represent the miles per gallon for four tests. The last row is the mean for each auto on the four tests and the mean for all three on the four tests. Auto C seems to have the best mileage of the three. However, these results come from random samples. How do we know that these differences are not attributable to chance? Of course, we do not know. Therefore, we must run tests of statistical significance.

Our first impulse, from chapter 7, is to run three t tests: auto A vs. B, auto A vs. C, and auto B vs. C. However, we cannot use this approach because proceeding to a series of two sample tests without first testing for the overall effect, greatly increases the probability of a type I error. Remember the additive nature of independent probabilities. If we set the alpha risk of all three tests at 0.05, these independent probabilities sum to an unacceptable (0.15) type I error probability.

We must look, instead, at the overall variance in the three samples. A statistically significant analysis of variance (AOV), allows us to proceed to a series of Z (T) or t tests. We start with the null, among three means and setting an alpha (a) risk:

$H_O: M_{autoA} = M_{autoB} = M_{autoC}$

$H_A: M_A \neq M_B \neq M_C$

$a = 0.05$

The null hypothesis as follows: Sample differences between the three cars are attributable to chance, and, therefore, actually equal. The alternative hypothesis is as follows: The differences are not attributable to chance, although, if the overall F test shows significance, we still have to run significance tests between each car to see which ones actually contribute to differences. We set the alpha risk at 0.05.

Analysis of variance uses three calculations. First, it computes the total sum of squares, signified by SS_{TV} or total variation (TV) for the entire data set.

Second, it calculates the within column sum of squares SS_{UV} or unexplained variation, UV (the degree to which each of the four samples differ from their sample mean or that the within car variance in miles per gallon is attributable to forces independent of the brand of automobile).

Finally, it calculates the between mean sum of squares, signified by SS_{EV} or explained variation (this is the amount of the total variance in MPG (TV) traceable to the maker of the automobile.

At no surprise, $SS_{EV} + SS_{UV} = SS_{TV}$. This equation suggests that we can break down (analyze) the total variance into two parts:

1. The variance attributable to the difference between the auto maker means (EV)

2. The variance attributable to random unmeasured forces (UV)

Here we show the details of three calculations:

1. $TV = SS_{TV} =$

$$\Sigma (X_i - M_{GRAND})^2$$

Equation 8.1

$$= \Sigma [(22\text{-}25)^2 + (21\text{-}25)^2 + (26\text{-}25)^2 + (23\text{-}25)^2 + (22\text{-}25)^2 + (25\text{-}25)^2$$
$$+ (24\text{-}25)^2 + (25\text{-}25)^2 + (25\text{-}25)^2 + (29\text{-}25)^2 + (28\text{-}25)^2 + (30\text{-}25)^2]$$
$$= SS_{TV} = 90$$

Equation 8.1 tells us to calculate the mean mileage of all three make of autos, the grand mean, $M_{GRAND} = 25$ mpg. Then we calculate the sum of the squared difference between each observation mileage and the grand mean, thus, the total mileage variation of the three cars is 90.

2. $SS_{UV} = \Sigma (X_{INCOLUMNS} - M_{COLUMNS})^2$ Equation 8.2

$$= (22\text{-}23)^2 + (21\text{-}23)^2 + (26\text{-}23)^2 + (23\text{-}23)^2 + (22\text{-}24)^2 + (25\text{-}24)^2 +$$
$$(24\text{-}24)^2 + (25\text{-}24)^2 + (25\text{-}28)^2 + (29\text{-}28)^2 + (28\text{-}28)^2 + (30\text{-}28)^2 =$$
$$SS_{UV} = 34$$

Equation 8.2) instructs us to calculate the squared differences between each observation in that column and each column mean and sum those squared differences. This yields the unexplained variance between the three cars, 34.

3. $SS_{EV} = \Sigma[n_{column} \times (m_{column} - m_{grand})^2]$ Equation 8.3

Equation 8.3) calculates the squared differences between each column mean and the grand mean, multiplies those squared differences by 4 and sums the set.

$$SS_{EV} = \{[4 \times (23\text{-}25)^2] + [4 \times (24\text{-}25)^2] + [4 \times (28\text{-}25)^2]\} = SS_{EV} = 56$$

$$\text{And } SS_{EV} + SS_{UV} = SS_{TV} = 56 + 34 = 90$$

Why should you multiply by four? The answer: because there are four trials. Had all the four trials been equal to their column mean, all of the variation would have been attributable to the auto maker.

These calculations produce exciting and valuable insights on their own and for subsequent discussion. Think about it. SS_{UV} describes that part of the variation in gas mileage not attributable to the differences in the carmaker's means, i.e., random variation, variation attributable to other unmeasured forces. If the observations in each column equaled their column means, then we could attribute all the variation (the total variation) in gas mileage to the differences in the cars. Car differences explaining all the variation in gas mileage means no random, unmeasured differences.

For the curious, divide SS_{EV} by SS_{TV}, $\dfrac{SS_{EV}}{SS_{TV}} = \dfrac{56}{90} \approx 0.622$ and observe that about 62.2% of the variation in car gas mileage is ascribable to the differences in the three cars. If there were no differences in the column means, then SS_{EV} would equal zero and all the car variation in recorded gas mileage is unexplained. Not incidentally, we signify the calculation $\dfrac{SS_{EV}}{SS_{TV}}$ by the symbol R^2 called the coefficient of determination. Because we found a relationship and described its strength, we call this calculation analytical. While, at this point, our interest concentrates on inference, not analysis, you can see the beginning of the golden thread that runs through the domains of description, inference, and analysis, integrating all three.

Now back to inferential testing. The null hypothesis is as follows: Mean differences attributable to chance, not different. How do we make that test?

So far, this series of calculations has resulted in two estimates of variation: SS_{EV}, based on the variation *between* the sample means and SS_{UV}, based on the variation *within* the samples. Since the first estimate, SS_{EV}, 56, somewhat larger than SS_{UV}, 34, it might be reasonable to assume that the large difference between the auto means cannot be a chance effect. However, we need to go three steps further.

To complete the test we make three additional computations. First, we compute an estimate of the variance of the explained variation, S^2_{EV}. Remember from chapter 7 that the sample variance equals

$$\frac{\Sigma D^2 = SS}{N-1} \quad \text{and, therefore } S^2_{EV} = \frac{SS_{EV}}{df_1 = K-1} = \frac{54}{2} = 27 .$$

Second, compute the within column sample variance, S^2_{UV}, an estimate of the variance based on unexplained variation.

$$s^2_{UV} = \frac{SS_{UV} = 36}{df_2 = (k(N-1) = (3*3) = degrees\, of\, freedom = 9} = 4$$

The denominator of this equation also represents degrees of freedom df_2. Our sample contains three sample column means with four observations in each column. Therefore we have degrees of freedom for unexplained variance, $df_2 = (N_{k1} -1) + (N_{k2} - 1) + (N_{k3} -1) = 9$, where N_k is the number of observations in each column. Dividing 36 by 9 yields the variance based on unexplained variation.

Third, calculate the F ratio, a statistic that compares explained variance to unexplained variance by dividing the variance due to explained variation by the variance based on unexplained variance.

$$F = \frac{S^2_{EV}}{S^2_{UV}} \qquad \text{(Equation 8.4)}$$

If our calculation yields a large F ratio, i.e., large explained variance (EV/df_1) relative to unexplained variance (UV/df_2), the probability that chance caused mean differences this large, must be low. How low?

The Sampling Distributions of F Ratios

To answer the question how low, we calculate the F ratio as follows:

$$F = F = \frac{S^2_{EV}}{S^2_{UV}} = \frac{27}{4} = 6.75.$$ As is the case with sampling distributions of means and proportion when testing hypotheses between two alternatives, there are also sampling distributions of F ratios that we can use to test null hypotheses among more than two choices. Note that we said sampling distributions because there are many. The shapes of these distributions depend on the degrees of freedom. In other words, there is a family of F ratio sampling distributions with varying degrees of freedom. Figure 8.1 shows two sampling distributions of F ratios, A for larger degrees of freedom and B for smaller.

Figure 8.1 Examples of Sampling Distributions of F for Larger and Smaller Degrees of Freedom

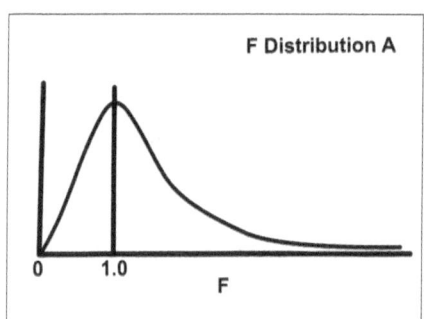
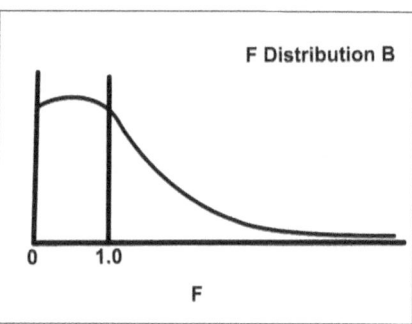

Distribution A with larger degrees of freedom has a smaller spread than B. There are as many sampling distributions of F as degrees of freedom, each different. The x-axes of these sampling distributions show F ratios; the Y-axes display the probabilities of obtaining an F ratio this large or small if the null hypothesis is true. Any sample F must fall on the x-axis of one of these distributions because they represent the entire sample Fs for df_1 and df_2.

To interpret our calculated F ratio, 6.75 we go to an F calculator on the Internet like *VassarStats*.[49] This calculator asks for the F ratio, df_1, and df_2. For those values, the calculator yields the probability of an F ratio this large or small purely by chance, if the null is true. The probability of obtaining an F ratio of 6.75 for df 2 and df 9 purely by chance is about 0.012. This probability, 0.012 means that for two and nine df, an F ratio this large, could occur purely by chance only about 12 times in a thousand. Figure 8.2 is a picture of the sampling distribution of F ratios for 2 and 9 degrees of freedom. We can easily see how the probabilities fall as F ratios increase,

Figure 8.2 F Distribution: df= 2, 9

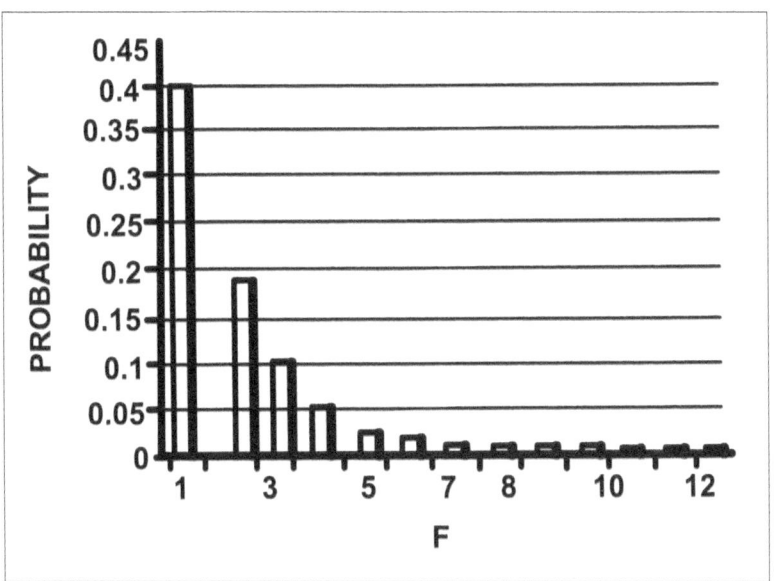

Based on this evidence at an alpha risk of 0.05, we would reject the null hypothesis and accept the alternative. The F test does not tell us which cars are significantly different, only that there are significant differences somewhere between the three. Rejection of the null means that the difference between the three means is probably real and allows (demands) proceeding to a series of two variable tests. These tests determine which of the tire differences

---

49. Richard Lowry, professor of Psychology Emeritus, Vassar College, Poughkeepsie, New York, USA, *VassarStats: Web site for Statistical Calculations*

produced the significance. Of course, statistical programs can do all this for you.

The data in tables 8.2 and 8.3 show how computer technology, running the same analysis, relieves us of calculation drudgery.

Table 8.2 Analysis of Variance for: Auto A, Auto B, and Auto C

Source	Df	EV, UV, and TV	S^2	F	P
Explained (EV)	2	56	28	7[52]	< 0.01
Unexplained (UV)	9	36	4		
Total (TV)	11	92			

Table 8.3 shows the mean and standard deviation for each of the three autos plus the grand mean.

Table 8.3 Statistics for the Three Autos

Variable	Mean	Standard Deviation	Sample Size	Degrees of Freedom
Auto A	23	2.1602	4	3
Auto B	24	1.4142	4	3
Auto C	28	2.1602	4	3
Grand Mean (All Three)	25	1.9437	12	11

We know from table 8.2, our overall test that significant differences exist among the means. Although the source of difference is self evident in this example, we still must formally test the means against each other. Most statistics programs have more than one algorithm for making tests

---

50. $S^2_{EV}/S^2_{UV} = F = 28/4 = 7$

of differences between the means. We used the test we consider most useful, the Tukey HSD test.[51] While the difference between the means of autos A and B does not produce significance, the differences among the means of auto C and autos A and B does produce significance. We made the tests by hand. Note: With two alternatives, we can make either two variable F tests or two variable t tests with the same result.

Tests of Significance between Two or More Sample Variances: Another F Test

In the Auto example given above, the sample variances, s^2 for the three cars are $(2.16)^2$, $(1.41)^2$, and $(2.16)^2$. Are the sample variances significantly different?

$$H_O: S^2_{CAR\,A} = S^2_{CAR\,B} = S^2_{CAR\,C}$$

$$H_A: S^2_{CAR\,A} \neq S^2_{CAR\,B} \neq S^2_{CAR\,C}$$

$a = 0.05$

The test is a simple F test. Take the largest variance, $S^2_{CAR\,A\,or\,C} = 4.67$ and the smallest variance, $S^2_{CAR\,B} = 1.99$ and divide the largest by the smallest. This division yields an F ratio of:

$$F = \frac{4.67}{1.99} = 2.35 \text{ for } df = n\text{-}1 \text{ for denominator, 3 and } df\ n\text{-}1 \text{ for numerator, 3.}$$

Using the VassarStats F calculator, the probability of a sample F of 2.35 for df_1 = three and df_2 = 3, purely by chance, is 0.2636. Thus, at an alpha risk of 0.05, we fail to reject the null hypothesis. Obviously, if the differences were significant, the variances for Autos A and C are the same, so the difference would have been between A, C and B.

Tests of Hypothesis among More Than Two Proportions

---

[51].   Lane, David M. *HyperStat Online Statistics Textbook*, 1993-2007.

(Nominal Data): The Chi Square Test

In chapter 7, we tested hypotheses for two proportions using T tests. For the same reasons as we discussed with more than two means and analysis of variance, we cannot execute a series of T tests with more than two proportions. In this case, as we will see with nominal data, we must use a test called chi-square (χ^2).

A Two by Two Example

Here is an example of a sample ($n=100$) taken randomly from a population of first year students majoring in economics. We measured two variables: gender and political preference. Table 8.1 shows the data.

Table 8.4 Comparison of Gender and Political Preference

	Male	Female	Total
Conservative	32	10	42
Liberal	8	40	48
Total	40	50	90

The observed counts or frequencies represent the actual breakdown of the sample. We have expressed the data as counts but we can easily convert them to percentages. For example, the sample shows that 80% of the males conservative while only 20% of the females have the same /political orientation. With females, it is exactly opposite. Based upon observation only, it would seem that a pattern exists. On the average, males seem to be more conservative than females.

Once again, however, we cannot be sure that this pattern (this relationship) is not attributable to chance. We cannot be sure, so we must make a test. When testing hypotheses concerning patterns of counts, frequencies, or percentages, we choose the test called chi-square (χ^2).

The null hypothesis is as follows: This pattern, relationship, or differences exhibited by the cross tabulation of gender and political preference is attributable to chance. The alternative hypothesis is as follows: A pattern probably exists. We show the formal statement of hypothesis and alpha risk below:

Ho: The cell frequencies (counts) follow no pattern
Ha: The cell frequencies (counts) follow a pattern
a risk: 0.05

As you can readily discern, pattern is a synonym for relationship. Again, we see the link between description, inference, and analysis. Indeed, that is what this test is about, i.e., the test for significance of the relationship between gender and political preference.

We make the χ^2 test as follows. For the four cells, we calculate the expected frequencies if no pattern existed. The percentage of conservatives in the sample is 47% (42/90). If no relationship existed between gender and political preference, the expected number of conservative males would be 0.47 × 40 or 19 and because there are 40 males, 40 - 19 or 21 would be the expected number of male liberals. The expected number of conservative females would be 23 with liberal females 27. The frequencies are the result of observing the totals for males and multiplying the percentage of conservatives in the sample by the number of males (0.47 × 40) yielding 19. Moreover, because we have only one degree of freedom in a 2 x 2 matrix, 19 automatically defines the other cells. We show observed and expected frequencies in table 8.2.

Table 8.5 Observed and Expected Frequencies

	Male	Female	Row Totals
Conservative	Observed = 32 Expected = 19	Observed = 10 Expected = 23	42
Liberal	Observed = 8 Expected = 21	Observed = 40 Expected = 27	48
Column Totals	40	50	90

The expected frequencies are what we would expect if no pattern existed, i.e., if the null hypothesis was true. Now we proceed to calculate the value of χ2. The first step centers on cell one, male conservatives. In cell one, we calculate the difference between the observed and expected, (32-19) = 13 and square that difference, 169. We then divide that squared difference by the expected, the ratio of squared observed to expected, 169/19 = 8.89, the x^2 value for cell one. Symbolically we represent that calculation as

$$\text{Cell 1 } x^2 = \frac{(O-E)^2}{E} = \frac{(32-19)^2}{19} = \frac{169}{19} = 8.89$$

We repeat the Cell one calculation for the other three cells.

$$\text{Cell 2 } x^2 = \frac{(8-21)^2}{21} = \frac{169}{21} = 8.05$$

$$\text{Cell 3 } x^2 = \frac{(10-23)^2}{23} = \frac{169}{23} = 7.35$$

$$\text{Cell 4 } x^2 = \frac{(40-27)^2}{27} = \frac{169}{27} = 6.26$$

We sum the cell x^2 values to yield 30.55, the x^2 value for the table and symbolically represent this calculation as

$$x^2 = \Sigma \, (\text{Cell } x^2) = 8.89 + 8.05 + 7.35 + 6.26 = 30.55$$

As is the case with F ratios, there is a family of sampling distributions of x^2's (chi-squares) for specific degrees of freedom. In this example, df equals one (1) because the row and column totals allow only one random choice, only one free cell entry. For example, entering 10 in cell one, given the row and column totals, defines the other cells. Technically df is equal to (k -1) × (row-1) where k equals the number of columns and row equals the number of rows. In this example df = (2-1) × (2-1) = 1.

Figure 8.3 shows the probability of H_o being true for x^2 = 30.55 with df = one or the probability that the observed pattern is caused by chance.

Table 8.3 shows the chi-distribution for one degree of freedom. The x-axis of figure 8.3 represents values of chi-square while the y-axis is probability. The boxes at the top of the bars shows the probability of realizing a x^2 of 30.55 for df = one, purely by chance if the null, H_o is true. For example, if X^2 => than 10, the probability of that happening by chance if H_o is true is approximately 0.0016. In our example, X^2 = 30.55 with one df. Given those values we know that the probability that the null is true is less than 0.002. Therefore, we reject the null hypothesis and accept the alternative, H_A that the pattern (the differences) was probably not caused by chance and therefore, the pattern is real. Figure 8.3 shows a chi-square distribution for one degree of freedom.

Figure 8.3 Sampling Distribution of Chi-square with One Degree of Freedom

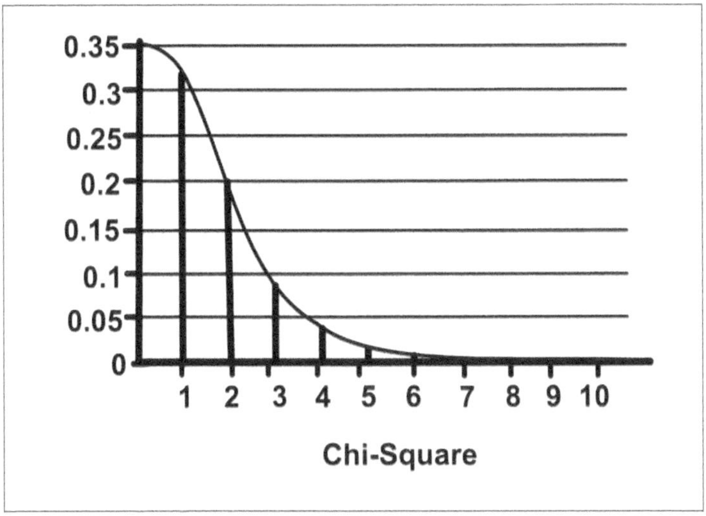

The x-axis depicts chi-square values, the y-axis probabilities if the null is true.

Another Example of a Chi-square Problem

Another example of a of a chi-square problem with a random sample (n = 100) taken from a population of retail stores selling like items or services in some hypothetical community. Table 8.6 shows the frequencies.

Table 8.6 Quality of Service of Retail Outlets in Some Hypothetical Community

Retail Stores Quality of Service	Poor Service	Mediocre Service	Good Service	Excellent Service
Frequencies	10 (25)	25 (25)	35 (25)	30 (25)

An expert retail quality-control team visited these stores and, based on a range of service criteria, judged the service poor, mediocre, good, and excellent. The question: How do we know that these count differences or proportions are not attributable to sample error? Of course, we do not know. Therefore, we must run a χ^2 test

The numbers in parenthesis represent the expected frequencies if there were no differences in service quality.

H_o: Observed difference ascribable to random error
H_A: Difference is real
Alpha risk = 0.05

We show the calculations below.

Cell 1 $\chi^2 = \dfrac{(10 - 25)^2}{25} = 9$

Cell 2 $\chi^2 = \dfrac{(25 - 25)^2}{25} = 0$

Cell 3 $\chi^2 = \dfrac{(25 - 35)^2}{25} = 4$

Cell 4 $\chi^2 = \dfrac{(25 - 30)^2}{25} = 1$

Total $\chi^2 = 14$

Df = Rows - 1 = 4 -1 = 3

In the calculation of *df*, there are four rows in the table with a total of 100. Therefore, we have three free row entries, the fourth prescribed by the total, or three degrees of freedom. Given the total, even if any one of the cells in the row was missing, we could restore the other cell values.

Probability of a type I error for $\chi^2 = 14$ and *df* =three is less than 0.001. Consequently, we reject the null and accept the alternative that there are service differences in the retail stores in some hypothetical community. We are now entitled (required) to make a series of tests between each level of service quality. One possibility is the test of significance between two sample proportions discussed in chapter 7.

Summary

The critical ideas in this chapter are the following:

1. The concept of analysis of variance

   Analysis of variance, a statistical method that tests hypotheses about the differences between three or more sample means. Our discussion of AOV also introduces part 3, "Analysis."

2. The calculation of the critical elements of Analysis of Variance

   AOV requires the calculation of an *F* ratio, a ratio comparing explained to unexplained variance.

3. Use and interpretation of the *F* test

   AOV's *F* ratio tests the significance of difference between two or more sample means using the sampling distribution of *F* ratios.

4. The notions of explained (EV) and unexplained (UV) variance

   EV and UV relate to the degree to which variable differences do and do not account for total variance (TV). Again, we can use the term *signal* as a surrogate for EV. UV defines the term noise, the residual variance.

5. The concept of chi-square (X^2)

    Chi-square, a statistical method used to test the significance of difference between two or more proportions. More generally, we also use chi-square to test statistical significance of nominal data across a broad problem range.

6. The calculation of the values of chi-square

    Using a technique analogous to F ratios, Chi-square tests the statistical significance of nominal data using squared differences between observed values and expected values if no real patterns in the data exist.

7. Interpretation of chi-square

    Chi-square tests use sampling distributions of chi-square, analogous to F distributions, in testing for statistical significance.

This chapter completes, for the moment, our discussion of statistical inference. As we proceed through the remainder of our book, the concept will occur repeatedly.

Chapter Eight

Appendix A

This data set of 25 observations is an unrestricted random sample consisting of three cardinal variables, $X1$, $X2$, and $X3$. Your task: Using hand calculations (spreadsheet), test the null hypotheses that the means and variances of the three variables are not significantly different. Check your findings using your computer program. Carefully explain your calculations and findings.

ID	X1	X2	X3
1	112.27	154.34	106.72
2	92.47	119.66	116.00
3	95.48	130.81	111.97
4	116.40	102.18	135.58
5	98.37	142.77	130.75

6	72.15	142.31	108.68
7	87.50	103.55	99.45
8	131.93	145.97	112.21
9	124.03	108.31	103.22
10	86.72	82.49	155.69
11	102.76	123.88	105.69
12	108.15	140.43	146.28
13	94.84	115.78	137.58
14	101.34	82.93	104.12
15	110.51	100.55	156.77
16	87.39	118.38	102.18
17	119.99	113.60	130.20
18	110.54	121.12	123.81
19	116.73	138.00	123.26
20	123.03	127.76	120.57
21	129.46	107.35	118.91
22	110.96	127.77	106.61
23	134.31	130.24	82.68
24	104.22	114.59	136.06
25	128.99	107.67	108.11

Appendix B

This data set is an unrestricted random sample of three nominal variables ($N = 25$). Using a hand calculation (spreadsheet), test the null hypothesis that the proportions of the variables are not significantly different. Explain your conclusions.

ID	X1	X2	X3
1	0	0	1
2	0	0	1
3	0	0	0
4	0	1	0
5	0	1	1
6	0	1	1
7	0	1	0
8	0	0	0
9	0	0	1
10	0	0	1
11	0	0	0

12	0	0	0
13	0	0	1
14	0	1	0
15	0	0	0
16	0	1	0
17	0	1	1
18	0	0	1
19	0	1	1
20	0	1	1
21	0	0	0
22	1	1	0
23	0	0	0
24	0	0	1
25	1	1	0

Analysis

Introduction

Part 3, Data Analysis, caps our presentation of the three dimensions of the statistical method that enable us to produce information from masses of uninformative data—or to again use the metaphor we've stressed throughout the book—to yield the clear signals from noise that make informed decision-making possible. In this sense, data analysis is the most important of the three because it links why-is and what-if questions to testable hypotheses emerging from problem settings, thereby producing the most information from the data at hand.

Chapters 9-12 open with a discussion of the different types of statistical relationships you will encounter as a data analyst, and introduces the most general and therefore most useful method for measuring those relationships, regression analysis. Statistically measurable relationships can exist between ratio, interval, ranked, and perhaps most significantly in management settings, nominal variables. Chapter 9, by sharpening and extending the concept of statistical relationships, gets to the core of "problem framing" by making the clear distinctions between these types of data, in their dual roles as observations on both dependent and independent variables, i.e., respectively, between those variables for which your questions seek answers, and those variables—reasons, factors, or forces—you believe will provide answers.

Chapter 10 discusses measures of relationships between pairs of variables: covariance and correlation coefficients between ratio, interval, ranked, and nominal variables, technically known as bivariate relationships. And in chapter 11, "Measuring relationships with regression analysis," we open the topic of regression analysis—and extend the case of bivariate relationships—with simple bivariate regression (one dependent and one independent variable), showing how it produces information (reduces ignorance) from data regardless of its type, i.e., nominal, ordinal, interval, or cardinal.

It is here that we begin to make our case, argued compellingly (we hope) throughout part 3, that regression analysis is the most general and useful tool for estimating the strength of statistical relationships, forecasting their bearing on future values of the dependent variable identified in the research question, and making inferences about the causes of those relationships. Chapter 12, "Measuring relationships with multiple regression analysis" moves bivariate regression into the realm of everyday reality, where stripping the noise from decision-making signals about the behavior of the dependent variable in question is almost always a matter of evaluating the contributions of more than one variable (factor, or force).

The remainder of part 3 focuses on assumptions of the OLS regression model, consequences of not meeting the assumptions, and methods of correcting for assumption failures.

CHAPTER 9

RELATIONSHIPS AND CAUSATION

Relationships

Two or more variables are related when they move together in the same or opposite directions. College freshmen grade point averages and high school grades, money supply and inflation, and germs and disease are just three examples of relationships.

Independent and Dependent Variables

To facilitate explanation, we introduce the notion of independent and dependent variables, X and Y. We usually assume that an X variable drives a Y variable $(X \rightarrow Y)$, but sometimes we label variables X and Y arbitrarily because of lack of certainty as to which drives which.

Relationships Defined

In producing bricks, we can define the relationship between labor costs and plant use employing the symbols X and Y. Table 9.1 shows a relationship in which we know that the X variable, plant use, partially drives Y, labor costs.

Table 9.1 Plant Use and Labor Costs

Plant Use, X measured in percentage of plant capacity.	20.00%	40.00%	50.00%	75.00%	90.00%
Labor Costs, Y measured in worker hours per 1,000 bricks of output.	95 hours	45 hours	30 hours	12 hours	4 hours

Table 9.1 shows labor costs decreasing as plant use increases. Figure 9.1 gives a picture of the relationship.

Figure 9.1 Labor Costs and Plant Use

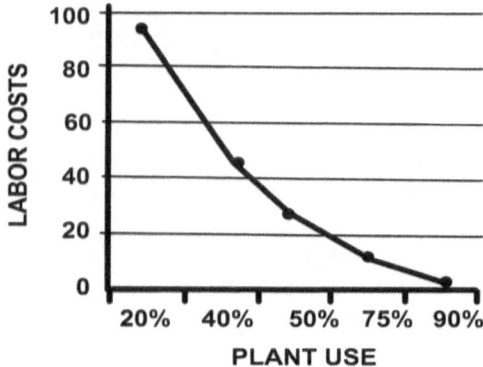

In figure 9.1 the two variables move together in opposite directions; as plant use increases, hours needed to produce 1,000 bricks declines. When two variables move in opposite directions, we call the relationship negative.

Another example, figure 9.2, shows the relationship between retail sales (Y) and income (X).

Figure 9.2 Income and Sales

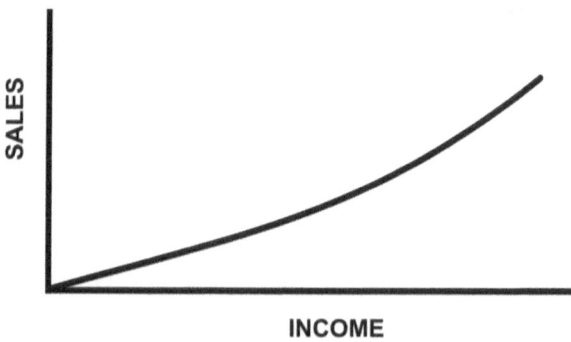

Figure 9.2 shows a positive relationship between sales and income because the two variables move together in the same direction. As we see in the upward sloping line, when income increases, sales rise and when income falls, so do sales.

Relationships Between Nominal, Cardinal, and Interval Variables

Relationship between Nominal and Cardinal Variables

Table 9.2 shows a relationship between gender (0 = male and 1 = female), a nominal variable and finger dexterity, an interval variable. The numbers in the body of the table represent finger dexterity.

Table 9.2 Gender and Finger Dexterity

	Males	Females
	3.85	8.83
	5.15	9.17
	6.85	10.35
	6.85	11.83
	8.15	13.17
Means	6.17	10.67

Table 9.2 captures this relationship by showing the male and female means; the female mean equals 10.67, the male mean equals 6.17. In this small sample, females have higher finger dexterity than males. Figure 9.3 pictures the relationship.

Figure 9.3 Gender and Finger Dexterity

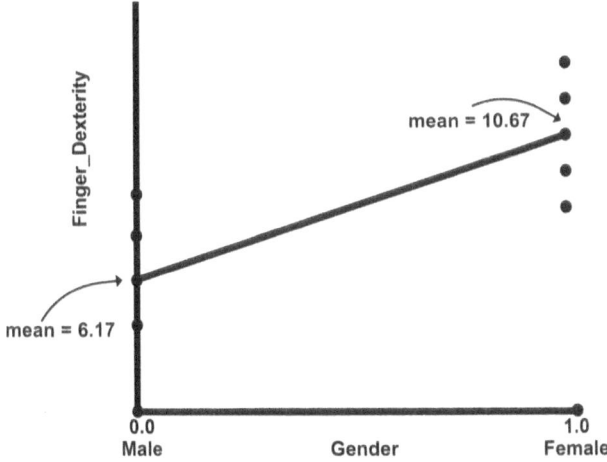

Figure 9.3, with the line plotted to the means, shows a positive relationship between the nominal independent variable, gender (females coded one; males coded zero) and the quantitative (interval) dependent variable, finger dexterity. If we reversed the gender coding the line would slope downward (negative). Whatever the coding of the nominal variable, the interpretation remains the same. Of course, it does not make sense to place finger dexterity on the X axis.

Relationship between Two Nominal Variables

Table 9.3 depicts a relationship between economic status and race. The numbers without parenthesis are the expected values if no relationship existed. The values not in parenthesis portray the observed or actual numbers (counts) of the economically advantaged and disadvantaged in each category. If no relationship existed, the observed frequencies in the cells would equal the expecteds. By looking closely, we can see a disproportionate number of disadvantaged nonwhites. Table 9.3 helps us see the relationship.

Table 9.3 Economic Status and Race

	White	Nonwhite	Total
Disadvantaged	(82) 60 41.4%	(63) 85 58.6%	145
Advantaged	(37) 59 72.8%	(44) 22 27.2%	81
Total	119 (56.25%)	107 (43.75%)	226

We arranged these data as a chi-square matrix, calculated the values in parenthesis by finding the percentage of white observations, 56.25% out of the total, 226, and multiplied this percentage, expressed as a decimal fraction (0.5625) by the total disadvantaged, 145 to give the expected number of disadvantaged whites, 82. Because we have only one degree of freedom, we calculated all the rest of the expected values by subtraction Table 9.3 shows nonwhites as disproportionately disadvantaged. Figure 9.4 pictures this pattern.

Figure 9.4 Economic Status and Race/Ethnicity

In figure 9.4 shows a bar graph of the relative disparity in economic status between whites and nonwhites. Whites are proportionally more advantaged than nonwhites. We devised this chart to show that even when both variables are nominal, we can graphically portray that relationship.

Forces Producing a Relationship

Chance

Some relationships occur by chance. An unrestricted random sample of fifty people from a list of all the registered voters in Dallas, Texas may show a relationship between physical beauty and intelligence purely by chance. Of course, as we have seen, we have methods for testing the probability of relationships occurring purely by chance.

Tertium Quids

A third variable, Z acting on X and Y can produce a relationship. We call these relationships tertium quids. For example, a random sample of ministerial salaries in the major cities of the country and the price of prostitutes will show a statistically significant relationship. This relationship is a tertium quid because the X and Y variables move together as average income in the cities varies.

Even if relationships are not causal, knowledge of stable relationships over time is valuable. For example, if a stable relationship (positive or negative) exists between soft-drink consumption (X) and academic performance (Y), a university registrar might consider an applicant's soft-drink consumption as one criterion for admission. Registrars could predict performance from soft drink consumption even though the relationship was not causal. In other words, if our statistical goals are solely predicting or forecasting,[52] stable spurious relationships are useful.

Cause

Finally, relationships can be causal. When, under controlled conditions or analysis, a change in X produces a change in Y, we say X is a cause of Y. We will explain this important idea in the next section.

Causation

Importance of the Distinction between Tertium Quid and Causation

Relationships are a necessary, but not sufficient condition for causal inference. That is, for two variables to be causal, they must be related. But relationship alone is not sufficient evidence of causation. Causal relationships provide the basis for ameliorative intervention. If Johnny cannot read, the cause(s) of his difficulty point to the best methods for

---

52.   We differentiate prediction and forecasting by the data used. While prediction and forecasting both make future projections. Predictions rely on cross-sectional data while forecasts rely on time series or panel data (cross sections combined with time series). Predictions stay within the measured constraints of the cross sectional data. For example, if we made a prediction of the probability of successfully completing a flight training program based on hand/eye dexterity, the prediction cannot extend beyond the lowest and highest measures of dexterity. Forecasts, on the other hand, project beyond the data.

teaching Johnny to read. Causation allows interventions that change what is to something better. Causal relationships also provide better foundations for predictions and forecasts.

A Note on Pseudo-Causal Semantics

To understand and use the term *cause*, we make a distinction between pseudo causal semantics and causation.

Circularities

When we habitually substitute grammar for meaning, we often assume that we are talking about causation when we are only making circular statements. For example, when children and parents habitually use the word *because* as an introduction to an explanation, they frequently use it alone in answering the question why.

The substitution of grammar for meaning leads to statements like the following.

> Question: Why does Millie commit crimes?
> Answer: Because Millie is evil.

This is a circular daisy chain. It seems to point to a causal connection between the two conditions, crime and evil, but merely relates the two expressions. In the case of Millie, if we ask, "What does it mean to be evil?" we find that criminal acts are the only criterion for defining evil Therefore, the statement says only "Millie is criminal because she is criminal." The circularity is obvious.

All the following assertions are circular.

1. John hates to work because he is lazy.
2. Grace does not learn because he is unmotivated.
3. Charlie speaks falsely because he is a liar.
4. Teachers do not belong to organizations that fight for teacher's rights because they are unprofessional.

All the assertions are at best naive or at worst designed to halt inquiry.

Temporal Sequences

Often, we erroneously think of an event as the sufficient cause of another if it temporally precedes the other.[53] For example, suppose that after a football game, (X), a race riot (Y) ensues. Some people assume that the football game, (X), caused the riot (Y) because it preceded it. This is a pseudo-causal temporal association. While time series analysis is a powerful tool for establishing causation, using temporal sequence alone as a sufficient criterion for causal assertions produces a false sense of cause because if we wait long enough, almost anything occurs after almost anything else. If, after a riot, a football game occurs, did the riot cause the football game?

Below are some additional examples of pseudo-causal temporal sequences:

1. An ant bite caused me to be sick.
2. It rained because the chief did a war dance.
3. We had a depression because we previously had a terrible inflation.
4. The train left the station because the little hand on the clock reached eight.
5. The assassination at Sarajevo caused World War I.

Our point: We cannot use mere time precedent as a sole criterion for causal inference.

Causation

We say that (X) causes (Y) when we manipulate X and observe a change in Y. The conditions surrounding manipulation take the form of an experiment characterized by randomization, controlled experimental intervention and experimental/ isolation. The diagram illustrates a classical experiment:

$$RO_1 \ X \ RO_2$$
$$RO_3 \quad RO_4$$

---

[53]. Of course, temporal sequence is a necessary, but not sufficient condition of causation.

The Rs represent random assignment of experimental research units (e.g., pigeons, rats, consumers, business enterprises) to at least two groups, experimental and control; O's represent observational measurements (e.g., heart rate, temperature, customer values, sales) taken before and after an experimental intrusion, X (electric shock, drug administration, increase in brand advertising). We call the group receiving the intrusion, experimental and the group not receiving the intrusion, control.

Depending upon the hypothesized direction of change, O_2 bigger or smaller than O_4 (assuming statistical significance), O_2 bigger or smaller than O_1, and O_2 bigger or smaller than O_3 enables causal inference. We can infer causation because random assignment prevents attributing the differences between O_1 and O_2, O_4 and O_2, O_2 and O_1 and no differences between O_3 and O_4 and O_1 and O_4, to the confounding of differential effects.[54]

By confounders, we mean a host of plausible alternative explanations for the changes in O. These include simultaneous irrelevant intrusion(s), maturation (e.g., older or wiser) of the research units, decay of the measurement instrument, (e.g., a weighing instrument using springs in which the springs stretch), systematic differences attributable to selection of the sample units, the effect on the research units of pre measurement (e.g., how a premeasurement might influence a post-measurement), or differential mortality (e.g., the X intrusion might result in sample units dying or dropping out of the experiment). Randomization assures equality of experimental and control groups and, together with experimental isolation, defeats the confounding effects attributable to plausible alternative explanations.

We call this process, controlled experimentation and, if our test is confirmed and replicated in independent studies, allows use of the word cause.

While classical experimentation, with quasi-experimental variations, represents the gold standard for causal inferences, ethical and practical considerations (particularly in institutional settings) often prevent its use. For example, ethical constraints prevent using human beings in experiments testing the causal links between smoking and

---

54.   For an elaboration of these ideas see CD, chapters 1 and 2.

lung cancer. In addition, limited generalizability, e.g., the degree to which experimental subjects may see themselves as special, often makes quasi-experimental designs or passive statistical analysis preferable. Due, in part, to these limitations and the critical need for effective interventions, passive statistical analysis has become important for drawing causal inferences.

In the chapters that follow, we show how to use statistical analysis to make causal inferences from passive data when classical or quasi experimentation is unethical or impossible. If theoretically plausible confounding effects can be controlled or eliminated by statistical analysis, causal inference becomes possible. We cited the problem of smoking and lung cancer as a condition in which classical experimentation was not possible. Medical researchers established the causal link by statistically eliminating all the plausible alternative causal explanations and still finding a relationship between smoking and lung cancer.

Summary

The critical ideas developed in this chapter are the following:

1. Relationships

    Two variables are related when they move together in the same or opposite directions.

2. Causation

    X causes Y when X and Y are related and when a controlled manipulation of X changes Y.

3. Experiment

    Controlled manipulation of X designed to produce a change in Y

4. Pseudo-causation
    a. Circularities: Mistaking word links with causation
    b. Temporal sequences: Mistaking time sequences for causation

5. Forces producing relationships

    Chance, Tertium Quid, and Causation

Stable, noncausal associations are useful as predictors or forecasters, but they take the world as it is. Causal associations produce guidelines for interventions that change the world.

CHAPTER 10

MEASURING RELATIONSHIPS: COVARIANCE AND THE COEFFICIENT OF CORRELATION

In this chapter, we focus on two variable (bivariate) analyses. In the chapters that follow, we extend the discussion to multiple regression analyses.

The Covariance

The Covariance (COV_{XY}) Defined

Assume we have a small random sample ($N = 6$) of families in some community and measure family income in thousands of dollars, labeled Y and years of schooling of the head of the household, labeled X. Table 10.1 shows the result of our sample.

Table 10.1 Years of Schooling and Family Income

Family	Years of Schooling	Family Income
	X	Y
A	12	50
B	12	49
C	15	53
D	14	52
E	15	56
F	16	58

We labelled schooling as the independent variable because our theory specifies that schooling affects income, the dependent variable (Y). If, on average, high values of X go with high values of Y and low values of X go with low values of Y, the variables have a positive covariance. If, on average, high values of X go with low values of Y, the variables have a negative covariance. If X tends to be neither higher nor lower for higher or lower values of Y, the variables have no relationship and zero covariance.

The Covariance Measured

We show the calculation of the covariance in equation 10.1.

$$COV_{XY} = \sum \frac{Dx \times Dy}{N-1}$$ (Equation 10.1)

The equation instructs us to calculate the differences (D) between the mean of X, and each individual value of X and between the mean of Y and each individual value of Y_i, multiply these differences to realize $D_X \times D_Y$, sum $\sum D_X \times D_Y$ and divide that sum by N - 1, the total number of observations in the sample minus one. Table 10.2 illustrates the calculation.

Table 10.2 Calculation of the Covariance

Family	Schooling	Income	DX	DY	DX × DY
A	12	50	-2	-3	6.00
B	12	49	-2	-4	8.00
C	15	53	1	0	0.00
D	14	52	0	-1	0.00
E	15	56	1	3	3.00
F	16	58	2	5	10.00
Totals	84	318	0	0	27.00
Means	14.00	53.00			5.40
Standard Deviations	1.67	3.46			
Variances	2.80	11.97			

$$COV_{XY} = \frac{\sum D_X * D_Y}{N-1} = \frac{27}{5} = 5.40$$

The COVxy, 5.40, show a positive relationship between X and Y. We graphically portray this relationship in figure 10.1 below.

Figure 10.1 Income and Schooling

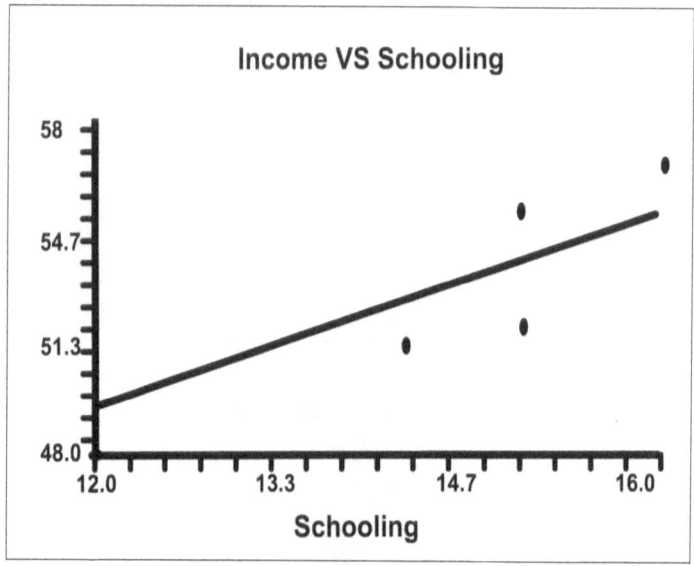

We constructed this scatter diagram, by plotting the pairs of X and Y values from table 10.1. The first plotted value, X = 12, Y = 50 shows the higher point on the figure's y-axis. The second plot, X = 12, Y = 49 falls just below the first. We continue plotting until we represent all six pairs (points). We have drawn a line through these data showing how the variables move together positively, high with high, low with low. You should understand that this chart also depicts the average relationship. We can interpret the line as a special kind of mean average running through the pairs of Xs and Ys. As education increases, income, on average, also increases. The degree to which the points do not fall on the line portrays the imperfection of the relationship, that is, the degree to which schooling does not explain income.

If the calculated COV_{XY} had been zero, indicating no relationship, the line would plot at the mean of Y, parallel to the x-axis (see below). If the COV_{XY} had been negative, the line would have sloped downward. Note: The COV_{XY}, an absolute measure of the relationship, tells us that a relationship exists, but tells nothing about the strength of the relationship because the covariance has the two variables units of measure embedded in it.

It is important to understand why the covariance has a positive algebraic sign. Looking back at table 10.2, you can see that all the values in the columns D_X and D_Y have algebraic signs. If high goes with high and low with low these signs match and their products will all be positive making the sum of $D_X \times D_Y$ positive. If the pairs of values in D_X and D_Y do not match at all, their products will be randomly plus and minus and their sum and covariance zero, no relationship. If low goes with high and high with low, the sum of $D_X \times D_Y$ and the covariance will register negative because their products are all negative, a negative relationship.

At this point, we cannot say anything about the statistical significance of the relationship because we have not tested it. Moreover, we cannot say that X causes Y. All we can say, from this sample, is that there may be a positive relationship between schooling and income.

Let us look at another example. Suppose we take a random sample of N = 6 from the same population and measure family income, X and number of children in the family, Y. Table 10.3 gives the data.

Table 10.3 Income and Number of Children

Family	Income	Number of Children	DX	DY	DX × DY
A	50	2	-3	0.67	-2
B	49	3	-4	1.67	-6.67
C	52	1	-1	-0.33	0.33
D	53	1	0	-0.33	0
E	56	1	3	-0.33	-1
F	58	0	5	-1.33	-6.67
Totals	318	8	0	0	-16
Means	53	1.33			COVXY = -16/5 = -3.2

The COV_{XY}, -3.2, shows a negative relationship; families with higher incomes, on average, have fewer children than families with lower incomes. Figure 10.2 pictures that relationship.

Figure 10.2 Children and Income

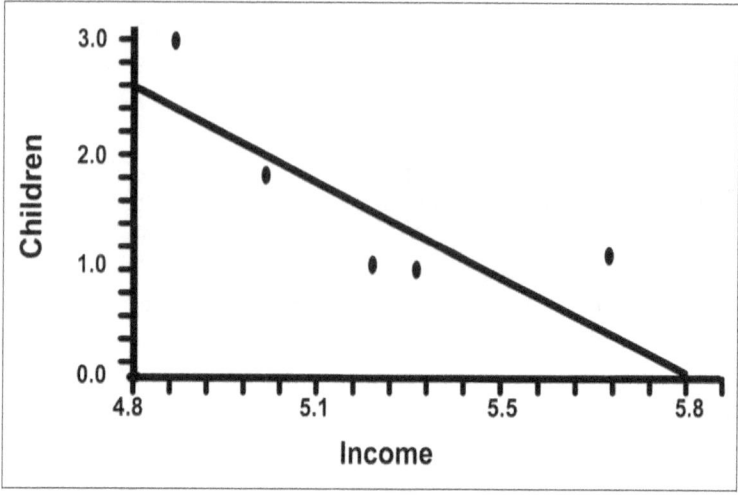

On average, the higher the income, the smaller the number of children.

Table 10.4 shows the result of another random sample (N =5) taken from the same population measuring creativity and intelligence.

Table 10.4 Intelligence and Creativity

Subject	Intelligence	Creativity	DX	DY	DX × DY
A	10	50	-20	10	-200
B	20	20	-10	-20	200
C	30	40	0	0	0
D	40	60	10	20	200
E	50	30	20	-10	-200
Totals	150	200	0	0	0
Means	30	40			0

The variables in this small sample have a covariance of zero (0), indicating no relationship between intelligence and creativity figure 10.4 plots this nonrelationship.

Figure 10.4 Creativity and Intelligence

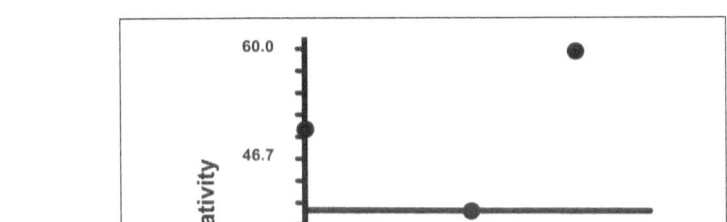

The straight line paralleling the x-axis plots to the mean of Y showing that creativity does not vary with intelligence. Note that the five data points (Y_i) vary randomly about the mean of Y. This also suggests that the nonrelationship tells us no more about creativity than its mean, that is, the relationship has not improved our signal.

We can extend the notion of the covariance to a nominal X variable and a cardinal Y variable. Consider the data in table 10.5. We coded the nominal variable, gender, 0 = male, 1 = female. We measured productivity with a single index bringing together several surrogates for this difficult to measure Y variable.

Table 10.5 Gender and Productivity

Individual	Gender (X)	Productivity (Y)	DX	DY	DX × DY
Jim	0	1	-0.375	-3	1.125
Bill	0	2	-0.375	-2	0.75
Frank	0	3	-0.375	-1	0.375
Ed	0	4	-0.375	0	0
Tom	0	5	-0.375	1	-0.375
Sue	1	3	0.625	-1	-0.625
Jill	1	4	0.625	0	0
Betty	1	5	0.625	1	0.625

Sum DX × DY					1.88
Covariance					0.268

The COVxy equals 0.268 indicating a positive relationship between gender and productivity; in this small sample; the covariance's positive sign indicates that females, on average, are more productive than males. Figure 10.5 shows this relationship.

Figure 10.5 Gender and Productivity

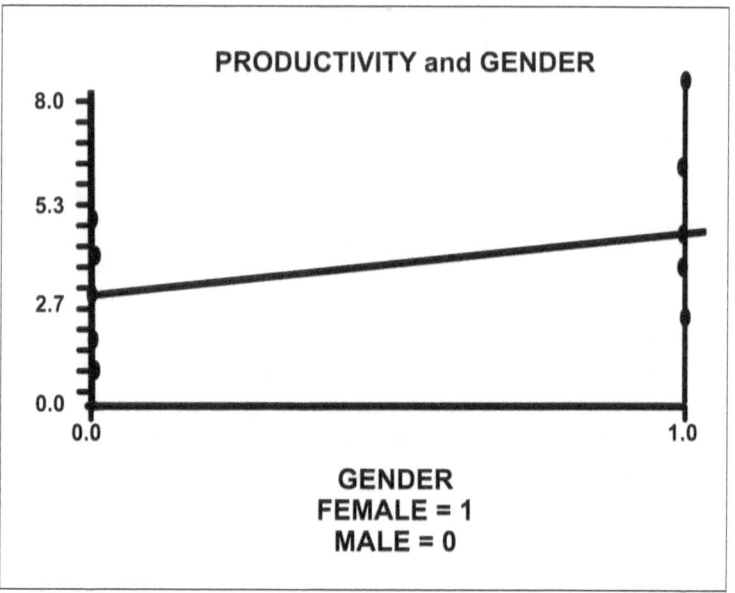

Notice that the line plots to intersect the mean productivity of males (3) and females (4). The mean difference in productivity between males and females defines the relationship. Reversing the coding of males and females turns the line's slope downward, but does not change its meaning.

We can also extend the idea of covariance to a condition in which X and Y are both nominal variables. The X variable signifies gender (1 = males); the Y variable, political preference (1 - liberal). In table 10.6 we coded females as zero, males as one; liberals as one, conservatives as zero.

Table 10.6 Gender and Political Preference

Individual	Gender (X)	Political Preference (Y)	DX	DY	DX × DY
A	1	1	0.500	0.3	0.15
B	1	1	0.500	0.3	0.15
C	1	1	0.500	0.3	0.15
D	1	0	0.500	-0.7	-0.35
E	1	0	0.500	-0.7	-0.35
F	0	1	-0.500	0.3	-0.15
G	0	1	-0.500	0.3	-0.15
H	0	1	-0.500	0.3	-0.15
I	0	1	-0.500	0.3	-0.15
J	0	0	-0.5	-0.7	0.35
Totals	5	7	0	0	-0.500
Means	0.5	0.7	0	0	-0.056
Standard Deviations	0.5	0.49			

The COV_{XY} equals -0.056 indicating a relationship between political preference and gender with females, on average, more liberal than males. Eighty percent (80%) of the females are liberal while sixty percent (60%) of the males are liberal. Figure 10.6 plots this relationship.

Figure 10.6 Gender and Political Preference

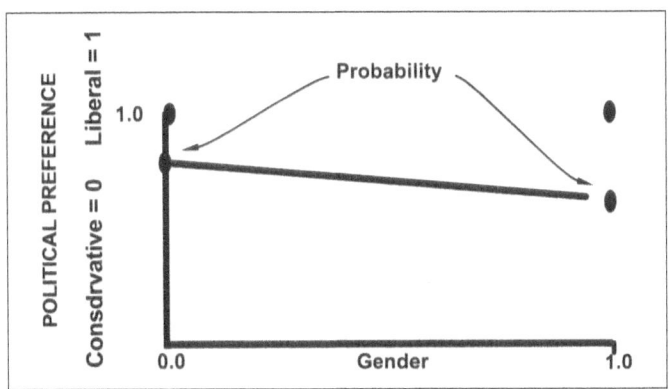

We have drawn the line so that it intersects the y-axis at the mean of Y for females (80%) and the x-axis at the mean of X for males (60%). Note: In this case, we did not use Chi-square, but calculated the covariance exactly as we calculated it for two cardinal variables.

The Coefficient of Correlation (R_{XY})

Coefficient of Correlation Defined

The coefficient of correlation measures the strength or degree of a relationship and, unlike the COV_{XY}, is a relative measure, i.e., stripped of the units of measure of X and Y. Thus, unlike the covariance, we can use this measure to compare two or more relationships. It is bounded by the values -1 to 1 with -1 a perfect negative relationship, 0 no relationship, and 1 a perfect positive relationship.

The Coefficient of Correlation Measured

The symbol for the symbol for the sample coefficient of correlation is R_{XY}. The formula for its calculation is given below:

$$R_{XY} = \frac{COV_{XY}}{S_X * S_Y}$$
(Equation 10.2)

The equation instructs us to divide the covariance, COV_{XY}, by the product of the standard deviation of X multiplied by the standard deviation of Y. Dividing by the product of two standard deviations strips the COV_{XY} of the units of measure of both X and Y. To illustrate this calculation we reproduce the data in table 10.2 as table 10.7 below.

Table 10.7 Schooling and Income

Family	Schooling (X)	Income (Y)	DX	DY	DX × DY
A	12.00	50.00	-2.00	-3.00	6.00
B	12.00	49.00	-2.00	-4.00	8.00
C	15.00	53.00	1.00	0.00	0.00
D	14.00	52.00	0.00	-1.00	0.00
E	15.00	56.00	1.00	3.00	3.00
F	16.00	58.00	2.00	5.00	10.00
Totals	84.00	318.00	0.00	0.00	27.00
Means	14.00	53.00			5.40
Standard Deviations	1.67	3.46			

The COV_{XY} = 5.40. The standard deviation of X equals 1.67, and the standard deviation of Y equals 3.46. Substituting into Equation 10.2 yields the coefficient of correlation:

$$R_{XY} = \frac{COV_{XY}}{S_X * S_Y} = \frac{5.40}{1.67322 * 3.464102} = 0.932.$$

By dividing the covariance by the product of the standard deviations of the two variables we create a standardized measure, i.e., we have removed the units of measure of the two variables, creating a *pure* measure useful for determining the strength of the relationship and comparing it to any other relationship. The coefficient of correlation for this sample is 0.932. The limits of this coefficient,-1 to 1, allows us to interpret this value as a strong relationship.

Another explanation of this calculation, reexpresses the data in table 10.7 as Z scores. Table 10.8 shows the calculation of the covariance of Z Scores.

Table 10.8 Z Schooling and Z Income

Family	Schooling	Income	Z Schooling	Z Income	Z Sch × Z Inc
A	12	50	-1.195229	-0.866	1.0353983
B	12	49	-1.195229	-1.1547	1.3801311
C	15	53	0.5976143	0	0
D	14	52	0	-0.2286	0
E	15	56	0.5976143	0.86602	0.5175492
F	16	58	1.952286	1.44337	1.7251639
Totals	84	318	0	0	4.658
Means	14	53			COVzxzy = R = 0.932

Reexpressing X and Y as Z scores, makes the covariance of the Z scores equal to the correlation coefficient of the original X and Y. Remember that we obtain the Z scores by calculating the difference between each value and its mean and dividing those differences by the variable's standard deviation. The covariance of the Z scores has already removed the units of measure from the data and, therefore, is the correlation coefficient. In table 10.9, we removed the unit of measure before we calculated the covariance. Thus

$$R_{XY} = COV_{ZXZY}$$

(Equation 10.3)

We note that equations 10.2 and 10.3 work no matter how we measure the variables, nominal, ordinal, interval, or cardinal. Older treatments of this topic gave different names and different formulae for the coefficient of correlations for nominal and ranked data, e.g., the point bi-serial coefficient. These different formulae all measured the correlation coefficient in the same way, but were developed for computational ease before computer technology. Modern software packages made these distinctions irrelevant.[55]

Please note that the coefficient of correlation measuring the strength of a relationship has in an important way increased our signal beyond knowledge of the arithmetic mean. In other words, we can make much better estimates of Y from X if we know that there is a relationship between X and Y.

Statistical Inference for the Coefficient of Correlation

Throughout this chapter, we have issued the disclaimer that none of our calculated coefficients had been tested for statistical significance. In this short section, we address that issue. Using the data on schooling and family income in which we calculated the R_{XY} as 0.93, suggesting a very strong relationship, we still do not know if the coefficient is statistically significantly different from zero. That is, we do not know if the R-value 0.93 occurred purely by chance.

---

[55].   JCC, p. 28.

In this circumstance, the shape of the sampling distribution of R is normal only when the population R = 0[56] If we want to test the null hypothesis that R equals 0, the standard error of R equals

$$S_{RSE} = \sqrt{\frac{1-R^2}{n-2}}$$ (Equation 10.4)

Using our convention, we can test the significance as follows.

H_O: R = 0
$H_{A:}$ R ≠ 0
Alpha risk = 0.05

There is a sampling distribution of R whose mean is zero and its standard deviation is the standard error of R. The distribution is normal when R equals zero and N is equal to or greater than 30. When N is less than 30, the sampling distribution of R is a t distribution.

The null hypothesis is that our sample R is a chance effect and, therefore the mean of the sampling distribution of R is zero (0). Our null, H_O says that the sample R departed from zero purely by chance. We know that our sample R must fall somewhere on the x-axis of the sampling distribution of R. Therefore, if we can estimate the standard deviation of the sampling distribution, S_{RXY}, we can make a test of statistical significance. So we now calculate an estimate of the standard error of R, S_{RSE}, as follows:

$$S_{RSE} = \sqrt{\frac{1-r^2}{n-2}} = \sqrt{\frac{1-0.869}{6-2}} = 0.181.$$

The standard error of R (S_{RSE}), our sample error, is 0.181. The t test is

$$t = \frac{R}{S_{RSE}} = \frac{0.932}{0.181} = 5.15.$$ Using a t table or a computer program for N -2

and df = 4 and a t value of 5.15, the probability of finding a sample R

---

56.   JCC, If we want to test any other null, e.g., R = 0.50, we must make a Fisher's Z′ prime transformation. All of our recommended programs will make that for you.

this large, 0.932, purely by chance if the true R is zero (0) is 0.0068. This means that if the population R is zero (0), finding a sample R of 0.932 could happen by chance only 68 times in 10,000. With a two-tailed test and an alpha risk of 0.05, we reject the null and accept the alternative that the true R is not zero.

We can also make a 95.45% confidence interval estimate that the true R is between +/- $2S_{RSE}$ of 0.932 or +/- 0.362 or 1.00 to 0.570. Because R cannot exceed +/- 1.00, we are 95.45% confident that the true R lies somewhere between 1.00 and 0.570.

Some Pitfalls in Calculating the Coefficient of Correlation[57]

Coefficients of Correlation with Truncated Data

Assume that we want to use intelligence as a criterion for hiring so we do a study of productivity and intelligence by taking a sample, n = 30 from a population frame of prospective employees on which we have productivity and intelligence data. Table 10.9 is an excerpt of those data.

Table 10.9 Unrestricted Sample

ID	Intelligence	Productivity
1	37.77	156.08
2	32.86	88.08
3	33.44	116.72
4	30.28	98.5
5	35.08	106.39
6	30.38	11.21
7	37.31	110.89
8	31.73	110.08
9	34.73	108.77
10	28.09	101.9

---

[57].   For a more elaborate discussion of these ideas see JCC, pp. 53-61.

We ran a correlation coefficient using the 30 applicants, with X equal to intelligence and Y equal to Productivity and tested the coefficient for statistical significance. We summarize the result in table 10.11 below.

Table 10.11 Correlation of Intelligence and Productivity

$(N = 30)$

$R = 0.4819$

Probability of a Type I Error	0.0107

From table 10.11 we reject the null, accept the alternative and conclude that the population R is probably not zero, closer to 0.4819, the sample R.

After some discussion, the analysts decided that the sample contained prospective employees with intelligence too low to perform the needed tasks adequately. Therefore, they decided to restrict the analysis to people in the "more normal range," say to people with intelligence of at least 100 but less than 127 thereby restricting intelligence to 101.9 to 126.64 rather than the full range, 88.06 to 156.08. Table 10.12 represents applicants within the truncated range.

Table 10.12 Truncated Range

ID	Productivity	Intelligence
3	33.44	116.72
5	35.08	106.39
6	30.38	111.21
7	37.31	110.89
8	31.73	110.08
9	34.73	108.77
10	28.09	101.90
11	32.19	107.63
13	32.02	109.83
14	31.12	116.89

15	30.16	124.66
16	32.08	126.64
17	21.96	103.26
19	34.09	110.21
21	33.42	105.39
22	21.05	122.95
24	24.68	113.69
25	33.25	116.33
26	34.38	123.24
29	28.13	123.49

We reran the correlation analysis using the truncated range of applicants, displaying the result in table 10.13. Using the data in table 10.13 we fail to reject the null (the sample R is not statistically significant). Table 10.13 shows the sample R falling to -0.08366. This decline was neither accidental nor did it result from sample error.

Table 10.13 Truncated Range Correlation Coefficient

$(N = 30)$

R	-0.08
Probability of a Type I Error	0.73

The correlation coefficient has fallen to -0.08 and statistical insignificance. When either the range of a variable is restricted deliberately or in data collection, R will be smaller in the truncated range than R for the full range of data.

Coefficients of Correlation when the Dependent Variable Contains the Sum of the Independent Variable plus Some Other Variable

Occasionally you will see a correlation coefficient computed between some variable X_1 and another variable X_2, in which X_2 is the sum of variables X_1 and X_3. Under these circumstances, a correlation can be expected between X_1 and X_2 because X_2 includes X_1, even when there is no correlation between X_1 and X_3. This is also true if X_1 is subtracted from X_3. Suppose, for example, from table 10.12, we added productivity to intelligence to create the variable productivity plus intelligence in table 10.14.

Table 10.14 New Variable Productivity Plus Intelligence

Productivity	Intelligence	Productivity plus Intelligence
33	117	150
35	106	141
30	111	142
37	111	148
32	110	142
35	109	144
28	102	130
32	108	140
32	110	142
31	117	148
30	125	155
32	127	159
33	72	105
21	102	123
25	89	114
33	83	116
34	89	123
28	95	123

The correlation analysis of productivity and productivity plus intelligence produces a statistically significant R of 0.3453. But this R is an artifact of the productivity plus intelligence variable containing the variable productivity. Thus, in a part-whole correlation in which the part of the dependent variable is in the independent variable, a relationship must exist.[58] If we rerun the analysis with intelligence and productivity, the R falls to 0.1039 and is statistically insignificant.

---

[58]   This seems to be intuitively obvious.

Correlations between Ratio or Index Variables and the Base

Ratio (index or rate) scores are those obtained by dividing one variable by another. When a ratio score, say X_1 is correlated with another ratio score having the same base, the resulting R may simply reflect relationship between X_1 and the base denominator. While correlating ratios reflecting a base denominator may be a legitimate analytical goal, the correlation obtained may be spurious, because what is correlated is the common base. If, for example, in table 10.15 we divided intelligence by age to create a new variable, a ratio of intelligence relative to age, and, if productivity was not related to intelligence but was related to age, a spurious correlation would result. We did exactly that with the following result:

Table 10.15 Correlation with Indices

ID	Productivity	Intelligence	age	Intelligence/age
3	33	117	26	4.49
5	35	106	48	2.22
6	30	111	39	2.85
7	37	111	27	4.11
8	32	110	21	5.24
9	35	109	40	2.72
10	28	102	47	2.17
11	32	108	22	4.89
13	32	110	37	2.97
14	31	117	33	3.54
15	30	125	39	3.20
16	32	127	36	3.52
17	22	103	38	2.72
19	34	110	24	4.59
21	33	105	47	2.24
22	21	123	32	3.84
24	25	114	49	2.32
25	33	116	31	3.75
26	34	123	42	2.93
29	28	123	33	3.74

Productivity, in this sample is not related to intelligence, but is negatively related to age ($R = -0.189$) and, therefore, productivity is positively

related to the index $= \dfrac{Intelligence}{age}$ ($R = 0.192$)

This correlation is spurious, representing the negative effect of age, the denominator in the index, on productivity. In this case the R between Productivity and the index reflects nothing except the relationship between Productivity and Age.

Summary

The critical notions discussed in this chapter are the following:

1. The notion of relationships

    Variables are related when they move together in the same or opposite direction.

2. The concept of the Covariance

    Covariance is an absolute measure of relationships.

3. The notion of the Coefficient of Correlation

    The coefficient of correlation is a relative measure of relationship showing its strength.

4. Statistical Inference and the coefficient of correlation

    A test of statistical significance when H_o: $R = 0$

5. Other conditions affecting the correlation coefficient

    a. Truncation
    b. Variable sums
    c. Ratio correlations.

Chapter 10

Appendix A

Appendix A has two cardinal variables. Your tasks are the following: (1) Using your spreadsheet, calculate the covariance and correlation coefficient, (2) what information can you derive from those two calculations, and (3) check your work by making the same calculations with your computer program.

ID	X1	X2
1	98.66	111.5
2	76.4	72.23
3	104.79	82.26
4	85.48	109.24
5	122.58	114.67
6	102.58	102.23
7	102.59	96.41
8	116.03	101.05
9	98.62	75.55
10	93.7	80.72
11	105.11	96.28
12	126.61	113.01
13	97.58	102.78
14	100.1	98.13
15	79.25	89.06
16	97.85	87.99
17	92.98	93.62
18	129.47	123.68
19	122.84	122.38
20	110.75	98.25
21	83.63	94.75
22	95.03	91.51
23	96.82	83.66
24	88.87	88.25
25	98.58	102.15
26	55.84	68.28
27	79.33	79.07
28	108.54	89.38
29	126.1	106.53
30	110.68	130.8

Appendix B

This data set has two nominal variables coded one and zero. Your tasks are the following: (1) Using your spreadsheet, calculate the covariance and correlation coefficient, (2) what information can you derive from those two calculations, and (3) Check your work by making the same calculations with your computer program.

ID	X1	X2
1	1	1
2	1	1
3	0	1
4	0	1
5	0	0
6	1	1
7	1	1
8	1	1
9	1	1
10	1	1
11	0	0
12	0	1
13	0	0
14	1	1
15	1	1
16	1	1
17	1	1
18	0	1
19	0	0
20	0	0
21	1	1
22	0	1
23	1	1
24	1	1
25	0	0
26	0	0
27	0	0
28	0	1
29	0	0
30	0	1

CHAPTER 11

BIVARIATE REGRESSION: ORDINARY LEAST SQUARES (OLS)

Our discussion of analysis has so far focused on the measurement of relationships: Covariance and correlation. This chapter extends the discussion to a second, more powerful tool called regression analysis. Regression, a general data analytic system, incorporates all the dimensions of inference, description, covariance, and correlation and produces, in its multiple forms, more useful methods for forecasting, predictions, and causal inferences.

Specifying a Relationship without a Functional Form

In statistics, when we wish to suggest (hypothesize) the existence of a relationship between two or more variables, we often use an expression that does not specify the functional form of the relationship, but only that X and Y are related. Such statements might look like the following:

1. $Y = f(X)$
   Or
2. $Y = f(X_1; X_2; X_3; X_4)$
   Or
3. $(Y_1; Y_2; Y_3; Y_4) = f(X_1; X_2; X_3; X_4)$

The first expression, a bivariate relationship (one Y and one X) reads: Y is a function of (or is related to) X without specifying an algorithm (an equation) defining the relationship; the second expresses a relationship with one Y and multiple X's; and the third, called multivariate, multiple Y's and multiple X's. We call the X variable(s), the independent variable(s). We also call the X variables exogenous because forces outside the model determine their values. We call the Y variable, the dependent variable. W also call the Y variable endogenous because the X variable(s) in the model determine Y. The X variable causes (or influences) Y and the direction of the effect is one-way, i.e., Y does not reciprocally cause or influence X. We call this one-way effect recursive and a two-way reciprocal effect, nonrecursive.

A Specific Functional Form: The Straight Line

In regression analysis, we must proceed from a nonfunctional form, e.g., $Y = f(X)$, to a statement of a hypothesized form of the relationship, i.e., a statement of the actual relationship defined by an equation. While the choice of functional forms involves many possibilities, in this section, we restrict the discussion to a straight line, the principal tool of OLS linear regression.

We define the equation for a straight line as follows:

$$Y = B_0 + B_1 X \qquad \text{(Equation 11.1)}$$

In this equation, we define the variables as X and Y and the constants, that we must calculate, as B_0 and B_1. B_0, called the Y intercept, specifies the value of Y when X equals zero (0). B_1, the regression coefficient, defines the effect of X on Y or the amount that Y changes, on average, for a unit change in X. B_0 and B_1, both expressed in Y's unit of measure, can have positive or negative signs.

Equation 11.1 defines a straight line as a perfect, idealized relationship. In other words, if we substitute values for the variable X and solve the equation for Y, the Y values form a straight line. Consider the data in table 11.1:

Table 11.1 Perfect Relationship

ID	X	Y
A	4	8
B	5	10
C	6	12
D	7	14
E	8	16
Sums	30	60
Means	6	12
Standard Deviations	1.41	2.82

This relationship, perfectly represented by the straight-line equation $Y = 0 + 2X$, does not allow Y to deviate from the line. Substituting the five values of X_i from table 11.1 we realize the values in the column headed Y. When $X = 4$, $Y = 8$, etc. Figure 11.1 plots that relationship.

Figure 11.1 A Perfect Straight Line Relationship

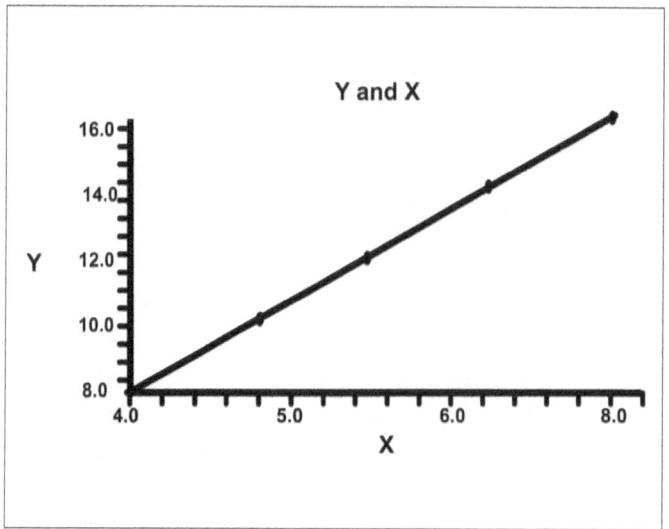

Figure 11.1 depicts a perfect relationship. When X changes by one (1), Y changes by 2; when $X = 5$, $Y = 10$; when $X = 6$, $Y = 12$; when $X = 7$, $Y = 14$; and when $X = 8$, $Y = 16$. We call this relationship deterministic, because of the perfect relationship between X and Y. Any estimate of Y from X contains no error.

Bivariate Linear Regression Analysis

Bivariate means two variables, X and Y. In bivariate linear regression analysis, we predict values of the dependent variable from a linear function of the form:

$$Y_c = B_0 + B_1X_i \qquad \text{(Equation 11.2)}$$

and

$$Y_i = B_0 + B_1X_i + \varepsilon_i \qquad \text{(Equation 11.3)}$$

In the first equation, Y_C represents the estimated value of the dependent variable or, more precisely, an estimate of the *mean* of the dependent variable, for each value of the independent variable. Therefore, each value of X has a corresponding value of Y_C. B_1, the slope of the line, also called the regression coefficient or the effects coefficient, represents the effect on Y when X changes by one. B_0, the regression constant, defines the value of Y_C when X is zero. We also call B_0 the autonomous value of Y.

In equation 11.3, Y_i, the actual value of Y, permits the calculation of ε_i, the residual or error term by obtaining the difference between Y_C, the estimated value or mean value of Y, and Y_i, the actual value of Y for a given value of X. The residual or error (ε_i), represented by the expression $Y_i - Y_C = \varepsilon_i$,[59] defines a new variable, $Y_{x'}$ with the effect of X removed or Y with the effect of X held constant.

Note: We can draw many straight lines through a set of X, Y variables. Bivariate regression analysis, however, calculates values of B_0 and B_1 producing a straight line that fits the data *best*, where we define best as a line that minimizes error, the sum of ε_i^2, i.e., the sum of the squared residuals (ε_i^2). By that, we mean a line that makes the sum of squared residuals smaller than the sum of squared residuals from any other possible line drawn through the data. The classical regression model (OLS), has this feature as its defining characteristic. The word *least* refers to the minimizing of the error term.

$$\Sigma\,(Y_i - Y_C)^2 = \varepsilon_i^2 = \text{Minimum} \qquad \text{(Equation 11.4)}$$

Calculation of B_0 and B_1

We can show (without proof) that the best-fit values for B_1 and B_0 are the following:

$$B_1 = \frac{COV_X}{S_X^2} \qquad \text{(Equation 11.5)}$$

---

[59]. This residual value is also assumed to be a risk coefficient, i.e., the degree to which any estimate using the X value as an estimator poses a risk of being in error. And, as we shall see, the residual is assumed to be normally distributed so that the probability of very large or small values is low. Therefore, all regression models assume calculable risk, not uncertainty. See chapter 3 for a more extensive discussion of this notion.

$$B_0 = M_Y - B_1 \times M_X \qquad \text{(Equation 11.6)}$$

Dividing the covariance[60] by the variance of X, Equation 11.5 eliminates from B_1 the unit of measure of X while retaining the unit of measure of Y. Since B_1 always takes on the unit of measure of Y, we also call B_1 an effects coefficient because it reflects the effect of a unit change in X on Y_C, the mean estimate of Y.

Multiplying the mean of X times the regression coefficient, B_1 and subtracting that product from the mean of Y, Equation 11.6, gives the regression constant, B_0, the unit of measure of which is Y. Note: If B_1 equals zero, then B_0 equals the mean of Y. This says no relationship between X and Y and the best estimate of Y is still the mean of Y. So if B_1 equals zero, our regression has not reduced noise. Table 11.2 represents a small data set (N = 5) on which we illustrate a bivariate regression analysis.

Table 11.2 Intelligence and Productivity (N = 5)

Individual	Intelligence (X)	Productivity (Y)	DX	DY	DX × DY
A	2	1	-4	-2	8
B	4	3	-2	0	0
C	6	2	0	-1	0
D	8	5	2	2	4
E	10	4	4	1	4
Totals	30	15	0	0	16
Means	6	3			$COV_{XY} = 4$
Standard Deviations	3.1622	1.581139			

---

60.   Remember that the covariance of X, Y is one measure of the relationship between X and Y.

Here are the steps:

First, calculate the $COV_{XY} = 4.0$

Second, calculate the variance of X, $S_X^2 = (3.162278)^2 \approx 10$

Third, calculate B_1, $\dfrac{COV_{XY}}{S_X^2} = \dfrac{4}{10} = 0.40$.

Fourth, calculate B_0, $B_0 = M_Y - (B_1 \times M_X) = 3 - (0.40 \times 6) = 3 - 2.4 = 0.60$

These coefficients produce the straight line, $Y_C = 0.60 + 0.40X_1$, the line that fits the data best because it minimizes the sum of the squared residuals, ε_i^2. When X increases by one, Y increases, on the average, by 0.40. When X is zero, Y has an autonomous value, 0.60. We summarize the data and the best-fit straight line table 11.3 and figure 3.2 below.

Table 11.3 Summary of the Regression Calculations $Y_C = 0.6 + 0.4X$

Col 1	Col 2	Col 3	Col 4	Col 5	Col 6	Col 7	Col 8	Col 9	Col 10
Individual	X	Y	Yc	Yi - Yc	Yc - My	Yi - My	$UV = (Yi - xc)^2$	$EV = (Yc - My)^2$	$TV = (Yi - My)^2$
A	2.00	1.00	1.40	-0.40	-1.60	-2.00	0.16	2.56	4.00
B	4.00	3.00	2.20	0.80	-0.80	0.00	0.64	0.64	0.00
C	6.00	2.00	3.00	-1.00	0.00	1.00	1.00	0.00	1.00
D	8.00	5.00	3.80	1.20	0.80	2.00	1.44	0.64	4.00
E	10.00	4.00	4.60	-0.60	1.60	1.00	0.36	2.56	1.00
Sums	30.00	15.00		0.00	0.00	0.00	3.60	6.40	10.00
Means	6.00	3.00							

Calculating Col 4, Y_C: When $X = 2$, $Y_C = 1.4$; when $X = 4$, $Y_C = 2.2$; when $X = 6$, $Y_C = 3.0$; when $X = 8$, $Y_C = 3.8$; when $X = 10$, $Y_C = 4.6$. Figure 11.2 plots the relationship between X and Y with the line of best fit.

Figure 11.2 Plot of Intelligence on Productivity

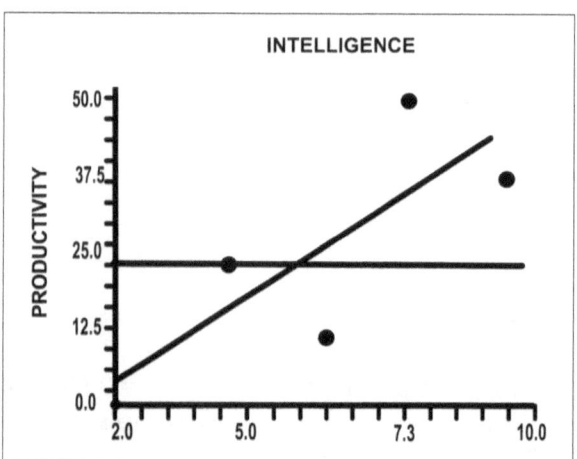

The sign of the regression coefficient, 0.40 shows the positive relationship between X and Y, intelligence and productivity. The regression constant, also positive equals 0.60. The correlation coefficient, R equals $\dfrac{COV_X}{S_X * S_Y}$ = $\dfrac{4}{5}$ = 0.80.[61] Look carefully at table 11.3, column (4). In this column, we calculated the points on the regression line, Y_C by substituting all the values of X in our data. These values represent the predicted values (on the average) of Y, Y_C, for each X_i.

You can visualize the line as a special kind of arithmetic mean running through the pairs of X's and Y's. Column (3) shows the actual values of Y_i represented by the points falling away from the regression line. From this figure, we can see that the relationship is not perfect, i.e., while X is a good predictor of Y, R equals 0.80, it is not perfect. We can also see that any prediction we make using this equation contains error as measured by the distance between the actual values of Y_i and points on the regression line, the predicted values, Y_C. The sum of Column (5) equals zero because the values of Y_C represent means running through the data. Remember from chapter 3, the sum of the deviations from any arithmetic mean always equals zero. Also, remember that we said

---

61.   At this point, we can offer another interpretation of R, the coefficient of correlation. If we reexpressed the X and Y variables as Z scores, and ran a bivariate regression on the Z scores, the regression coefficient for the Z scores is R, the correlation coefficient. Therefore, R represents how much, on the average, ZY changes per unit change in ZX.

that these ε_i values represent noise, the degree to which we have not explained the variation in productivity. But we have sharpened our signal by reducing noise because the sum of the regression squared residuals is smaller than the sum of the squared residuals from the mean of Y.

In column (8), we simply squared column (7) calling the sum, 3.6, unexplained variance (noise), UV. In column (9), we calculated the sum of the squared differences between the mean of Y, and the points on the regression line, Y_c and called that sum, 6.4, explained variance (Signal), EV. The horizontal line running across figure 11.2 is the mean of Y, M_Y. Finally, we calculated, in column (10), the total variance of Y, TV, as the sum of the squared differences between each actual value of Y, Y_i, and the mean of Y, M_Y. Thus, TV =10.

It is no accident that the sum of independently calculated UV and EV equals TV, i.e., if we add unexplained variance, 3.6 and explained variance, 6.4; we get total variance, 10. Then by dividing EV, 6.4 by 10,

$\dfrac{EV}{TV}$ we realize a new measure, $R^2 = 0.64,$[62] called the coefficient of

determination, the percentage of variation in productivity attributable to intelligence. Conversely, K^2, the coefficient of nondetermination, equals $1 - R^2 = 36\%$, measures the percentage of variation in productivity not attributable to intelligence. Figure 11.4, Venn diagram helps understand the interpretation of R^2.

Figure 11.4 Venn Diagram of X, Y, Y_c, R^2, and K^2

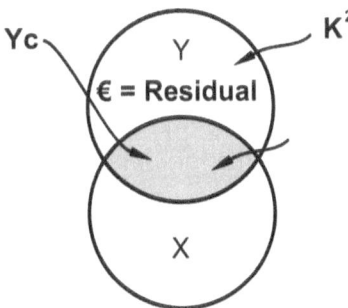

The shaded area of figure 11.4 defines the R^2.

---

62.   Refer back to chapter 8 and our discussion of R^2 in the context of analysis of variance to again see the golden thread running through description, inference and analysis.

Note: The area in figure 11.4 labeled ϵ occupies the same space as K^2 and equals the residuals. Its interpretation conceptually transcends these two labels because we also interpret ϵ as a new value of Y or Y with the effect of X removed. In statistics we call that process partialling or Y with the effect of X removed.

Prediction and the Standard Deviation of Regression, SD_{XY}.

At this stage, it should be clear that prediction defines one important use of regression analysis. In our example of intelligence and productivity, if we know the intelligence of a prospective worker, we can predict within a certain range of error, the productivity of that potential worker and we can use this prediction as one criterion in a hiring decision. A caveat: the prediction must stay within the range of intelligence used in the regression analysis. For example, if a prospective employee has a measured intelligence of 3.3, we can predict, on the average, that her productivity will be $Y_c = 0.60 + 0.40(3.3) = 1.92$. This value, 1.92 is an "on the average" estimate also called a point estimate. We cannot, however, use this equation to predict the productivity of a person with an intelligence of 15 or 2 because those two measures of intelligence lie outside the range of our analysis.

How much error is in this estimation? We can calculate a standard deviation of regression and if we assume that the sampling distribution of mean estimates is normally distributed, we can construct a confidence interval estimate of productivity. We give the equation for the standard deviation of regression below:

$$S_{YC} = \sqrt{\frac{\sum (Y_i - Y_C)^2}{N - 2}}$$

(Equation 11.6)

The equation instructs us to take the difference between each actual value of Y_i and Y_C, square and sum these differences, divide that sum by N - 2, the number of pairs minus 2, and take the square root of that quotient. We placed he sum of squared differences between Y_i and Y_C, 3.6, in column 5 of table 11.3. Dividing 3.6 by 5 - 2 and taking the square root, we find the standard deviation of regression.

$$S_{YC} = \sqrt{\frac{3.6}{3}} = 1.09$$

Given our assumption of the normality of the sampling distribution of predicted values of Y, we can build a 95.45% confidence interval estimate of the applicant's productivity, 1.92 ± 2 × 1.09 = 1.92 ± 2.18 or 4.10 to -1.16. This range of error is obviously too large so we need to increase the size of the sample, N to reduce the error

Testing the Statistical Significance of the Regression Coefficient, B_1.

In this context, how do we know that B_1 does not equal zero, i.e., how do we know B_1 of the population from which this sample was taken is not zero and that our sample B_1 0.40, is not probably attributable to chance? Of course, we do not know (and cannot know with certainty), but we can make a test of statistical significance. For a bivariate regression, two tests are possible, a T (Z) or (t) test and an F test. We use the T (Z) test when $n => 30$ and the t test when $n < 30$.

The T or t test

To make the T test we assume the existence of a normal sampling distribution of B_1 whose mean is zero. Because the sampling distribution of B_1 represents all the possible B_1's taken from samples size N, our sample B_1 must fall somewhere on the X axis of the sampling distribution of B_1. For a large sample, $N => 30$, we assume normality. For a small sample, $N < 30$ we assume a t distribution, a symmetrical distribution that is symmetrical but with larger standard deviation (errors). To test the null hypothesis that B_1 equals zero. We estimate the standard error of B_1 using the following equation:

$$Se\ B_1 = \frac{S_Y}{S_X}\sqrt{\frac{1-R^2}{N-2}} \qquad \text{(Equation 11.8)}$$

This equation says (1) divide the sample standard deviation of Y by the sample standard deviation of X, then (2) calculate the square root of the quotient of the unexplained variance and sample size - 2, and (3) multiply that quotient by step 1.

$$Se\ B_1 = \frac{1.581}{3.162}\sqrt{\frac{0.36}{3}} = 0.5\sqrt{0.12} = 0.173205$$

And because our sample size N is small, 5 and $df = N - 2$, or 3 our test is t.

H_o: $B_1 = 0$
H_A: $B_1 \neq 0$ (a two tail test)
Alpha risk = 0.05

$$\text{The } t \text{ test} = \frac{0 - B_1}{Se_{B1}} = \frac{0.40}{0.173205} = 2.31$$

The numerator of this t test equals zero, the null, minus our sample regression coefficient. Our null hypothesis is that the true population B is zero. Looking up the t value (2.31 two tail) for alpha risk = 0.05, $df = 3$, and $Se_{B1} = 0.173205$, the probability of a B_1 being as large s 0.40 purely by chance if the true value is zero, is 0.1041. Therefore, at an alpha risk of 0.05, we cannot reject the null, i.e., concluding that probably no relationship exists in the population from which the sample was taken. If the relationship had been statistically significant, we could make an interval estimate of B_1 between which we are 95% confident that the true B_1 is 0.40 +/- 2 × 0.1.73205 or about 0.74 to 0.06, again, too large for any useful application.

Bivariate OLS Regression with a Nominal Independent Variable

In our previous example, we used two variables that were both cardinal. OLS regression is also applicable with nominal independent variables. For example a regression with productivity as Y and gender as X. Table 11.4 is an excerpt (10) of a random sample ($n = 100$) of workers in a hypothetical firm. We coded the nominal variable as zero equal male, one equal female.

Table 11.4 Productivity and Gender

ID	Productivity	Gender
A	98	1
B	70	0
C	88	0
D	82	1
E	69	0
AA	73	1
BB	79	0
CC	82	1
DD	91	1
EE	70	0

Table 11.5 summarizes an OLS Bivariate regression using these data.

Table 11.5 Regression of Productivity and Gender

Variable	Regression Coefficient	Standard Error	T(Z) Value	Probability
B_0	72.2	0.7108	101.579	0.0000
B_1	16	1.0053	15.916	0.0000
R^2	0.72			

The coefficients, B_0 and B_1 are statistically significant at alpha < 0.0001. We reject the null and accept the alternative. Remember that the coefficients B_0 and B_1 are always scaled in the unit of measure of Y. In this example, the regression coefficient, B_1 equals 16 and B_0 equals 72.2. Solving the equation for $X = 1$, females, we estimate productivity as 88.2. Solving for $X = 0$, we have a mean estimate for male productivity of 72.2, the regression constant. Thus, with a nominal independent variable, the Regression coefficient, B_1, equals the mean difference between the two groups and the B_0 coefficient equals the mean of the group coded 0. We can also make a 95% interval estimate of B_1: $16 \pm 2 \times 1.005 =$ about 14 to 18, a strong estimate.

Summary

The critical ideas discussed in this chapter are the following:

1. The regression constant, B_0 and the regression slope coefficient, B_1

   The regression constant, B_0 is the value of Y_C when X is zero (0). This value is also called the autonomous value of Y. The regression slope coefficient, B_1 is the average change in Y, Y_C for a unit change in X. Please note that the unit of measure of B is always the unit of measure of Y. If Y is measured in pounds, the unit of measure of B_0 and B_1 is also in pounds. B_1 can be positive, negative, or zero. If it is positive, the relationship is positive, i.e., when X increases, Y_C increases. When B_1 is negative, the relationship is negative, i.e., when X increases, Y_C decreases. When B_1 is zero, there is no relationship between X and Y, B_0 is the mean of Y and in the scatter diagram the regression line runs parallel to the X axis.

2. The error term

The regression error term is the squared difference between the actual value of Y_1 and the estimated value of Y_c. An important point: Taken alone we see the error term is a completely new variable, Y with the effect of the X variable removed or partialled.

3. The coefficients of determination, R^2 and nondetermination, K^2

The coefficient of determination measures how much of the variation in Y is explained by X, the degree to which our regression analysis effectively explains the variation in Y. The coefficient of nondetermination, K^2 measures the independence of Y from X., the percentage of the total variation in Y not attributable to X.

4. The standard deviation of regression, S_{Yc}

The standard deviation of regression measures prediction error.

5. The standard error of B_1

The standard error of B_1, the denominator in our T (Z) or t test, represents the key inferential statistic in bivariate OLS regression.

Appendix A

Chapter 11

These variables, X_1 and X_2 are both cardinal data. Your tasks: Using your spreadsheet calculate the regression slope, B_1 and constant B_0. Explain the meaning of these calculations; 2) Test the null hypothesis that B_1 equals zero. Explain the meaning of that test; and 3) Make a 95% interval estimate of the mean of Y using X equal 52. Check your calculations using your statistics program.

ID	X1	X2
1	111.69	76.65
2	91.86	25.44
3	98.26	38.83
4	101.23	50.31
5	87.32	24.90

6	106.49	65.24
7	103.20	55.09
8	78.51	20.00
9	126.43	80.00
10	112.28	80.00
11	93.00	26.25
12	101.28	49.28
13	96.69	37.99
14	102.16	51.61
15	114.31	80.00
16	91.56	26.29
17	99.39	47.75
18	87.34	20.00
19	119.24	80.00
20	102.49	57.49
21	90.54	33.70
22	99.87	44.45
23	106.97	65.75
24	102.53	55.58
25	108.86	71.95
26	113.11	80.00
27	97.26	39.36
28	102.22	54.17
29	90.78	22.41
30	98.03	36.74

Appendix B

Chapter 11

This data set contains a nominal independent variable. Your task: Using your spreadsheet calculate the regression slope, B_1 and constant B_0. Explain the meaning of these calculations; 2) Test the null hypothesis that B_1 equals zero. Explain the meaning of that test; and 3) Make two 95% interval estimates of the mean of Y using X equal one and X equal zero.

ID	Y	X1
1	90.56	0
2	102.19	1
3	85.06	0
4	97.89	0

5	116.70	1
6	111.38	1
7	79.01	0
8	98.78	0
9	100.79	0
10	94.56	1
11	106.94	0
12	105.80	1
13	103.55	1
14	101.39	0
15	112.41	1
16	94.65	0
17	86.90	0
18	110.62	1
19	97.49	0
20	118.83	1
21	86.22	0
22	105.28	1
23	86.20	0
24	98.61	0
25	100.37	0
26	101.15	0
27	95.13	0
28	107.02	1
29	105.23	1
30	127.80	1

CHAPTER 12

UNIVARIATE[63] MULTIPLE REGRESSION

In this chapter, we begin a discussion of univariate multiple regressions as the principal tool for estimating, forecasting, and causal inference. Bivariate regression used one dependent variable and one independent variable. Univariate multiple regression is defined by one dependent variable and multiple independent variables. With multiple independent variables, we enter one of the most exciting parts of our book.

Univariate Multiple Regression

Returning to the problem posed in chapter 11, an effort to explain and predict productivity of potential employees by relating productivity to intelligence, in this chapter, we expand our analysis to two independent variables, intelligence and energy. Limiting the discussion to two independent variables makes multiple regression easier to explain, but all our calculations extend to any number of independent variables, although the principle of "less is more" holds. Assume that we can deduce from a theory that adding a second independent variable, energy level, will improve our understanding and prediction of productive performance. To test this theory we take a random sample of 30 from our hypothetical employee pool, measure productivity, intelligence and, energy level, recording the result in table 12.1.

---

[63]. In part 6, we address multivariate techniques using multiple dependent and multiple independent variables.

Table 12.1 Productivity, Energy, and Intelligence

ID	Prod	Intel	Energy	ID	Prod	Intel	Energy	ID	Prod	Intel	Energy
1	89.77	104.48	48.56	11	84.78	101.67	48.78	21	82.93	118.95	48.8
2	76	88.72	43.74	12	87.87	101.45	59.42	22	99.3	118.64	62.54
3	65.39	102.59	50.43	13	73.36	95.73	53.74	23	81.67	114.95	71.43
4	101.92	128.95	44.34	14	98.27	110.45	60.29	24	95.82	99.4	65.93
5	55.44	106.59	39.47	15	68.37	78.03	52.14	25	94.7	105.76	54.7
6	63.19	101.24	42.43	16	95.21	126.15	53.42	26	58.95	91.11	43.61
7	54.46	82.58	36.23	17	70.05	79.63	46.44	27	64.09	89.25	44.3
8	101.87	130.19	49.08	18	73.73	86.9	47.47	28	63.98	102.27	41.13
9	83.68	109.67	51.99	19	82.3	112.57	50.49	29	81.38	88.35	49.35
10	68.35	95.59	47.52	20	67.96	88.87	58.71	30	77.02	109.62	55.64

Analyzing these data takes the form of

$$Y_i = B_0 + X_1 B_1 + X_2 B_2 + \varepsilon_i \text{ or} \qquad \text{(Equation 12.1)}$$

$$Y_C = B_0 + X_1 B_1 + X_2 B_2 \qquad \text{(Equation 12.2)}$$

Comparing this equation to the bivariate equation in chapter 11, you can see that adding one additional term, energy expands our analysis. The multiple regression analysis of these data proceeds as follows.

The Correlation Matrix

Step 1 estimates the simple (zero order) correlation matrix. This matrix shows the correlation coefficients of all the variables in the analysis. Table 12.2 shows the matrix.

Table 12.2 Correlation Matrix for Energy, IQ, and Productivity

	Energy	Intel	Productivity
Energy	1.000	0.272	0.554
Intel	0.272	1.000	0.674
Productivity	0.554	0.674	1.000

The diagonal of the correlation matrix in table 12.2, all ones, simply shows the correlation coefficients of a variable with itself. The lower triangle shows the correlation coefficients of each variable with the other variables in the analysis. The upper triangle mirrors the lower triangle. The table's initial triangular value, 0.272, portrays the relationship between the independent variables, energy and intelligence. The apex value of the triangles, 0.554, displays the relationship between energy and productivity. The last element in the triangle, 0.674 exhibits the relationship between intelligence and productivity. Obviously, the two triangles are mirror images of each other.

We call these correlations zero-order coefficients because the coefficients reflect relationships without considering the other variables in the matrix. Correlation matrices give us initial clues as to how the multiple regression analysis may turn out.

The Coefficient of Multiple Determination, MR^2

Step 2 estimates the Coefficient of Multiple Determination, MR^2, the percentage of the total variation of productivity attributable to the independent variables in the analysis, in our example, intelligence and energy. Intuitively, we should see that, unless energy and intelligence are completely independent ($R_{x1,x2}$ equals zero), we cannot simply calculate the bivariate, zero-order $R^2{}_s$ for intelligence and energy and add them together, because that sum double-counts the variation shared by the two independent variables. Figure 12.1 reveals the effect of double counting. If we ran two squared bivariate correlations and added the two together, the result would count the O value twice. Figure 12.1 helps us understand why two bivariate R^2 added together, double counts the joint variance.

Figure 12.1 Multiple Regression Diagram

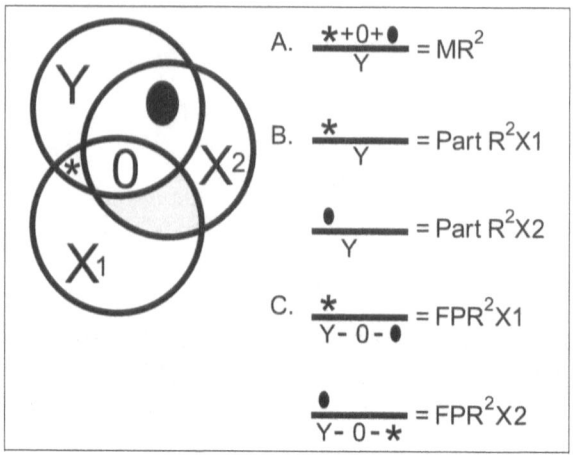

A. $\dfrac{*+0+\bullet}{Y} = MR^2$

B. $\dfrac{*}{Y} = \text{Part } R^2 X1$

$\dfrac{\bullet}{Y} = \text{Part } R^2 X2$

C. $\dfrac{*}{Y-0-\bullet} = FPR^2 X1$

$\dfrac{\bullet}{Y-0-*} = FPR^2 X2$

From figure 12.1, we see that simply running two bivariate regressions would double count the area O.

To avoid double counting, we first calculate a bivariate R^2 between intelligence (X_1) and productivity (Y). That value is $(0.6746)^2$ or 0.4451.[64] This yields the areas * and O infigure 12.1. Then we run a regression between energy (X_2) as the dependent variable and intelligence as the independent variable realizing the residuals. These residuals represent a new variable, energy, $X_{2.X1}$ with the effect of intelligence removed.[65] We call this calculation partialling and it lies at the heart of multiple regression. Figure 12.2 portrays the residual as a new variable.

Figure 12.2 The Residual as a New Variable

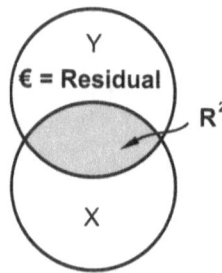

64.   We can, of course, simply square the value from the correlation matrix generated from step one.
65.   See that discussion in chapter 11.

Figure 12.2 shows the partialling effect. The regression of X_2 on X_1 produces a residual that is a new variable $X_{2.X1}$ with the effect of X_1 removed.

Then, we run a regression between productivity, Y and this residual, energy with intelligence held constant, $X_{2.X1}$, to determine the effect of energy on productivity. The $R^2_{Y.X2.X1}$ of that regression equals 0.1488. This yields the area • in figure 12.1. Compare this value to the value of the bivariate R^2 of Y on unpartialled X_2 $(0.5548)^2 = 0.3078$ in table 12.1. This process of partialling assigned the joint variance between intelligence and energy to the intelligence variable, the variable used first. Thus, the multiple effect (MR^2) of X_1, intelligence and, $X_{2.X1}$, energy with intelligence held constant, on productivity equals $0.4451 + 0.1488 = 0.5939$. Figure 12.1 shows MR^2 as the sum of × + O + • avoiding the double count of the cross-hatched part of the figure.

Intelligence and energy explain almost 60% of the total variation in productivity. If our analysis contained a third variable, e.g., gender, we would partial X_1 and X_2 from X_3, save the residuals as $X_{3.X1.X2}$, now gender with intelligence and energy partialled. Then regress Y on $X_{3.X1X2}$ and add the R^2 to 60%. At this point, we would run a protected F test, discussed subsequently, to determine overall model statistical significance.

The OLS equation

If our F test was statistically significant, we would proceed to step 3. Here, we estimate the OLS coefficients in the multiple regression equation, B_0, B_1 and B_2, so as to minimize the error term, $\Sigma(Y_i - Y_c)^2$, ε^2, the squared sum of the actual Y_i values minus the predicted Y_c values. B_1 is the partial regression coefficient for intelligence, with the effect of energy held constant and B_2 is the partial regression coefficient of energy with the effect of intelligence held constant.

To estimate B_1 we run a bivariate regression between intelligence and energy with intelligence as the dependent variable and realize the residuals. We repeat: these residuals represent a new value of intelligence with the effect of energy partialled or held constant. Then we run a bivariate regression between productivity, Y and the intelligence variable with the effect of energy removed (held constant), $X_{1.X2}$ to realize B_1. B_1 is the mean change in Y, per unit change in $X_{1.X2}$.

We estimate B_2 by regressing energy on intelligence and realize the residuals as a new energy variable, $X_{2.X1}$, with the effect of intelligence held constant. Then we run the regression Y on $X_{2.X1}$ to realize B_2.

The values of the fully partialled regression coefficients, B_1 and B_2 are 0.5702 and 0.7179 respectively. These values tell us how much, on the average, productivity changes with a unit change in $X_{1.X2}$, holding constant the effect of intelligence on energy and energy on intelligence.

Finally, we calculate B_0 using the following equation.

$$B_0 = M_Y - B_1 \times M_{X1} - B_2 \times M_{X2} \qquad \text{(Equation 12.3)}$$

The regression constant (intercept) equals the mean of Y minus the products of B_1 times the mean of X_1 and B_2 times the mean of X_2. B_0, the result of that calculation equals - 16.0507. Using these constants, we unveil the OLS multiple regression equation: $Y_C = -16.0507 + 0.5702 \times X_1 + 0.7179 \times X_2$. This multiple regression equation gives the fully partialled effects of intelligence and energy on productivity. Substituting values of intelligence and energy produces predictions of mean values of productivity.

The Part Coefficient of Determination, PTR^2

Step 4 estimates the Part Coefficients of Determination, PTR^2 for the two independent variables. We define PTR^2 as the proportion of the total variation in Y explained by each of the independent variables with the other independent variables in the equation held constant. In other words, we calculate the coefficient of correlation between productivity and intelligence, with energy held constant and square it. That value is $PTR^2_{Y,X1.X2.} = 0.2961$. Then we calculate $PTR^2_{Y,X2.X1}$ using the same strategy, regressing productivity on energy with intelligence held constant. That value $PTR^2_{Y,X2.X1}$ equals 0.1488; energy, independent of intelligence, explains about 15% of the total variance in Y.

Figure 12.1 clarifies: $PTR^2_{Y,X1.X2}$ between productivity and intelligence (with energy held constant) equals * divided by Y. $PTR^2_{Y,X2.X1}$ between energy and productivity, holding intelligence constant equals ● divided by Y. We interpret these calculations and the diagram to mean that intelligence, (independent of energy) and energy (independent of intelligence) explain about 30% and 15% respectively of the variance in Productivity. Note that the sum of the two PTR^2s, equals about 45%, a value less than the MR^2 of about 60% because the PTR^2 ignores the common variance, O in figure 12.1, between intelligence and energy. It is safe to point out that the joint variance between energy and intelligence explains an additional 15% (60% - 45%) of the variation in productivity.

Figure 12.1 MR² PTR² FPR²

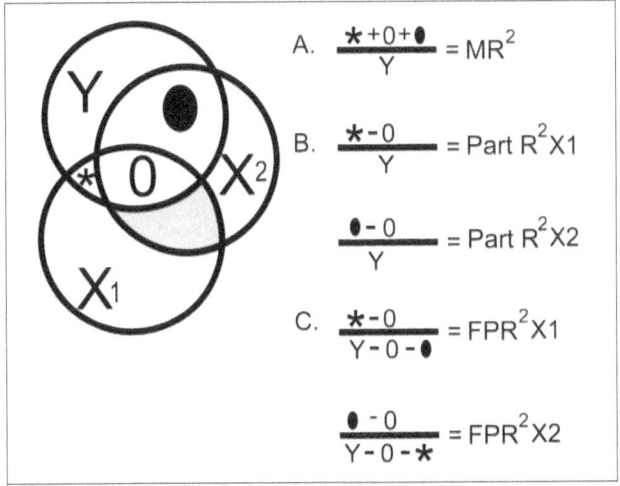

$$A. \quad \frac{* + 0 + \bullet}{Y} = MR^2$$

$$B. \quad \frac{* - 0}{Y} = \text{Part } R^2 X1$$

$$\frac{\bullet - 0}{Y} = \text{Part } R^2 X2$$

$$C. \quad \frac{* - 0}{Y - 0 - \bullet} = FPR^2 X1$$

$$\frac{\bullet - 0}{Y - 0 - *} = FPR^2 X2$$

The Fully Partialled Coefficient of Determination, *FPR²*

The proportion of variation in productivity explained by intelligence with energy partialled from both productivity and intelligence, defines the fully partialled coefficient of determination.

To realize the FPR^2 of productivity on intelligence, we partial energy from intelligence and productivity.

In our example, the FPR² of productivity on intelligence with energy held constant is about 42%. Again, referring to figure 12.1 (C), *FPR²* of productivity on intelligence with energy held constant equals × divided by Y - O - •. You can see that most of the time *FPR²* will exceed *PTR²* because we have partialled (removed) the effect of all the other independent variables from Y as well as X_1.

It is critical to note that in this example of a fully partialled OLS multiple regression, the effect of the joint variation between the independent variables on the dependent variable has been thrown into the regression intercept (see equations 12.1 and 12.3). Therefore, it has not been quantitatively lost, but lost interpretatively; the only interpretation in this analysis of the effect of this joint variation in the X variables on Y is to subtract from the multiple coefficient of determination (MR^2) the part coefficients of determination (PTR^2s).

Inferential Statistics with OLS Multiple regression

In step four, because our data represents a sample, size $N = 30$, we must calculate the appropriate inferential statistics.

The F Test

An F test specifies our first inferential assessment, testing the null hypothesis that the overall relationship, the coefficient of determination, MR^2 of the population probably equals zero (H_o: $MR^2 = 0$). The total variance of productivity, TV equals 5854.116. Figure 12.2 shows how we calculated TV. TV is the sum of the squared deviations of each individual Y_1 observations from the mean of Y, M_y. EV, the explained variance in Y attributable to X_1 and X_2, equals 3535.139. Figure 12.2 shows how we calculated EV. EV is the sum of the squared deviations of the predicted values of Y (Y_c) from the mean of Y. UV, unexplained variance, is that part of Y not attributable to the independent variables. UV equals 5854.139 - 3535.116 = 2318.978. In this example, we calculated UV by subtracting EV from TV. Figure 12.2 shows how UV is calculated independently: UV equals the squared sum of the deviations of each observations of Y_1 from the corresponding points on the regression line Y_c.

The estimate of the variance   based on EV is

$$S^2_{EV} = \frac{EV}{df} = \frac{3535.139}{2} = 1767.569.$$ The estimate of the variance based on

UV is $S^2_{UV} = \dfrac{UV}{df} = \dfrac{2318}{27} = 85.888.$ The F ratio is $\dfrac{S^2_{EV}}{S^2_{UV}} = \dfrac{1767.569}{85.888} = 20.580.$

Figure 12.2 TV, EV, and UV

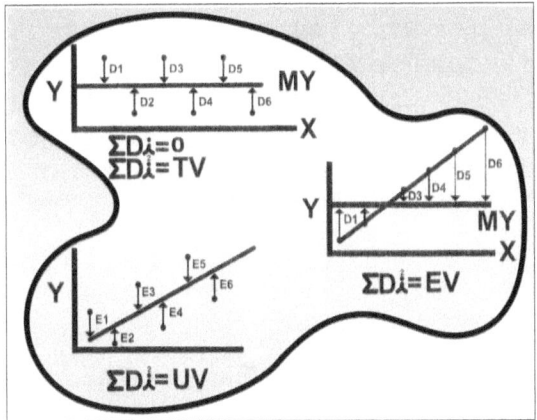

Figure 12.2 shows how TV, EV, and UV are calculated.

The probability of finding an F ratio this large, 20.580 for $df = 2, 27$, purely by chance if the population MR^2 equals 0, is less than 0.00001. Therefore, we reject the null (H_O: $MR^2 = 0$) and accept the alternative, H_A, that MR^2 in the population is greater than 0 and a relationship between productivity and intelligence and energy probably exists. Note: Again, we call this test a protected F test because it protect us from a type I error. If the equation does not pass the F test we are not entitled to proceed because a series of T or t tests without a significant F have additive errors. For example, if an analysis using three independent variables, each with an alpha risk of 0.05, fails the F test but proceeds to three T or t tests on each of the regression coefficients, an additive type I error has a probability of 0.15.

T (Z) or t tests of the Statistical Significance of the Regression Coefficients

We are entitled (required) to make these tests only if the F test is statistically significant. It (the F test) tells us that at least one of the independent variables is significant. But which of the independent variables are causing the significance of the F test?

To test the statistical significance of each of the regression coefficients, B_0, B_1, and B_2, we assume the existence of a normal sampling distribution for each coefficient whose mean is the true coefficient and its standard deviation is its standard error. We will not repeat the equations for the estimates of the standard errors of the coefficients as they parallel the estimate of the standard error of a bivariate relationship. In the case of our three coefficients, B_0, B_1 and B_2, the standard errors SE are respectively 14.8840, 0.2254, and 0.1269. In the case of each of the coefficients the null hypothesis is $H_0: B = 0$ and the alternative H_A is $B \neq 0$, a two tailed test or $B >$ or $<$ than zero, one tail tests.

Using the standard error of B_0, SE_{B0}, $T(Z)$ for $B_0 = \dfrac{-16.0507}{14.8840} = -1.078$.

The probability of a T value this large, if the null is true is, 0.2904,[66] larger than our alpha risk of 0.05. Therefore we cannot reject the null hypothesis that B_0 equals 0 . The standard error of $B_1 = SE\ B_1 = 0.2254$.

The T(Z) test for $B_1 = \dfrac{0.5702}{0.2254} = 2.530$. The probability of a T value this large if $B_1 = 0$ is less than 0.01. Therefore we reject the null hypothesis and accept the alternative that intelligence and productivity are related in the population.

Using the standard error for B_2, $SE\ B_2 = 0.1269$ we can make a 95% interval estimate for B_1: 0.3097 to 0.8306. Note: The interval estimate yields additional information beyond the T test; its interval estimate tells us how much random error is in our estimate.

For B_2, $T = \dfrac{0.7179}{0.1269} = 5.657$. The probability of a T value this large, if B_2 = 0, is less than 0.001. Therefore, we reject the null and accept the alternative that energy and productivity are positively related. A 95% interval estimate for B_2 equals 0.2554 to 1.1804. These interval estimates may suggest more error than we can tolerate in which case we must take a larger sample.

The Action Usefulness of OLS Multiple Regression

Prediction

Clearly, the fully partialled OLS regression model is an excellent device for making predictions. Our prediction equation, $Y_c = -16.0507 + 0.5702$

[66.]   We looked up this value from an F table. See chapter 7.

× X_1 + 0.7179 × X_2 enables predicting the productivity of potential employees by measuring intelligence and energy levels, substituting those values for X_1 and X_2 and solving for Y_C. Then we can calculate the standard error of estimate. The standard deviation of regression is the standard deviation of the error variable, the residuals, $\varepsilon_i = Yi - Y_C$.

We compute the standard deviation of regression ($SD_{YX1,X2}$) as

$$\sqrt{\frac{\Sigma(Y_i - Y_C)^2}{N - k - 1}}$$ (Equation 12.4)

N = sample size; k = number of predictors.

For our analysis, $SD_{YX1X2} = \sqrt{\dfrac{\Sigma(Y_i - Y_C)^2}{n - k - 1}} = \sqrt{\dfrac{470.61}{30 - 2 - 1}} = 4.17$

For an applicant with an energy score of 49 and an intelligence score of 110, we make a point estimate of her productivity by substituting 49 and 110 into the regression equation. That substitution yields a point estimate of 81.84. Then we can make a 95% confidence interval prediction of her productivity as 81.84 ± 1.96 standard deviations of regression or 81.84 x 4.17 = ± 1.96 = 90.01 to 73.67. A personnel officer, using this equation as one hiring criterion can, with 95% confidence, estimate the productivity of a potential employee with an intelligence of 110 and an energy level of 49.

Causal Inference

The following is a brief reminder of the requisite conditions for making causal inferences. If X causes Y:

X must precede Y in time,

X and Y must be related,

Some theoretical system (theory) or justifiable confederacy of hunches (soft theory) must link X to Y,

A manipulation of X must result in a change in Y in the hypothesized direction. Obviously, in many circumstances, we cannot meet this condition. Sometimes the manipulation is impossible or impractical. Or, sometimes the manipulation is unethical.

The effect of X on Y must involve some process, mechanism, or design eliminating, controlling, or isolating plausible alternative explanations for the change in Y. One such design is the classical experiment discussed in Chapter 9

Because we often cannot run classical experiments for reasons also discussed in chapter 9, we often must rely on passive data analysis to draw causal inferences. One method for causal inference with passive data uses OLS multiple regression to analyze nonexperimental data for establishing causal effects. In subsequent chapters we discuss other methods. For now there is one situation in which the fully partialled OLS multiple regression model works well.

Suppose we have a theory from which we deduce the hypothesis that a training method, X has an important causal effect on productivity. Our theory also tells us the other variables that affect productivity: intelligence, education, energy level, and personality. Assuming that we cannot run any kind of controlled experiment because it is impractical, to test the causal hypothesis, we can run a fully partialled multiple regression. Assume we take a random sample of 100 of our employees, 50 of whom have taken the training and 50 have not. We have all the variables in our data base so we can run an OLS multiple regression with Productivity as the dependent variable and intelligence, energy level, education, personality, and employee participation in the special training as independent variables. The equation for our analysis is as follows:

$$Y_{Productivity} = B_0 + B_1 X_{intelligence} + B_2 X_{energy} + B_3 X_{education} + B_4 X_{personality} + B_5 X_{method} + \varepsilon$$

If our analysis has met the following conditions: 1) we have not omitted a critical variable; 2) our measurements are valid and reliable; 3) the overall multiple regression is atavistically significant; 4) and the B_{method} coefficient is statistically significant with the hypothesized sign, we can logically infer the possibility of a causal effect of training. Think for a moment about what we did.

First, in this simplified model, we identified the variables affecting productivity that must be controlled. In this causal analysis, we call those variables confounders, because without controlling or eliminating their effect on productivity, we cannot claim the causal effect of training because we have not eliminated the confounding effects possible alternative explanations.

Second, we ran the multiple regression as specified above so that we partialled from training (coded 1 if the worker had the training, 0 if the worker did not) the effects of the other plausible causal variables. If training has a positive sign with statistical significance and with replicated verification in other places and other times and the training effect holds, we can argue that this training causes an increase in productivity.

Summary

The most important ideas discussed in this chapter are the following:

1. The correlation matrix

    The correlation matrix shows the zero order correlation coefficients among all the variables in the multiple regression.

2. The coefficient of multiple determination

    The coefficient of multiple regression represents the total percent of the variance in the dependent variable explained by the independent variables in the multiple regression.

3. The residual as a new variable

    If we regress Y on X_1, the residuals represent a new variable, Y with the effect of X_1 removed or partialled.

4. The multiple regression equation

    The multiple regression equation shows the regression constant plus fully partialled regression coefficients of the independent variable on Y, the dependent variable.

5. The multiple regression coefficients

   a. The regression constant (B_0)

   b. The regression coefficients $(B_1 \ldots B_N)$.

   The multiple regression coefficients show the effect, on average, of a unit change in each independent variable on Y with all the other independent variables partialled (held constant).

6. The part coefficient of determination (PTR^2)

   The part coefficients of determination show the percentage of the Y variable explained by each of the independent variables with the other independent variables held constant (partialled).

7. The fully partialled coefficient of determination $(FPR^{2)}$

   The fully partialled coefficients of determinations show the percentage of Y, with the other independent variables partialled from Y and each other, explained by each independent variable.

8. The protected F test

   Protected F tests the hypothesis that MR^2, the percentage of variation in Y, explained by all the independent variables in the analysis, equals zero. This test protects the regression against the additive probabilities of each independent variable.

9. T (Z) or t tests on the regression coefficients

   T (Z) or t tests the probability of each regression coefficient equaling zero.

10. Estimation using multiple regression

   We can use multiple regression as a predictive tool by finding independent variables (predictors) with strong relationships with the dependent (predicted) variable.

11, Causal inference using multiple regression

We can use multiple regression to test the causal effect of one variable (the causal agent) on a dependent variable (the caused variable) if we can identify and partial all the other plausible causal agents from the caused variable.

In the next chapter, our discussion focuses on how we find and discuss how to correct (or ameliorate) regression problems and failures to meet assumptions of the OLS regression model.

Chapter 12

Appendix

In the data set given below, we have six variables in a data set where $N = 30$. Your tasks are the following:

1. Put your imagination to work and assign names to Y and the Xs. Using your statistical program, run a univariate multiple regression and interpret your results.
2. In your interpretation, discuss the degree to which you might feel uneasy about your interpretation of the regression coefficients.
3. For the more intrepid, using a spreadsheet, calculate the values of B_0 and B_1, test the null hypothesis that B_1 equals zero, and interpret your findings.
4. Using your statistical program, make a 95% interval estimate of Y for an individual with X scores of 39.2, 41.6, 23.9, 4.06, and 18.22.

ID	Y	X1	X2	X3	X4	X5
1	91.14	43	38.50	23.04	19	4.16
2	86.77	33	32.55	20.93	18	4.35
3	113.54	55	84.70	28.30	20	4.20
4	90.71	42	51.23	22.18	17	4.71
5	109.15	52	61.32	27.22	20	4.24
6	83.94	45	80.45	22.25	10	4.38
7	100.53	50	75.62	27.35	13	4.91
8	87.23	33	51.23	20.27	10	4.57
9	82.31	33	62.67	28.55	11	4.66
10	92.20	39	40.34	22.70	13	4.01

11	118.76	55	99.96	30.65	22	4.59
12	113.89	55	79.09	28.12	23	4.16
13	106.34	46	49.48	25.57	17	4.37
14	95.72	43	41.66	25.52	13	4.69
15	127.1	55	100.73	36.42	20	5.83
16	87.43	36	48.27	23.39	10	4.61
17	97.39	43	48.90	23.84	16	5.22
18	102.55	55	51.50	28.46	16	5.07
19	90.83	50	40.54	22.75	15	4.02
20	113.11	51	68.39	27.79	20	4.17
21	83.66	33	64.22	21.42	17	4.30
22	79.21	33	43.46	25.23	18	4.51
23	105.40	55	72.33	28.54	19	4.80
24	86.76	43	47.50	21.53	10	4.66
25	114.47	55	92.30	29.22	20	4.40
26	102.33	51	43.35	26.72	14	4.56
27	112.48	55	74.71	27.56	20	4.09
28	74.22	41	40.56	24.89	10	4.92
29	75.14	36	62.29	20.23	13	4.39
30	104.44	45	42.90	25.49	15	4.05

CHAPTER 13

ASSUMPTIONS OF THE OLS REGRESSION MODEL

The Classical Regression Model, whether bivariate or multiple, makes certain assumptions, the violation of which often have dire consequences to the usefulness of an analysis. This chapter concentrates on the assumptions and the consequences of their violation. The chapters that follow address how to correct for violations of the assumptions. For simplicity, our explanations often use bivariate examples, but the ideas apply with equal force to multiple regressions

Correct Model Specification

OLS regression assumes that we have correctly specified our regression model. Correct model specification involves three elements: (1) linearity, (2) reliable and valid measurement, and (3) correct independent variables in the model. We discuss each of these below.

Linearity

This assumption relates to the specification of the form of relationships between the independent and dependent variables. Calculations of regression coefficients, tests of significance, and confidence intervals assume a linear relationship between the variables X and Y. If a straight line does not approximate the relationship between the variables, i.e., the actual form approximates curvilinear concave, convex, U or J shaped distributions, the result will yield biased estimates of R^2, B's, F, and T's. Table 13.1 shows an excerpt of a larger data set relating intelligence and productivity for five individuals.

Table 13.1 Excerpts of Productivity and Intelligence

ID	Productivity	Intelligence
1	44.7	106.17
2	43.96	100.64
3	44.34	110.51
4	51.49	104.35
5	67.7	130.02

Figure 13.1 plots the relationship between intelligence and productivity with a linear regression line and a Loess[67] plot using the complete data set.

Figure 13.1 The Relationship between Intelligence and Productivity

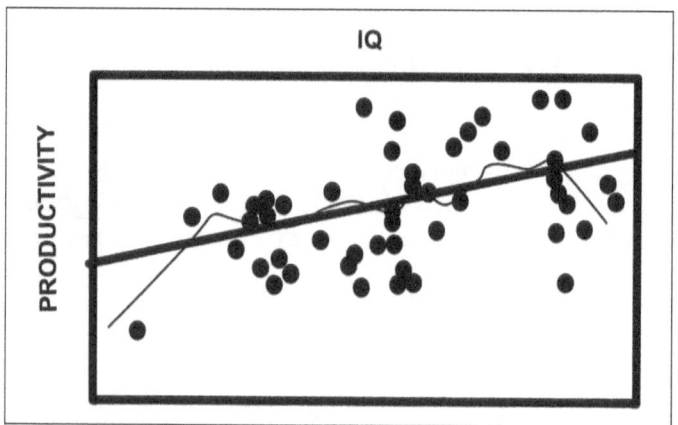

The loess plot suggests that some transformation might be appropriate, but we tried a linear fit anyway. For the linear regression, the statistically significant R^2 equaled 0.2211. The regression coefficients, B_0 and B_1, both statistically significant at alpha = 0.05, equaled 28.525 and 0.2061 respectively. Although statistically significant, the linear fit was not optimal. Some kind of transformation (we'll deal with that in

---

[67]   A loess plot developed by W. S. Cleveland in 1979, weights and smoothes original data.

chapter14) would increase MR^2, change the regression coefficients, and the F ratio.

Reliable and Valid Measurement[68]

Reliability (Revisited)[69]

Reliability refers to the consistency of a measurement, or whether a measured variable produces approximately the same value each time we use it. Thus, repeatability and consistency define reliability. We consider a measure reliable if it produces approximately the same value in repeated measurements of the same person or object. We do not measure true reliability, we estimate it. For example, a simple bathroom scale is designed to measure weight. One way to estimate the scale's reliability might measure the weight of a random sample of 100 people, reweighing them a second time without significant delay or intrusions that might affect weight, and correlating the two sets of weights. We call the correlation coefficient a reliability coefficient, R_{xx}. That reliability coefficient will probably never be 1.00 because no measure is perfectly reliable. The difference between the estimated coefficient and 1.00 is an estimate of measurement error. That is, each measured observation is assumed to consist of the true measure X_T plus $X\varepsilon$, a random error, so that a measured variable, $X_{Meas} = X_T + X\varepsilon$. Thus, we may think of any measured variable, X as composed of some *true* value plus random error. This random error is assumed to have a mean of zero and a zero correlation with the true score. If a measurement has a reliability coefficient of zero, then it measures nothing and will not correlate with any other variable. The lower the reliability of a measurement, the lower is the correlation with any other measure.

We can also estimate the reliability of a measuring instrument, e.g., a test with multiple items by measuring its internal consistency, i.e., by correlating each item on the test with every other item and averaging the correlations. Cronbach's alpha is one of the tests that uses this method; most computer programs have this routine.

---

68.   For much more extensive treatment of this issue see ZC, chapters 3 and 4 and KP, chapter 9.
69.   In chapter 1, we introduced this topic. This material is an elaboration of those ideas.

The point of this discussion is that unreliable measurements mask relationships, increasing the probability of a type II error. If you use variables with low reliability you probably won't find an effect even if it actually exists. Therefore, before you begin an analysis, you should always try to establish some estimate of a variable's reliability.

Validity (Revisited)

Reliability should not be confused with validity. A reliable variable measures something but not necessarily the something in which you have an interest. Thus, we can say a variable's reliability is a necessary condition for its validity. While a variable can be reliable but invalid, the variable cannot be valid and unreliable. We establish a variable's validity by correlating it with some other variable that reflects what we want to measure. To illustrate what we mean, let's ask why some persons regularly weigh themselves on a bathroom scale? One reason is the belief that weight correlates with health. So it is in this context that health concerns define the bathroom scales validity. Thus, to establish the scale's validity we would use a health score as a dependent variable, measure say 100 men of a certain age range and socio/economic circumstance, weigh the men on a bathroom scale, and run a correlation analysis between weight and health. The resulting R^2 becomes a validity coefficient.

Thus, we measure a variable's validity by correlating it with some process or pattern of behavior external to the variable. An opinion scale designed to measure attitudes toward pure food and drug laws, represents another example of measuring validity. One way to establish the scale's validity might use Young Democrats and Young Republicans on university campuses testing whether the scale significantly discriminates between the two groups. The correlation coefficient between the attitude test and the two groups measures its validity.

To put this discussion of validity in more formal terms, we say that there are three types of validity: Face validity, criterion validity, and construct validity. Examples of these types of validity provide their definition. Suppose we prepared a test of arithmetic skill: addition, subtraction, multiplication, and division. Then we asked a group of experts to examine our test to see if we have adequately measured the domain of arithmetic skills. If the experts affirm that the test adequately measures the domain, we say the test is face valid. Our example of the bathroom scale, weight, and health defines criterion validity, That is, a variable is criterion valid if it correlates with an appropriate variable that we measure independent of the variable on which we want to test validity.

With respect to construct validity, in chapter 1, we alluded to a complex socio/psychological concept, alienation. Using theory, sociologists developed a test designed to measure this concept and then proceeded to test its validity. One test used a large group of high school students measuring their participation in the high school community. Their validity hypothesis was that the alienation scale should discriminate between those students who actively participated from those who did not. Thus, what distinguishes criterion from construct validity is the theoretical basis for the measurement.

Correct Independent Variables in the Model

This assumption relates to the theory that guides the choice of independent variables used in a model. Excluding plausible variables has two consequences: First, the effects of the excluded variables are not recognized and second, their effects are not partialled from the included variables in the equation. The effect of violation is biased coefficient estimates and tests of statistical significance.

More formally, we say that excluding an independent variable, X_3 will cause the other included independent variables in the model, X_1 and X_2, if they are related to X_3, to be correlated with the residuals because the residuals contain X_3, the unmeasured effect. Thus, the effect of excluding X_3 biases the estimates of the included variables. Therefore, if the variables intelligence and energy influence productivity, the omission of energy biases the estimate of the effect of intelligence on productivity.

Including irrelevant variables that should not be in the model can also bias regression coefficients and significance tests. For example, in our regression of productivity on intelligence and energy, if we add a third independent variable related to intelligence and/or energy, but irrelevantly related to productivity, the probability of a type II error increases because we lose degrees of freedom and, because of the partialling of the wrong variable, reduce the effects of the included independent variables.

Correct model specification, then involves linearity, reliability, validity, and correct independent variables in the model. Violation of any of these assumptions results in biased regression coefficients and tests of statistical significance.

Constant Variance of Residual (Homoscedasticity)

The OLS regression model assumes that the variance of its residuals is constant throughput the total range of X values or, in the case of multiple regression, the Y_c (predicted) values that are a linear combination of the independent variables. If our model adheres to this assumption, we call it homoscedastic; if it fails to meet the assumption, we call it heteroscedastic. Violation of the assumption (heteroscedasticity) causes incorrect standard errors and significance tests, but does not bias regression coefficients. In practice, OLS regressions are robust in the face of violations of this assumption, but become problematic when the ratio of unequal variances within an independent variable set exceeds ten. Figure 13.2 is an example of heteroscedastic residuals.

Figure 13.3 Heteroscedastic Residuals

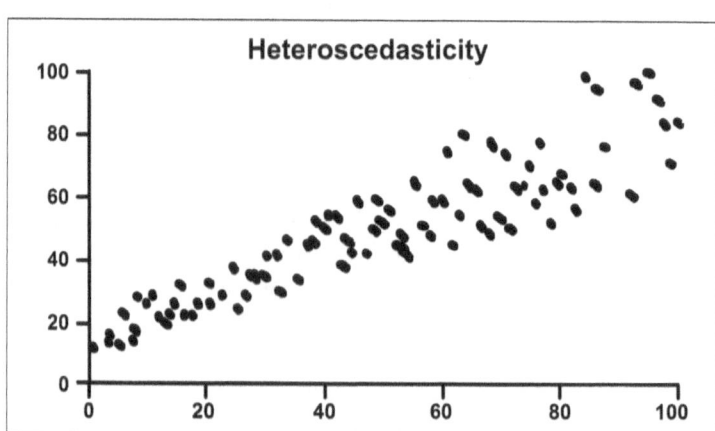

From figure 13.3, we can see the variance of residuals becoming larger as X increases. Residual variance can also decrease as X or Y_c increases or bulge in the middle, i.e., increase to a point and decline.

Uncorrelated Residuals

We call the condition of correlated residuals, autocorrelation. Autocorrelation most often occurs with time series for reasons we explain in part 5. However, violations can also occur with cross sectional data and, in that context, we call the condition serial correlation. Clustered or temporarily linked data are often serially correlated. For example, a random sample of people taken from adjacent neighborhoods will often result in serial correlation because people within the same or adjacent socio/economic neighborhood(s) are more alike than people from a different socio/economic neighborhoods. Autocorrelation causes incorrect standard errors and significance tests, but does not bias coefficient estimates. If, however, we are analyzing time series instead of cross sectional data and the residuals are autocorrelated (which is highly likely), our regression effects may only reflect the autocorrelation. In part 5 we elaborate that notion. Autocorrelation also signals non random error terms putting in doubt the assumption that we are dealing with risk.

Figure 13.4 depicts positive and negative autocorrelations. With positive autocorrelation, if one residual goes down, the next one goes down or if one goes up the next goes up. With negative autocorrelation, if one residual goes up, the next goes down; if the next goes down, that which follows goes up.

Figure 13.4 Plots of Autocorrelated Residuals

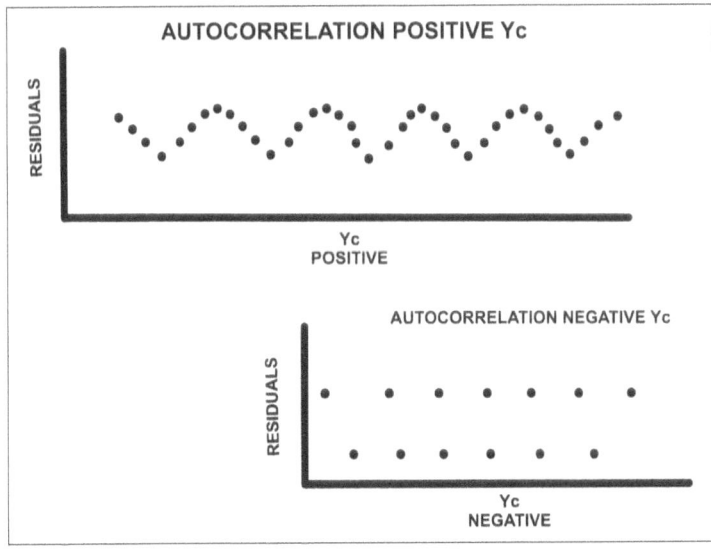

We can usually detect autocorrelations with plots like figure 13 4. Lagged residual correlations also give strong evidence of autocorrelation. By lagged correlation we mean correlating observation one against observation two, observation two against three, and so on through the rest of the series.

Independent Variables Uncorrelated with Residuals

In OLS regression, the dependent variable, Y will correlate with the residuals because the residuals represent the amount of the variance in Y that is not explained by the independent variable(s). But, the independent variables should not correlate with the residuals. If they do, it means that there are unmeasured effects not included in the equation that correlate with the independent variable(s) (see the discussion of correct model specification).

In addition, reciprocal causation also causes a correlation between the independent variables and the residuals. If we run a regression $Y = f(X)$ and then run a another regression $X = f(Y)$, the independent variables in these two equations will correlate with the residuals. To illustrate, consider equations 1 and 2 below.

1. $X_3 = B_{03} + B_{13}X_4 + \varepsilon_3$
2. $X_4 = B_{04} + B_{14}X_3 + \varepsilon_4$

In these two equations the two residuals, ε_3 and ε_4 correlate with X_3 and X_4.[70] This means that the two independent variables correlate with their respective residuals. We call this model reciprocally causal and its effect biases the regression coefficients.

Normally Distributed Residuals

OLS assumes that the residuals approximate a normal distribution. With large samples, violations of normality do not affect standard errors or confidence intervals. Non-normal residuals do signal, however, serious problems of model misspecification. Again, treating the regression as a risk model is based on a pattern of residuals that allows the calculation of the occurrence of very large or very small values, e.g., the normal distribution.

---

[70]   If you have difficulty seeing why this is true, we elaborate in chapter 20.

OTHER FORCES AFFECTING THE OLS
MULTIPLE REGRESSION MODEL

Multicollinearity

If the independent variables in a multiple regression relate to one another, we say that the variables are multicollinear. Multicollinearity can be small, moderate, or severe. Any collineraity will make the regression coefficients smaller than the zero-order coefficients in a bivariate analysis because of partialling the joint variance between all the independent variables. This partialling removes the joint variance from the independent variables and dumps it into the regression constant, B_0. Thus, the fully partialled OLS multiple regression model will not allow a complete interpretation of the effects of the common variance between the independent variables.

You can see this condition by looking again at the equation for b_0, the regression constant.

$$b_0 = m_y - b_1 \times m_{x1} - b_2 \times m_{x2}$$

If the x variables are collinear, the regression coefficients will be smaller than their bivariate coefficients because of the partialling and, because they are smaller, the value of b_0 will be larger. The larger the collinearity between the independent variables, the smaller is the partialled coefficients, and the larger are the regression constants, b_{0s}.

If our research interest focuses solely on prediction, modest collinearity does not affect predictive outcomes. However, in the face of severe multicollinearity, bias infects the entire multiple regression output, because severe multicollinearity can produce variables with little or no variance so that almost any regression line fits equally well. Figure 13.5 depicts this phenomenon.

But if our research interests lie in estimating the effect of any of the independent variables, even when mr^2 is statistically significant and substantial, modest[71] multicollinearity casts serious doubts on analytical interpretation. The partialling effect of the analysis will not only reduce the size of the estimates of the regression coefficients, but may also fail

---

[71]. In chapter 15, we will show you how to distinguish degrees of collinearity.

to show statistically significant coefficient effects of any independent variable. Such a circumstance defies good sense because something must be responsible for the significant mr^2.

Figure 13.5 The Effect of Severe Multicollinearity

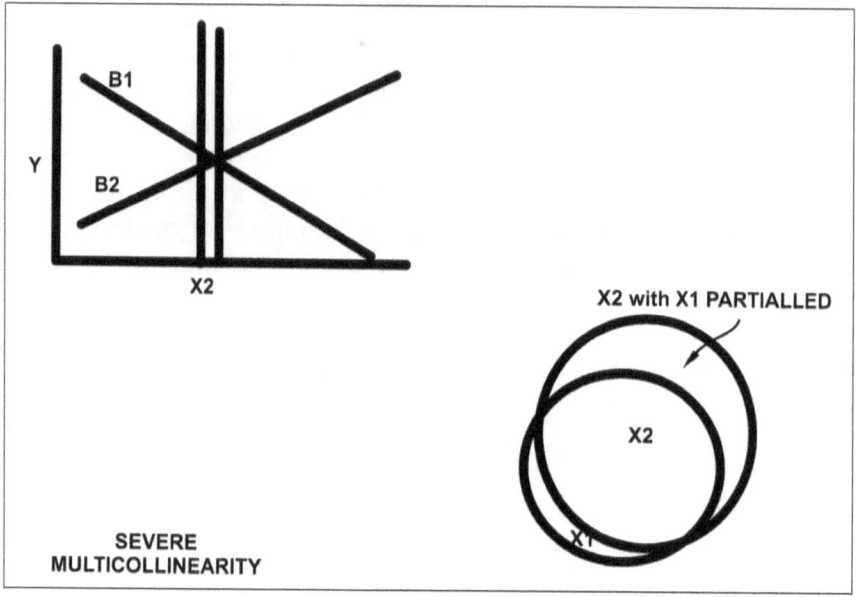

You can see from figure 13.5 that after partialling X_1 from X_2, X2 has little variation left in it, almost any regression line and regression coefficient will fit equally well.

Outliers

We define outliers as extreme values that do not fit the rest of the data. Outliers can affect only a single variable (univariate) or affect the analytical output of bivariate or multiple regressions. Outliers have three causes: (1) simple errors in data collection and recording, (2) the inclusion of an accurate but rare case, e.g., genius teachers in a study of teaching effectiveness, or (3) anomalies not anticipated by theory. Whatever the cause, outliers can substantially affect description, inference, and analysis of data. In chapter 3, we showed how an outlier could affect description, i.e., the mean. Here we show how an outlier can affect a bivariate regression. Figure 13.6 plots a regression of Y on X without an outlier.

Figure 13.6 Regression Without an Outlier.

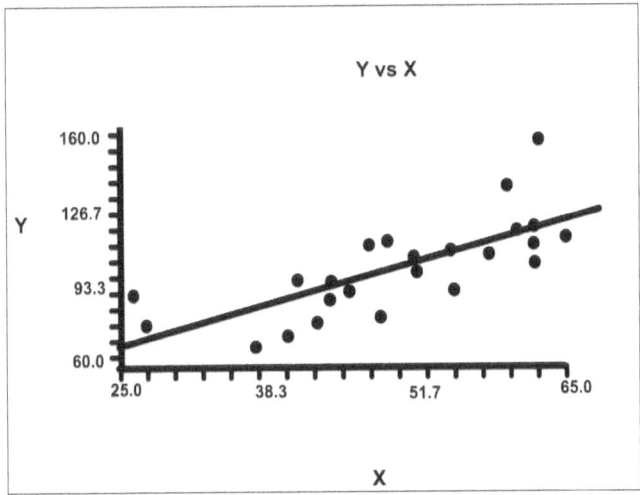

In figure 13.6 you can see the regression line running through the data without an outlier. The statistically significant R^2 equals 0.447. Figure 13.7 Plots a regression of the same data with an outlier inserted.

Figure 13.7 Regression with an Outlier

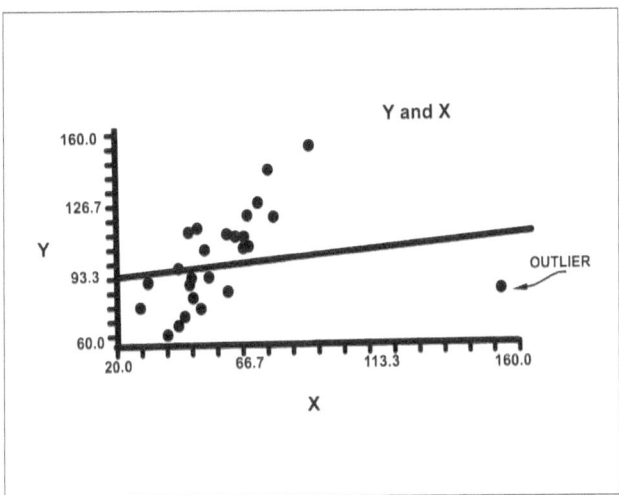

Now you can see the effect of the outlier as OLS seeks to minimize the sum of the squared residuals, $\Sigma\epsilon^2$, pulling the regression line with a large slope down and to the right almost horizontal to the x-axis. For

OLS to minimize the sum of ε^2, it must follow the outlier, lowering R^2, and making the regression statistically insignificant. The R^2 declines from 0.45 in figure 13.6 to 0.03 in figure 13.7.

Missing Data

Missing data have multiple causes: careless recording, persons refusing to answer certain questions on a questionnaire or test items on surveys, confidentially issues, and other undefined problems. Some data go missing randomly. In other cases, the missings are systematic, e.g., censorship. Taking data from chapter 12, we randomly deleted four observations from the independent variable and ran a bivariate regression. We show the results in table 13.8.

Table 13.8 Regression with Missing Data

R^2	0.343
B_1	1.160
T	4.312
Probability of a Type I Error	0.005

Then, we reran the regression with the missings replaced and summarize the results in table 13.9.

Table 13.9 Regression without Missing Data

R^2	0.447
B_1	1.259
T	4.315
Probability of a Type I Error	0.00026

You can see that even randomly missing data can affect all the critical estimates in a regression analysis. In chapter 15, we address useful methods for handling missing data.

Summary

The critical concepts discussed in this chapter are the following:

1. Assumptions of the OLS univariate regression model. These assumptions are the following:

   a. Correct Model Specification. Correct model specification includes

      i. Linearity
      ii. Reliable and Valid Measurement
      iii. Correct independent variables in the model

   b. Homoscedastic Residuals
   c. Uncorrelated Residuals
   d. Normally distributed residuals

2. Other forces affecting OLS multiple regression

   a. Multicollinearity
   b. Outliers
   c. Missing Data

In chapters 14, 15, and 16, our discussion focuses on how we find and correct (or ameliorate) failures to meet assumptions and other regression issues pertaining to the OLS multiple regression model.

CHAPTER 14

CURVILINEARITY IN OLS REGRESSION

In this chapter we discuss methods by which we detect and remedy nonlinearity. We note that before running a regression (or any other kind of analysis), you should always make a complete description of all the variables in the data set. Minimally, your descriptions should include the mean, median, standard deviation, skewness, kurtosis, scatter diagrams between each of the independent variables with the other independent variables as well as the independent variables and the dependent variable. You should also show graphics for each variable including normal plots, scatter plots, stem-leaf plots, box-whisker plots, and histograms.

These descriptions function as alerts to potential problems that may arise in analysis, e.g., univariate outliers, missing data, severe data gaps, nonlinear relationships, and non-normal distributions.

Monotonic and Non-monotonic Curvilinearity

A monotonic nonlinear relationship preserves the data's order. That is, if the relationship between X and Y is positive, it remains positive throughout all values of Y for any given value of X. A non-monotonic, nonlinear relationship does not preserve the order. In a nonmonotonic relationship between X and Y, the function may be positive (or negative) as X increases, and opposite throughout the subsequent range of X. A U shaped relationship is an example of a non-monotonic function. Figure 14.1 illustrates these nonlinear forms.

Figure 14.1 Monotonic and Non-monotonic Forms

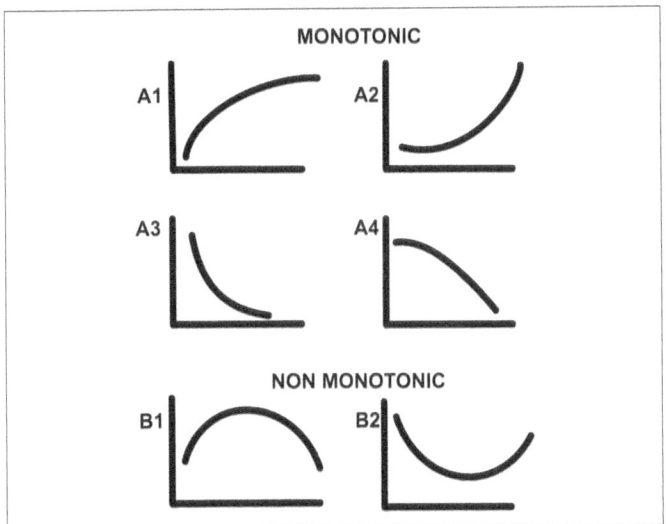

In figure 14.1 we see in figures 14.1.A1 through 14.1.A4 monotonic relationships. In figures 14.1. B_1 and 14.1. B_2, we show nonmonotonic curvilinear relationships.

Detecting Non-Linearity

In business, sociological, historical, and educational research, we generally lack theoretical guidance in explaining nonlinear relationships. Lack of theoretical guidance forces empirical solutions. One empirical solution uses a series of scatter diagrams with lowess plots between the dependent variable and each independent variable. If these plots suggest the presence of nonlinearity, we run a series of simple regressions between the dependent variable and each of the independent variables realizing the residuals from each regression. Using these residuals as surrogates for the dependent variable, we plot scatter diagrams and lowess curves with the independent variable on the X axis. These plots confirm or disconfirm nonlinearity and suggest a remedy.

In these discussions we use a hypothetical random sample, N = 50, with productivity as the dependent variable and intelligence, energy level, and gender (female = 0) as independent variables. We begin the search by constructing a scatter diagrams and lowess plots of the dependent variable on the independent variables.

Figure 14.2 shows the plot of productivity on intelligence with a linear regression line and a lowess plot. Using a moving regression and a precise set of rules specifying how to solve for an appropriate value of Y_c for given a value of X, the lowess plot shows a relatively smooth curve representing the relationship between X and Y. The algorithm defining the lowess plot forces it to run through the middle of the data, summarizing the relationship between X and Y.

Figure 14.2 Productivity and Intelligence with Lowess Plot

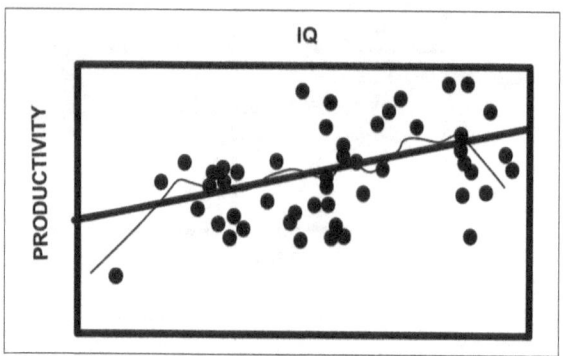

The lowess plot suggests a non-monotonic, nonlinear relationship. As X increases Y increases, until, at the higher end of X, Y begins to fall,

To confirm the nonlinearity suggested by the lowess plot we run a bivariate linear regression between productivity and intelligence, save the residuals, and plot the residuals against intelligence. We show this plot in figure 14.3.

Figure 14.3 Residuals of Productivity Plotted on Intelligence with Lowess Plot

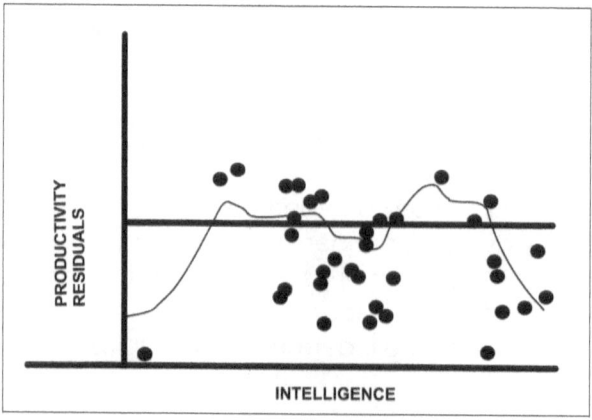

Partialling intelligence from productivity, shows, as expected, the flat regression line. The lowess fit of productivity with intelligence held constant, confirms our previous plot of an inverted bowl shaped relationship between intelligence and productivity, suggesting the use of a nonlinear fit. The point of using the residuals in figure 14.3, centers on removing the linear effect and checking to see if some non linear pattern persists.

This plot suggests a nonmonotonic, inverted bowl-shaped relationship between energy and productivity.

Solutions to Nonlinearity

As a first step in correcting for nonlinearity, we ran a bivariate linear regression of productivity on intelligence, with the results summarized in table 14.1.

Table 14.1 Regression of Productivity on Intelligence

	Regression Coefficient	Standard Error	T (Z) Value	Probability of Type I Error
Intercept	54.261	13.31	4.074	< 0.01
Intelligence	0.4284	0.131	3.287	< 0.01

Statistically significant at $P < 0.01$, the relationship yields a regression coefficient (B_1) of 0.4284, telling us that a unit increase in intelligence produces an average increase in productivity of about 0.43. This regression, although interesting, misses, as suggested by the lowess plot, the possible nonmonotonic curvilinear relationship between productivity and intelligence.

Can we do better by making a transformation that gives a clearer picture and more information? Indeed, we can do better by running a quadratic polynomial and testing the null of a nonmonotonic relationship.

The Quadratic Polynomial

One method for fitting nonlinear relationships of almost any shape uses a power polynomial. The quadratic equation provides an example.

Quadratic equations take the form of $Y_C = B_0 + B_1X + B_2X^2$. Looking closely at the equation, we see that the quadratic fit simply creates a new X variable, intelligence-squared, the purpose of which models the curved portion of the hypothesized relationship.

Before creating the quadratic variable, X-squared, we reexpress the X variables by centering,[72] i.e., subtracting the mean from each observation in the series. The purpose of centering is to lower the dependence of X and X-squared. Intuitively, we sense that if we add an independent variable, squared intelligence to our equation, we introduce a high degree of collinearity between X and X-squared. Centering X reduces the collinearity.[73]

The reason why centering reduces collinearity is that the arithmetic mean of X and X^2 partly determines the covariance between X and X^2. Algebraically, the Covariance of Centered X, and Centered X^2 equals zero if the two variables are symmetrically distributed and their means equal zero. In other words, two types of collinearity obtain: (1) artificial collinearity produced by scaling (as in squaring a variable or multiplying two variables together) and (2) real, collinearity produced by skewness.

Because centering reduces collinearity, the reexpression increases our chances (lowers the probability of a type II error) of finding a significant nonlinearity in the squared variable.

We now run a centered multiple regression of $Y_{CPRODUCTIVITY} = B_0 + B_1 CIntelligence + B_2 CIntelligence^2$.

Table 14.2 summarizes that regression.

---

[72]   This idea of centering was borrowed from JCC, pp. 34 and 201. Cohen does not cite a reference so we assume it was his idea to apply this mathematical notion to regression or he may have thought that it was intuitively obvious. In any case, the concept is very useful. Notice that we used the term reexpress to describe centering because it does not change a variable's structure.

[73]   If X is perfectly symmetrical, centering X before squaring will make the correlation coefficient between x and x-squared zero.

Table 14.2 Regression of Productivity on Centered Intelligence and Centered Intelligence Squared

	Coefficients	St. Error	T (Z) Ratio	Probability of a Type I Error
Intercept	100.36	2.4237	41.41	0.0000
CIntel	0.44	0.1273	3.48	0.0011
CIntel²	-0.012	0.0066	-1.97	0.0307
MR²	0.24			

The overall model and the two independent variables are statistically significant. How do we interpret this regression? Certainly, we have improved the fit. The R^2 jumped from 0.18 to 0.24 and the significant curvilinear regression coefficient has a negative sign. A simple interpretation of the curvilinear effect calculates the predicted values of productivity and plots the predicted values against intelligence. Figure 14.5 shows this plot.

Figure 14.4 Predicted Values of Productivity and Intelligence

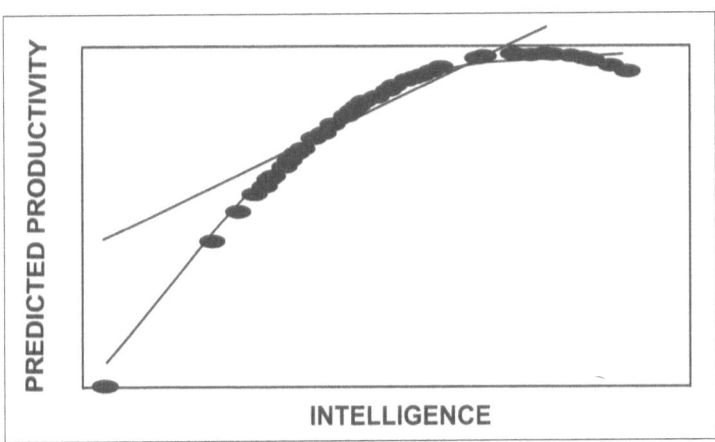

As intelligence increases, productivity increases up to a maximum point, then declines suggesting that the job requires intelligent persons, but not too intelligent. We can actually calculate this intelligence maximum point. For the quadratic, we have one bend at a value of intelligence that maximizes average productivity, $Y_{C\ MAX}$. We call that value the maxima, and calculate X_{MAX}, with the following equation:[74]

$$X_{MAX} = \frac{-B1_{IQ}}{2B2} = \frac{-.4437}{2*0.0124} = \frac{-.4437}{0.0248} = -17.48. \qquad \text{(Equation 14.1)}$$

The equation says calculate the value of I_{MAX} (the intelligence value that will maximize productivity) by dividing the negative value of B_1 by 2 times B_2. Thus, IX_{MAX} equals 17.48.

Because we used the centered scores in the regression, we now restore our metric by adding the mean of intelligence to this value, reexpressing back to the original metric, an intelligence score of 83.60,[75] at which productivity is maximized. For any personnel manager, this calculation yields an applicant's intelligence that maximizes productivity. We can also calculate $Y_{C\,MAX}$ directly using equation 14.2 below.

$$Y_{C\,MAX} = \frac{4(B_2)(B_0) - B_1^2}{4(B_2)} =$$

$$\frac{4(-0.0124)100.3568 - 0.4437^2}{4(-0.0124)} = \frac{-0.0496 * 100.3568 - 0.1969}{-0.0496} = 99.58 \text{ (Equation 14.2)}$$

Equation 14.2 says that, on average, productivity reaches its maximum at $Y_{C\,Max} = 99.58$, after which it declines. So a person with an intelligence score of about 84 has a predicted productivity of about 100. It certainly behooves a personnel manager to hire people, everything else being equal, with intelligence scores of around 84.

We ran the same analysis, this time using productivity and energy, but the squared value of energy was not statistically significant, causing us to conclude that its relationship was probably linear. We then ran the full multiple regression of productivity on intelligence, energy, and

---

74.   See JCC, pp. 205-206. In using this equation, you are doing a little applied calculus. While the equation looks formidable, do not be spooked by it. Simple arithmetic will do it.

75.   We note that this value is not an Intelligence Quotient on which 100 is the mean normal, but a raw score.

gender: $Y_C = B_0 + B_1 CIntel + B_2 CIntel^2 + B_3 Energy + B_4 Gender$. Table 14.3 summarizes that regression.

Table 14.3 Productivity on Energy, IQ, and Gender

Independent Variable	Coefficient	Standard Error	T(Z) Value	Probability of a Type I Error
Intercept	104.76	2.3133	45.285	0.0000
CEnergy	-0.5747	0.1233	-4.661	0.0000
CIntel	1.1474	0.132	8.695	0.0000
CIntel2	-0.0119	0.0047	-2.559	0.0139
(Gender=1)	-10.7386	4.0883	-2.627	0.0117

$MR^2 = 0.6474$

The model's MR^2 (0.6474) was statistically significant at $P < 0.05$ as were all the independent variables. Energy is negatively related to productivity, that is, the lower the energy level, the higher the productivity, a counterintuitive finding. We interpret the negative coefficient for gender, coded zero = female; one = male, to mean that females on average produce about 11 points more than males. To find the maxima for productivity on intelligence, we simply repeat the calculation in Equation 14.2 with the partialled coefficients. The statistical significance of squared intelligence confirms our test of curvilinearity.

Power Transforming X and Y Variables as a Solution to Monotonic Nonlinearity[76]

To explain this idea, we assume that a plot of Y on X reveals a nonlinear monotonic relationship with one bend. Figure 14.5 serves as a reminder of the look of a nonlinear monotonic relationship.

---

[76] Once again, we borrowed the essence of these ideas for this discussion from JCC, pp. 233-236.

Figure 14.5 A Nonlinear Monotonic Relationship

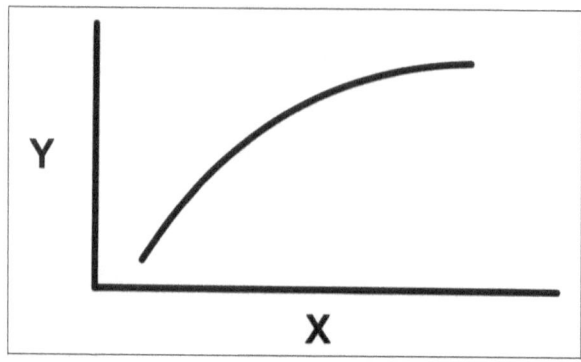

An effective way to solve monotonic nonlinearity uses a power transformation. Power transformations denote a set: reciprocals, $\frac{1}{X}$, logarithms, lnX, roots, $\sqrt{X}$, and powers, X^2, While it may not be obvious, all these transformations are power transformations represented by $X \rightarrow X^\lambda$ where λ is the power to which X must be raised to achieve linearity. For example, we can designate a reciprocal transformation, $\frac{1}{X}$, by a negative power, X^{-1}, a root transformation, e.g., the square root, by a fractional power, $X^{0.5}$, and by a power greater than one, X^2. In addition, a log transformation, lnX by a zero power, X^0. How do these power transformations achieve linearity in monotonic nonlinear relationships? We achieve linearity by either contracting or stretching the tail of either the X or Y variable. Roots, logarithmic, or reciprocal transformations contract (pull in) tails; transformations with roots greater than one expand (push out) tails. If we can fix nonlinearity by pulling in or pushing out the tails of variables, then we see that it is skewness in X or Y that causes nonlinearity in all monotonic relationships.

In table 14.6, X and Y, both symmetrical, are increasing by a constant amount ploting as a staright line. Figure 14.6 shows that plot.

Table 14.6 Two Variables Increasing by a Constant Amount

Y:	2	4	6	8	10	12	14	16	18	20
X:	1	2	3	4	5	6	7	8	9	10

Figure 14.6 Plot of Two Symmetrical Variables

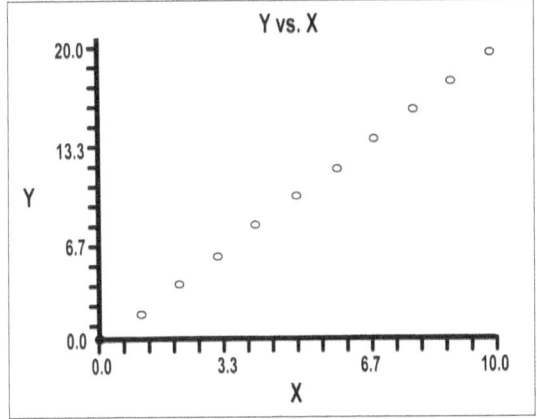

Figure 14.6 shows the linearity of a relationship between two symmetrical variables, in this case both variables increasing by a constant amount

In table 14.7, the Y variable doubles (skewed) (increases at a constant rate) as X increases by a constant amount (one) (symmetrical).

Table 14.7 A Non-linear Relationship

```
Y:  2  4  8  16 32 64   128256
X:  1  2  3  4  5  6    7  8
```

Figure 14.7 shows this relationship.

Figure 14.7 Plot of a Symmetrical X and Skewed Y

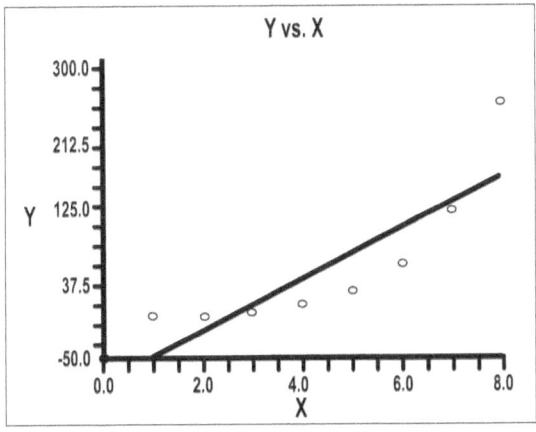

The plot shows this relationship as curvilinear monotonic. A log transformation of Y will make this relationship linear by pulling in the tail of Y and make it symmetrical because a log transformed variable that is increasing at a constant rate will plot linearly. A power transformation greater than 1 of X would expand its tail to make it also skewed. Obviously pulling in the tail of Y is more appropriate because it is the skewed variable. So let us try the natural log of Y. Table 14.8 shows the log transformed Y variable.

Table 14.8

LnY:	0.301	0.602	0.903	1.204	1.505	1.806	2.107	2.408
X:	1	2	3	4	5	6	7	8

In table 14.8 the natural log of Y causes Y to increase by a constant making it linear with X. The log transformation of any variable increasing by a constant rate will make that variable symmetrical. In this case, Y is doubling each time. Figure 14.8 shows this relationship.

Figure 14.8 Shows LnY plotted against X

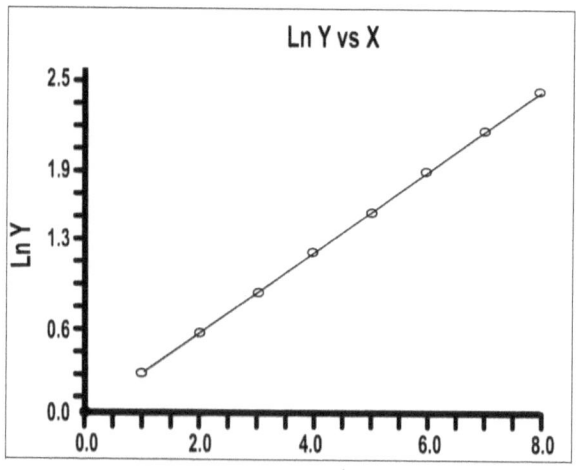

By making Y increase by a constant amount, i.e., pulling in its tail, we have made the relationship linear. We repeat: we can also achieve linearity by transforming X, i.e., by pushing-out its tail to correspond to the skewed Y. But in this case, with a skewed Y and symmetrical X, that doesn't make much sense

Summary

The critical ideas discussed in this chapter were:

1. Monotonic nonlinearity

    A curvilinear relationship that preserves the data's order, i.e., a positive relationship between X and Y remains positive throughout all values of Y for any given value of X.

2. Nonmonotonic nonlinearity

    Non-monotonic nonlinearity is a nonlinear relationship that does not preserve the order. In a non monotonic relationship between X and Y, the function may be positive (or negative) as X increases, and opposite throughout the subsequent range of X.

3. Quadratic transformations with nonmonotonic curvilinearity

    One method for fitting nonlinear relationships of almost any shape uses a power polynomial. The quadratic equation provides an example. Quadratic equations take the form of $Y_C = B_0 + B_1X + B_2X^2$. Looking closely at the equation, we see that the quadratic fit simply creates a new X variable, IQ squared, IQ^2, the purpose of which models the curved portion of the relationship.

4. Centering to reduce collinearity with quadratics

    Before creating the quadratic variable, X^2, we reexpress the X variables by centering, i.e., subtracting the mean from each observation in the series. This reexpression does not change the variable's structure; it simply causes the means to equal zero (0). Amazingly, centering X lowers the dependence of centered X and centered X squared. Intuitively, we sense that if we add an independent variable, intelligence[2] to our equation, we introduce a high degree of collinearity between X and X^2. Centering X reduces the collinearity between X and X^2. Because centering reduces collinearity, the reexpression increases our chances (lowers the probability of a type II error) of finding a significant non linearity in the squared variable.

5. Power transformations to achieve linearity

    Power transformations denote a set: reciprocals, $\dfrac{1}{X}$, logarithms, InX, roots, $\sqrt{X}$, and powers, X^2. While it may not be obvious, all these transformations are power transformations represented by $X \rightarrow X^\lambda$ where λ is the power to which X must be raised to achieve linearity. For example, we can designate a reciprocal transformation, $\dfrac{1}{X}$, by a negative power, X^{-1}, a root transformation, e.g., the square root, by a fractional power, $X^{0.5}$, a fractional power, a squared transformation, of course, by a power greater than one, X^2. In addition, a log transformation, InX by a zero power, X^0. How do these power transformations achieve linearity in monotonic nonlinear relationships? We achieve linearity by either contracting or stretching the tail of the X or Y variable. Roots, logarithmic, or reciprocal transformations contract (pull in) tails; transformations with roots greater than one expand (push out) tails.

6. The role of skewness in creating nonlinearity

    If we can fix nonlinearity by pulling in or pushing out the tails of variables, then skewness in X or Y causes nonlinearity in all monotonic relationships.

Chapter 14

Appendix

This data set represents a curvilinear relationship between Y and X. Your tasks are the following: (1) determine the nature of the curvilinearity, (2) make the appropriate transformation, and (3) explain your work.

Y	X
134	125
117	124
98	48
90	50
95	47
103	73

104	121
82	51
118	125
117	122
134	128
92	46
120	125
120	123
89	52
108	108
109	105
98	58
122	129
116	115
134	127
80	45
90	53
97	48
85	49
97	59
112	116
111	107
121	121
93	44
93	66
115	130
100	65
119	125
99	42
112	112
133	120
119	126
96	47
123	124
102	120
132	125
109	45
93	46
101	123
101	45
98	47
113	110
105	51
84	54

CHAPTER 15

MISSING DATA AND OUTLIERS

In chapter 12 we discussed how missing data and outliers affect regression coefficients and probabilistic inferences. In this chapter, we show how to detect and ameliorate the effects of these problems.

Missing Data

All the statistical programs we cited will detect and identify missing data; some will provide solutions.

Missing data have several causes.

1. Recording error
2. Incomplete data collection
3. Non response to questionnaires
4. Privacy issues
5. Privileged information

You must make every effort to retrieve missing data, using the corrective methods discussed here only after making strong efforts to retrieve the missings. While mathematical statisticians have made major theoretical progress in ameliorating the effect of missing data, there are no substitutes for the real thing. If irretrievable, you must try to understand why the observations are missing, because the answer to why affects your action decision. For example, if privileged information causes a large number of missings in a single variable or throughout the entire data set, you should consider dropping the variable or the entire data set or collecting a new sample.

Missing data can be random or systematic. When a variable has completely random missings, the missings will not be related to the other variables in the analysis and this randomness makes estimating methods unbiased. Nonrandom (systematic) missing data will be related to variables in the model or to variables that affect the outcome of analysis. Large samples with small percentages of missings, i.e., < 3%, probably permit ignoring the problem. With small data sets or data sets with a high percentage of missings, we must take action.

There are multiple methods for dealing with missing data. To illustrate and explain how to use these methods, we generated a hypothetical random sample with multiple missing observations. Table 15.1 shows an abbreviation of that data set.

Table 15.1 Data With Missing Observations, M = Missing (N = 200)
Only a fraction of the data is shown.

ID	Productivity	Intelligence	Energy	Gender 1= Female; 0 = Male
1	57.77	107.52	236.72	M
2	38.51	75.12	202.83	1
3	41.12	M	194.19	1
4	47.31	79.17	200.45	0
5	43.68	89.95	M	0
6	44.05	44.38	198.73	1
7	44.76	90.74	193.02	0
8	M	112.42	266.13	1
9	53.13	74.56	269.41	0
10	45.67	92.34	225.91	0
11	48	93.69	196.79	1
12	39.26	82.12	210.56	0
13	45.27	100.64	244.89	1
14	41.6	94.88	187.37	0
15	42.35	70.5	228.77	0
16	61.39	83.54	235.18	1
17	56.63	86.25	238.18	1
18	M	103.92	185.95	0
19	49.18	86.16	220.62	1
20	40.37	60.27	192.71	0

********	********	********	********	********
********	********	********	********	********
********	********	********	********	********
187	M	56.42	206.39	1
188	38.63	85.86	179.97	0
189	52.51	76.01	222.91	0
190	45.16	96.32	219.59	0
191	58.57	95.31	309.9	1
192	49.21	93.24	215.31	1
193	56.16	M	232.77	1
194	52.26	84.94	247.47	0
195	43.84	84.64	189.28	0
196	43.22	105.87	222.11	0
197	44.79	91.97	222.29	1
198	44.98	105.39	213.75	0
199	M	92.31	218.77	0
200	51.24	105.44	253.04	1

The symbol M in the table indicates missing data representing 13% of the sample. Stars show omitted portions of the table.

Full-Row Deletion

The simplest way to deal with missing data is full-row deletion. Using this method, if a missing value occurs, the analysis ignores the entire row in the regression. If you choose to ignore the problem, your statistical program will probably use this method. Using real as opposed to estimated data describes this method's singular advantage; the obvious weakness of full-row deletion centers on using only a subset of the sample, i.e., in this example, missing observations equal 26 out of 200. Omitting this many rows is probably fatal to any analysis, particularly if the missings are systematic, e.g., all females or all low energy. Full-row deletion also loses degrees of freedom.

Mean Substitution

Mean substitution replaces all missing observations in a variable by the mean of that variable. While this method preserves existing data, it reduces variation in the regression variables and the standard errors thus increasing the probability of a TYPE II error. While most computer programs offer this method as an option, most experts reject it. The reasons for rejection center on the failure to compensate for systematic missings, reducing the variance, distorting the mean, and misrepresenting correlations with other variables.

Pairwise deletion

Understanding pairwise deletion requires a brief discussion of correlation and covariance matrices. In chapter 12, we introduced the notion of correlation matrices. Here we repeat, as table 15.2, the table used in that earlier discussion.

Table 15.2 Correlation Matrix for Energy, IQ, and Productivity (from chapter 12)

	Energy	Intelligence	Productivity
Energy	1.0000	0.2721	0.5548
Intelligence	0.2721	1.0000	0.6746
Productivity	0.5548	0.6746	1.0000

Without an elaborate proof, we can use this matrix to calculate the regression coefficients expressed as Z scores, the fully partialled and part correlation coefficients, and the multiple coefficient of determination. By adding the means and variances of each variable to the matrix, we can create a covariance matrix that restore the original units of measure. In other words, if you have a correlation matrix and the mean and variance of each variable, you can calculate the full regression analysis. In chapter 12, we did not use matrices; we used instead, simple arithmetic and algebraic calculations.

With that brief aside, here is the way pairwise deletion works. Using all the available data, pairwise deletion constructs new correlation and covariance matrices using only the actual values, and calculates a new regression analysis with its total array of output. The matrix consists of correlation coefficients and covariances estimated from variables with different size Ns because our estimate uses variables with different numbers of missing data.

While pairwise deletion has an aura of sophistication, it suffers from essentially the same weaknesses as mean substitution.

Mean Substitution with Companion Dummy Variables

Pioneered by Jacob Cohen,[77] this method takes advantage of information "conveyed by the missingness." The method of companion dummy variables substitutes the arithmetic mean for the missing observations, creates a companion dummy variable, coding the missings as one (1) and the actual data as zero (0), and runs the regression with the new companion missing data as an independent variable. Again, many experts (those who know) have rejected out of hand Cohen's method.

From our perspective, the inclusion of a missing dummy variable, in part, addresses the objections raised to mean substitution alone. As Cohen argues, the correlation between the missing dummy and other variables in the analysis enables identification of systemic problems in the missing data. Including the missing dummy also adjusts for the differences between the mean of X and the estimated value of X.

For example, substituting the mean of energy for the missings and creating the companion dummy realizes a regression coefficient, $B_{MISSINGS}$ showing the difference in productivity of persons with a mean level of energy. Thus, despite its disfavor and admitting that it has not found widespread use, Cohen continued to argue that using a dummy missing variable with mean substitution has several advantages:

1. It is simple.
2. It uses new information.
3. It aids in understanding the why of missings.
4. And it uses actual data.

On this, as well as many other issues, we stand with Cohen.

Missing Value Imputation Using OLS Regression[78]

A more sophisticated method for coping with missing data uses a technique called imputation. One version of imputation (OLS) estimates missings by using OLS and the variable containing the missing observations as a dependent variable with one or more variables either

---

77. JCC, pp. 442-451. We are not sure whether Cohen invented the idea or borrowed it from someone else. Our sense is that he invented it.
78. An even stronger method for imputing missings lies in maximum likelihood estimation. For the interested reader, we refer you to Cohen cited in the previous footnote.

in or out of the model as independents that correlate strongly with the variable sporting the missings. This technique uses the predicted values from the regression as substitutes for the missings. While the procedure makes good estimates of the missings, it has three limitations: (1) it assumes a multivariate normal distribution which may not be accurate; (2) like mean substitution, it tends to lower estimates of the variance, and (3) it relies on the relationships between the variable with missing values and the other variables. If these relationships are weak, it makes the estimates of missings unreliable.

A variation of this method also uses a dummy missing variable in the analysis to form the estimating equation. Such an addition strengthens the method as described above. In this variation, the procedure estimates an equation using full-row deletion and employs the equation to estimate the missing observations including the dependent variable.

Comparison of Methods for Coping with Missing Observations

To make comparisons, we randomly eliminated 26 observations to create the missings in table 15.1. Then we used the data with missing values, and ran regressions using each of the methods discussed in this section. We replaced the missing values in table 15.1, and reran a regression with no missing values.

Table 15.3 summarizes these regressions.

Table 15.3 Comparison of Methods

Method	R^2	B_0 (Intercept)	B_1 (Energy)	B_2 (Gender)	B_3 (Intelligence)
No Missing Data	0.71	3.2510 (0.2219)	0.127 (0.0000)	2.382 (0.0005)	0.190 (0.0000)
Rowwise Deletion	0.68	4.4700 (0.1650)	0.129 (0.0000)	2.304 0.0034)	0.177 (0.0000)
Mean Substitution	0.67	0.4660 (0.8689)	0.129 (0.0000)	2.080 (0.0068)	0.191 (0.0000)
Mean Substitution with Dummies	0.68	-7.66 (0.0850)	0.130 (0.0000)	2.126 (0.0058)	0.185 (0.0000)
Regression Imputation	0.71	3.322 (0.2088)	0.128 (0.0000)	2.128 (0.0004)	0.187 (0.0000)

In our hypothetical sample, all the methods did a reasonable job of compensating for missing observations. The values in the body of the table showing parentheses are probabilities of type I error. Coefficient estimates were close and all the independent variables were statistically significant. As a rule of thumb, we use either mean substitution with dummy variables or OLS imputations with missing dummy variables.

Outliers

Outliers, unusual observations, can occur in a single variable (we call these outliers univariate) or in a regression analysis (calling these outliers multivariate) or both. We look for multivariate outliers in the residuals.

Univariate outliers distort a variable's descriptives. Multivariate outliers can distort an OLS fit by pulling the fit too far in the direction of the unusual observations, influencing regression coefficients, MR^2, and inferential statistics, e.g., standard errors, T (Z), t, and F tests.

Two conditions cause outliers: (1) error and (2) accurate but unusual observations. A false data entry, e.g., 150 instead of 15, represents an example of error. The solution to error finds the right entry and replaces the erroneous entry or if we cannot find the correct entry, eliminates the observation or treats it as missing.

The second type of outlier, a true but unusual observation, e.g., the intelligence of a genius child in a sample of *normal* fourth graders or an observation outside the range allowed by theory (if you have a theory). As suggested by Jacob Cohen, in the theory domain, unusual observations may contain valuable information suggesting a different theoretical approach to the problem. Before considering the possible elimination of outliers from data, one should try to understand why they appeared and whether it is likely that similar values will continue to appear in additional samples

Detecting Univariate Outliers

Box-whisker plots described in chapter 4 embody an excellent device for detecting univariate outliers. Table 15.4, a hypothetical random sample ($N = 25$), shows a variable, productivity, that contains suspicious, but not definite outliers.

Table 15.4

ID	Productivity	Intelligence	Energy	Gender
1	63.81	118.6	111.67	0
2	70.57	89.34	129.47	1
3	64.84	102.92	124.35	1
4	75.27	106.65	156.76	0
5	55.26	121.47	98.58	1
6	69.01	113.26	120.35	1
7	69.53	88.70	174.08	1
8	66.78	92.95	148.62	0
9	76.62	95.24	142.23	1
10	59.56	84.62	118.84	0
11	66.44	97.34	145.17	1
12	59.35	91.06	123.72	1
13	54.41	84.23	105.33	1
14	60.92	90.10	198.93	1
15	65.58	101.97	155.27	1
16	57.95	87.57	132.64	1
17	55.77	80.42	138.43	0
18	95.99	127.32	155.26	1
19	61.54	112.80	111.07	0
20	55.09	94.10	125.74	0
21	56.87	96.29	138.02	1
22	91.26	134.83	156.84	0
23	65.34	130.88	101.92	0
24	66.54	128.74	139.59	0
25	55.19	85.71	126.14	1

Figure 15.1 depicts a box plot of the variable productivity.

Figure 15.1 Box Plot of Productivity

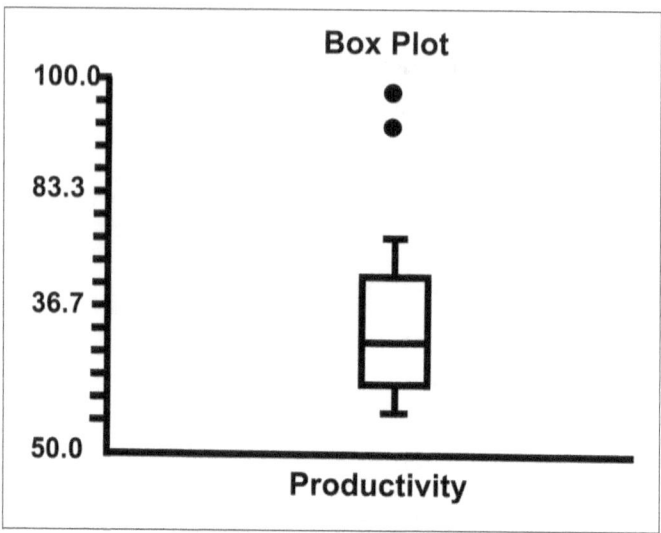

As a reminder, the vertical box ends represent the 25th and 75th percentiles or the InterQuartile range. The horizontal line near the middle of the box shows the median The T-shaped lines extending vertically from each end of the box depict the whiskers. The upper whisker pictures a value 1.5 times the InterQuartile range above the 75th percentile. The lower whisker pictures a value 1.5 times the interquartile range below the 25th percentile. We label observations beyond three Inter Quartile Ranges below and above the 25th and 75th percentiles as unusual and, therefore, highly probable outliers. We label observations beyond 2 IQRs, as suspicious. In figure 15.1, the two points outside the upper *T* values in the plot lay more than two InterQuartile Ranges outside the 75th percentile demanding our attention. We use points to show suspicious outliers and stars to show probable outliers.

If the two suspicious observations 18 and 22 (91.26 and 95.99) represent correct values, we probably do not wish to discard them. While not confirmed outliers, the plot alerts us to the possibility that these suspects may affect our regression analysis or may be of theoretical interest. The two independent variables, intelligence and energy, show no univariate outliers.

Detecting Multivariate Outliers

So, we proceed to run an OLS multiple regression summarized in table 15.5: Productivity on intelligence, energy, and gender, ignoring the effect of the potential univariate outliers in productivity.

Table 15.5 regression of productivity on energy, gender, and intelligence

Independent Variables	Regression Coefficients	Standard Errors	T(Z) ratios	Prob
Intercept	-6.014	14.0706	-0.427	0.6734
Energy	0.2243	0.0664	3.377	0.0028
(Gender=0)	-1.7894	3.2293	-0.554	0.5854
Intelligence	0.4105	0.0974	4.214	0.0004
MR^2	0.56			

In this regression, MR^2, energy and intelligence are statistically significant. We saved and plotted (another box plot) the residuals from this regression shown in figure 15.2.

Figure 15.2 Box Plot of Residuals from Regression in Table 15.5

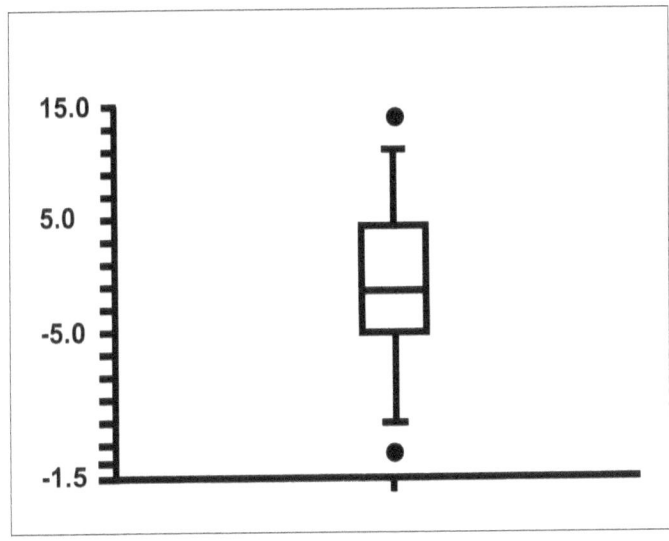

Figure 15.2, shows two suspicious positive and negative outliers in the regression residuals, a signal that must be investigated because they could have influenced the regression.

Detecting Multivariate Outliers in the Independent Variable(s): Leverage and Hat Diagonals

While the box plot puts us on alert, these suspicious outliers require a more complex set of calculations.

The first calculation, called Leverage, identifies outliers for one independent variable. We calculate leverage scores for each observation (h_i is its symbol). Equation 15.1 shows the calculation of h_i.

$$h_i = \frac{1}{N} + \frac{(X_i - M_X)}{\Sigma D^2},$$
(Equation 15.1)

where h_i is the leverage for each case, N the sample size or number of observations in the sample, M_X, the mean of the independent variable X, and ΣD^2, the sum of the squared deviations of X from its mean. Observations far from the mean, yield large h_i values, reaching a maximum of one. Large values of h_i signal potential outliers for one independent variable.

Calculating leverage h_i's for more than one independent variable combines the independent variables into a single component by capturing the common variance of the set with a technique called principal components (discussed in chapter 25). This calculation uses matrix algebra, so we'll simply let our computer program do it for us. Many programs call leverage values using more than one independent variable, Hat Diagonals.

For our data set, our program calculates the hat diagonals. Note: Equation 15.1 still holds for Hat Diagonals, the only difference is we calculate the hat diagonals from the group of independent variables transformed into one. Hat Diagonals greater than two times the number of regression coefficients in the model divided by the number of observations have high leverage (i.e., $H_i > 2 * \frac{P}{N}$). In our example 2 times (3 regression coefficients, B_1, B_2, B_3 divided by 25) gives 0.24. We identify a potential X set outlier if it has a Hat Diagonal value greater than 0.24. Using that criterion, observations number 8, 14, 15, and 18 with leverage values > 0.24, define potential outliers in the X domain. A box plot of Hat Diagonals suggests that, while the observations apparently do not contain outliers, they do contribute to the right skew in the leverage values. Figure 15.3 shows that skewness. .

Figure 15.3 Box Plot of Hat Diagonals (Leverage)

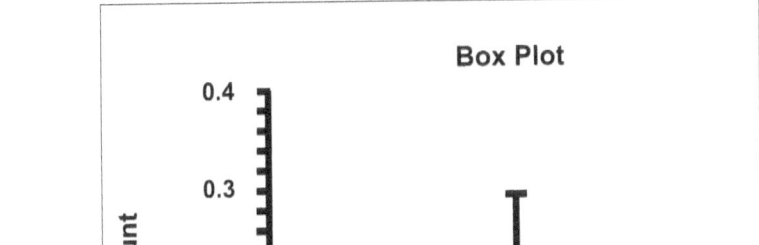

Our box plot of Leverage (Hat Diagonals), does not suggest an outlier problem in the X variables. Using the 0.24 criterion, however, still makes us suspicious.

Detecting Outliers in the Dependent Variable: Discrepancy and Rstudent

The second calculation measures outliers in the dependent variable, called Discrepancy. Discrepancy measures the distance (D) from predicted to actual values of Y. A measure called the external standardized residual represents the principal device for measuring discrepancy. We calculate discrepancy by running a series of regressions (in this case 25) with one observation removed. Using the 25 observations, we recalculate Y_C and $Y_i - Y_C$ (D_i) and standardize (create a Z score) each of these observed D_is. This measures the possibility of a Y outlier with a single observation removed. Recalculating discrepancy with one observation removed keeps the dependent variable outlier search honest by preventing an outlier from hiding as it pulls the equation toward itself. Finally, we divide the standardized residuals by their standard errors, producing a T (Z) ratio called Rstudent.

If the regression assumptions of normality are valid, values of Rstudent have a T (Z) distribution with n-k-1 degrees of freedom. The easiest and most conservative way to detect an outlier in the Y variable looks at a box plot of Rstudent. Figure 15.4 shows no outlier but perhaps potential at observation 14. Another way of checking looks at the T (Z) values of Rstudent. If normality characterizes the Rstudent distribution and if Rstudent equals 2.00 or greater, we probably have an outlier in Y. An observation with a T (Z) value greater than 3.00, almost certainly signals an outlier in Y. Even if Rstudent signals an outlier, do we label the observation influential? The answer is no! We need to make at least one other calculation.

Figure 15.4 Box-plot of Rstudent

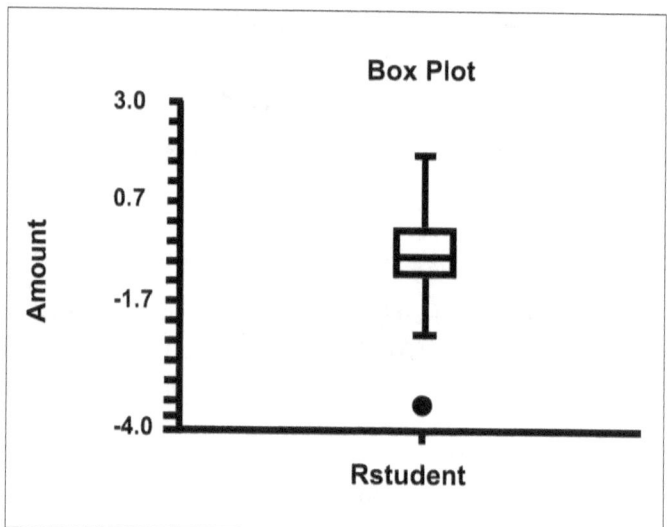

This plot shows one suspicious observation in Y, case number 14. This duplicates the finding in our univariate search discussed above.

Detecting Outliers that Affect the Total Regression Equation: Influence and DFFITS

We measure total influence using a quantity called DFFITS, a technique that combines leverage and discrepancy to show how the regression equation would change when we omit a given observation from the regression.

$$\text{Influence} = DFFITS_i = \textbf{\textit{Rstudent}}_i * \sqrt{\frac{h_i}{1 - h_i}} \ .$$ (Equation 15.2)

We calculate DFFITS for all the observations by multiplying Rstudent by the square root of h_1 divided by $1 - h_1$. DFFITS has a minimum of zero and grows larger as Rstudent and h increase. We box plot the DFFITS to identify observations that potentially modify the entire regression equation.

Figure 15.5 is a box plot of DFFITS. Figure 15.5 Box Plot of DFFITS

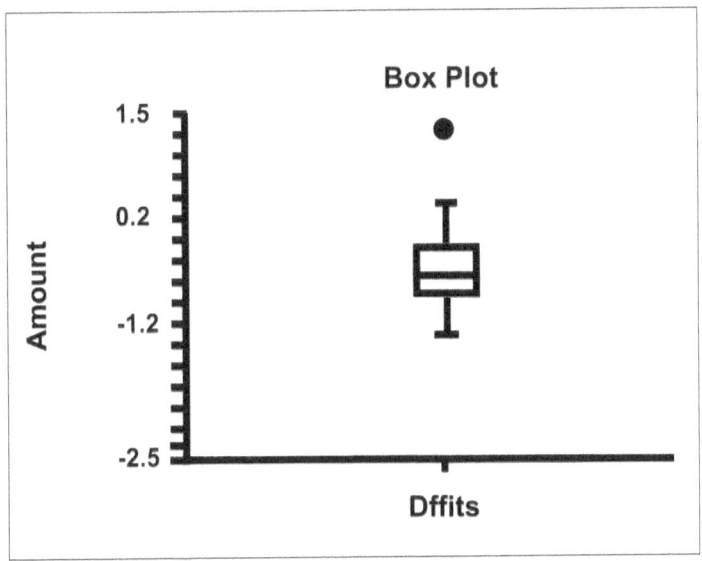

Observations 14 (DFFITS = -2.1214), a suspicious outlier identified with a solid circle, probably affects the total regression. So the univariate possible outliers identified earlier apparently do not affect the equation. Perhaps you can get a better picture of the influence statistic if we say that outlying values of Y combined with high leverage values of X have the greatest influence. Looking back at the data you can see that the problem probably lies with the outlier of 60.92 in productivity combined with an outlier in energy of 198.93, although our original box plot of energy did not convey that information.

Detecting Outliers Affecting a Specific Regression Coefficient: DFBETAS

If our interest lies in detecting an outlier effect on a single regression coefficient, we use a device called DFBETAS. For example, with the variables in our hypothetical data set, our inquiry might center on the effect of intelligence on productivity holding energy and gender constant. We calculate DFBETAS for any independent variable by dropping the first observation and generating a regression coefficient for the set with that observation dropped. We then drop the second observation and run another regression realizing its coefficient. We repeat this process throughout the data set until we have regression coefficients for each dropped observation. Then we subtract the regression coefficient with all observations from the new coefficients without the observations $(B_{dropped i} - B_{total})$ and standardize these differences to produce DFBETAS. DFBETAS show how much a coefficient would change if we dropped that observation from the data. Many experts suggest that we should look carefully at DFBETA values exceeding $\dfrac{2}{\sqrt{N}}$. In our example, two divided by five $(\dfrac{2}{5}) = 0.40$.

If the DFBETA shows the independent variable coefficient changing for any dropped observation, it means that relieved of that outlier, the regression line probably does not require a different slope to minimize the squared residuals and accommodate the outlier. In our example, figure 15.6 shows our box plot of the DFBETAS for intelligence.

Figure 15.6 box plots of dfbetas for intelligence

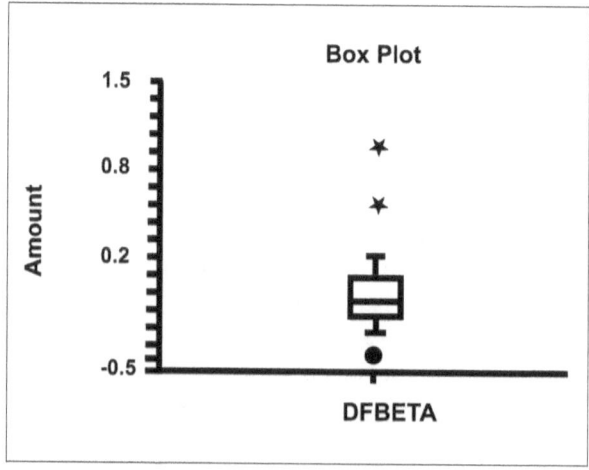

Here we see two confirmed outliers at observations 18 and 22 that affect the regression coefficient for intelligence and certainly requires our attention.

In table 15.5 we summarizes all the possible outlier threats

Table 15.5 Outlier Threats

Row	RStudent	Hat Diagonal	DFFITS	DFBETAS (Intelligence)
14	-2.8345	0.3590	-2.2436	0.9846
18				-1.0790
22		0.2882	-0.6592	

Summarizing table 15.5, Rstudent tells us that we probably have an outlier in the dependent variable at observation 14. Hat Diagonals (Leverage) shows two possible outliers in the independent variable set at observations 14 and 22. DFFITS tells us that an outlier at observation 14 probably affects the entire regression equation. Finally, observations 14 and 18 probably affect the regression coefficient for intelligence.

Correcting for the Effect of Outliers

If not a mistake, outliers that represent no research concern to us, e.g., the genius child in a sample in which our interest is in *normal* children, we delete the observation. On the other hand, if an outlier is not a mistake, but represents a true unusual indigenous element of our data set, we have two methods available for dealing with it.

Transformations

Often we can minimize the affect of outliers by one of the transformations discussed in chapter 4, e.g., a Log or square root transformation, if outliers in the dependent variable produce a significant right skew. In our example, we tried a natural log transformation of productivity and reran the regression with negligible results. Then we tried natural log transformations on all the cardinal variables, productivity, intelligence, and energy. Table 15.6 summarizes that regression.

Table 15.6 Regression with Log Transformed Productivity, Energy, and Intelligence

Independent Variable	Regression Coefficient	Standard Error	T(Z) ratio	Probability
Intercept	-0.9573	0.933	-1.026	0.3166
Gender = 0	-0.0196	0.0433	-0.453	0.6552
LnEnergy	0.475	0.1241	3.827	0.001
LnIntelligence	0.6095	0.1384	4.405	0.0002
R^2	0.5916			

Compare these results with the regression summarized in table 15.5. The log transformations of Y and the Xs produced a better fit (R^2 = 0.59) than the regression that paid no attention to outliers (R^2 = 0.55). All other transformations produced no better results.

Robust Regression

Some statistical programs have a module that uses a robust weighted regression. Employing leverage and discrepancy statistics, a robust regression weights each observation in terms of its affect on the regression. This method assigns the weight of one (1) to non-outlier observations weights of less than one to those with outlier effects. The greater the outlier effect, the smaller the weights that diminish outlier effects.

With NCSS, we ran a robust regression using the untransformed Y value. Table 13.7 summarizes the result of that analysis.

Table 15.7 Robust regression with Untransformed Y and Xs

Independent Variable	Regression Coefficient	Standard Error	T Value to test	Prob
Intercept	-5.0127	12.5942	-0.398	0.6946
Energy	0.2435	0.0604	4.032	0.0006
(Gender=0)	-0.6557	2.6958	-0.243	0.8102
Intel	0.3676	0.0853	4.312	0.0003
	R^2 = 0.5946			

Comparing the results in table 15.7 with table 15.6, the robust regression was no better than the log transformation.

Then we tried the robust regression using log transformations of productivity, energy, and intelligence. Table 15.8 summarizes that robust regression.

Table 15.8 Robust regression with all Cardinal Variables Log Transformed

Independent Variable	Regression Coefficient	Standard Error	T ratio	Prob
Intercept	-283.8827	60.0935	-4.724	0.0001
Gender	-0.495	2.5711	-0.193	0.8492
lnEng	34.3371	7.8559	4.371	0.0003
lnIntel	39.2953	8.6471	4.544	0.0002
R^2		0.6178		

Comparing tables 15.8 with 15.7, we see the robust regression with all variables log transformed was a slightly better fit. R^2 jumped to 0.62. But neither the transformations nor the robust regressions worked better than dropping the outlier observation.

A Note on Dropping Outliers

Assuming, for the sake of argument, we decide to drop observation 14 and rerun the regression without it. Table 15.6 summarizes that regression.

Table 15.6 Regression without Observation 14

Independent Variable	Regression Coefficient	Standard Error	T Value to test	Prob
Intercept	-18.0046	12.8918	-1.397	0.1778
Energy	0.3323	0.0690	4.819	0.0001
Gender	-2.4135	2.8036	-0.861	0.3995
Intelligence	0.3965	0.0845	4.694	0.0001

$R^2 = 0.6818$

Comparing table 15.6 to table 15.5 we see the difference between retaining and dropping the outliers. First, looking at the R^2s between retained ($R^2 = 0.5578$) and dropped ($R^2 = 0.6818$) shows a large difference. Second, look at the regression coefficients. In the analysis in which we retain the outliers the coefficients for energy (0.2243) and intelligence (0.4105) show large differences from the analysis in which we dropped the outliers, energy (0.3323) and intelligence (0.3965). In the dropped analysis the coefficient for energy was much bigger; the coefficient for intelligence was slightly smaller.

We emphasize that good practice shuns thoughtless dropping of outliers. A decision to drop outliers requires strong justification. As explained earlier, if we have based our analysis on strong theory, an indigenous outlier may signal the need for theory adjustment. Dropping an indigenous outlier requires explaining that the outlier observation was either outside your research interests or such an anomaly that its retention would substantially reduce the generalizability of your findings.

Summary

The major concepts discussed in this chapter were the following:

1. Missing data

    Missing data have several causes.

    a. Recording error
    b. Incomplete data collection
    c. Non-response to questionnaires
    d. Privacy issues e. Privileged information

    You must make every effort to retrieve missing data, using the methods discussed here only if, after careful searching, you cannot retrieve the missings.

2. Observations missing at random

    When our data have completely random missings, this randomness makes estimating methods unbiased.

3. Observations missing systematically

    We attribute nonrandom missing data to missing observations systematically related to variables in the model or to variables that affect the outcome of analysis. Large data sets with small percentages of missings (< 3%) probably permit ignoring the problem. With small data sets or data sets with a high percentage of missings, we must take action.

4. Methods for estimating missing data,

    a. Full-row deletion

        The simplest way to deal with missing data is full-row deletion. Using this method, if a missing value occurs, the analysis ignores the entire row in the regression. If you choose to ignore the problem, your statistical program will probably use this method.

    b. Pairwise deletion

        Using all the available data, pairwise deletion constructs two new correlation and covariance matrices, and calculates a new regression analysis with its total array of output. The matrix consists of correlation coefficients and covariances estimated from variables with different size Ns because our estimate uses variables with different numbers of missing data.
        While pairwise deletion has an advanced aura, it suffers from essentially the same weaknesses as mean substitution.

    c. Mean substitution

        Mean substitution replaces all missing observations in a variable by the mean of that variable. While this method preserves existing data, it reduces variation in the regression variables and the standard errors in regression thus increasing the probability of a TYPE II error.

    d. Mean substitution with dummy variables

        Using a dummy variable, the method of companion dummy variables substitutes the arithmetic mean for the missing observations, creates a companion dummy variable coding the missings as one (1) and the actual data as zero (0), and runs the regression with the new companion missing data as an independent variable(s).

e. Imputation

The method of imputation (OLS) estimates missings by using the variable containing the missing observations as a dependent variable and one or more independent variables either in or out of the model that correlate strongly with the variable sporting missings. This technique uses the predicted values from the regression as estimates of the missings.

5. Outliers

Outliers, unusual observations in a single variable can occur in a single variable (we call these outliers univariate); in a regression analysis, we call these outliers multivariate. We look for multivariate outliers in the residuals.

Univariate outliers distort a variables descriptives. Multivariate outliers distort an OLS fit by pulling the fit too far in the direction of the unusual observations, influencing regression coefficients, MR^2, and inferential statistics, e.g., standard errors, T (Z), and F tests.

a. Outliers due to error

Two conditions cause outliers: (1) error and (2) accurate but unusual observations. A false data entry, e.g., 150 instead of 15, represents an example of error. The solution to error finds the right entry and replaces the erroneous entry or if we cannot find the correct entry, eliminates the observation.

b. Indigenous outliers

The second type of outlier, a true but unusual observation, e.g., the intelligence of a genius child in a sample of *normal* fourth graders or an observation outside the range allowed by theory (if you have a real theory). We should not discard these outliers without clear justification.

c. Detecting univariate outliers

Box-whisker plots described in chapter 4 embody an excellent device for detecting univariate outliers

d. Detecting multivariate outliners

i. Outliers in the X domain: leverage and hat diagonals

Large values of leverage signal potential outliers for one independent variable. Large values of hat diagonals signal potential outliers for the independent variable set.

ii. Outliers in Y: discrepancy and Rstudent

A measure called the external standardized residual represents the principal device for measuring discrepancy. We calculate discrepancy by running a series of regressions (in this case 25) with one observation removed. Using the 25 equations, we recalculate Y_C and $Y_i - Y_C (D_i)$ and standardize the D_i values (create a Z score). We divide the standardized residuals by their standard errors, producing a T (Z) ratio that many programs call Rstudent

iii. Outliers affecting the regression equation

We measure total influence using a quantity called DFFITS, a technique that combines leverage and discrepancy to show how the regression equation would change when we omit a given observation from the regression.

$$\text{Influence} = DFFITS_i = \textbf{\textit{Rstudent}}_i * \sqrt{\frac{h_i}{1 - h_i}} .$$

iv. Outliers affecting specific regression coefficients

If our interest lies in detecting an outlier effect on a single regression coefficient, we use a device called DFBETAS. For example, with the variables in our hypothetical data set, our inquiry might center on the effect of intelligence on productivity holding energy and gender constant.

e. Methods for dealing with outliers

   i. Transformations

       Often we can minimize the affect of outliers by one of the transformations discussed earlier, e.g., the quadratic or power transformation.

   ii. Robust regression

       A robust regression weights each observation in terms of its affect on the regression, incorporating leverage and discrepancy statistics. This method assigns the weight of one (1) to non-outlier observations weights of less than one to those with outlier effects. The greater the outlier effect, the smaller the weights with smaller weights diminishing outlier effects.

   iii. Dropping Outliers

       We emphasize that good practice shuns thoughtless dropping of outliers. A decision to drop outliers requires strong justification.

Appendix A

Chapter 15

The data set given below contains missing observations. Your tasks are the following:

1. Give the variables names and identify the dependent variable.

2. For each of the variables, use four methods for estimating the missings. At least two methods must include mean substitution with a dummy variable and imputation.

3. Run five regressions, one analysis using full-row deletion, the other four using the methods you chose.

4. Compare the results of the five regressions.

1	87.09	57.57	204.01	265.00
2	98.76	50.20	199.41	270.39
3	113.88	39.35	215.00	220.31
4	106.55	49.08	243.58	M
5	106.21	49.55	M	307.49
6	84.01	41.22	220.31	280.53
7	132.95	58.24	247.45	347.10
8	95.32	45.21	153.19	282.82
9	131.64	59.96	218.89	411.26
10	78.66	43.40	158.78	221.25
11	106.15	53.21	198.15	257.87
12	107.29	59.78	171.95	268.05
13	107.62	54.64	153.39	279.77
14	78.48	37.21	169.47	241.28
15	120.65	54.61	237.41	362.77
16	122.57	62.96	193.65	341.71
17	107.88	52.15	173.81	298.68
18	113.60	M	228.37	365.44
19	145.08	70.57	280.51	423.84
20	112.06	52.22	230.68	387.92
21	101.57	50.78	193.05	296.62
22	103.60	45.73	209.46	258.82
23	118.72	61.39	219.76	385.21
24	96.51	45.53	179.38	238.45
25	111.96	49.55	201.02	358.23
26	122.56	51.51	256.88	402.49
27	M	49.92	222.77	304.78
28	M	41.16	214.33	284.88
29	114.29	49.04	239.07	384.00
30	84.94	39.13	179.51	311.12

Appendix B

Chapter 15

The appendix B data set contains outliers. Your tasks are the following:

1. Run an analysis that detects the univariate outliers and identifies the suspicious.

2. Give the variables names and identify the dependent variable.

3. Run a multiple regression and identify outlier observations that could affect the regression.

4. Run two regressions, one with the outliers retained, the other with outliers dropped. If the results differ, explain why they differ.

5. Assuming that we should retain the outliers, run one or more analyses that might ameliorate outlier effects.

	X1	X2	X3
1	110.86	55.5	355.88
2	96.58	53.61	182.51
3	116.4	64.55	228.94
4	111.26	58.46	231.47
5	89.25	58.11	208.32
6	101.22	49.9	247.09
7	81.89	57.06	188.74
8	81.2	43.85	168.58
9	152.69	65.72	230.37
10	87.25	48.32	181.2
11	94.67	45.38	262.62
12	81.84	44.97	203.18
13	112.6	62.73	166.45
14	78.54	60.53	196.01
15	123.44	63.54	245.08
16	89.28	46.59	159.83
17	100.06	56.6	162.51

18	76.18	57.48	246.92
19	110.31	61.06	247.21
20	99.45	63.47	186.13
21	100.56	58.9	252.93
22	86.03	51.87	350.95
23	86.74	60.45	211.64
24	99.02	80.36	259.31
25	95.72	51.64	219.13

CHAPTER 16

AUTOCORRELATION, HETEROSCEDASTICITY, MULTICOLLINEARITY, MODEL SPECIFICATION, AND NORMALITY

In chapter 13, we discussed the assumptions of OLS regressions and the problems arising from assumption violations. In this chapter, we discuss detecting and correcting assumption violations.

Autocorrelation

OLS regression assumes uncorrelated residuals. Usually, violation of residual independence does not bias the regression coefficients, but affects probabilistic inference by making standard errors smaller and increasing the probability of type II errors. However, in certain contexts, e.g., time series, autocorrelation can dangerously affect our analytical interpretation. We have much more to say about autocorrelation and its effect in our discussion of time series analysis.

Not surprisingly, we measure autocorrelation with a correlation coefficient, not between two variables, but between two values of the same variable at times X_T and X_{T-N} or at sequence X_1, X_N. When using cross sectional data, we call the correlation of residuals serial correlation. With time series, we call the problem autocorrelation. We will have much more to say about autocorrelation and time series in part 5.

An easy to understand example of serial correlation (in cross sectional data) involves a situation in which a teacher asks her students to participate in a political telephone survey. Each student is assigned 50 calls. Many of the students find the process tediously onerous: people are rude, insulting, or difficult to reach. As a result, some students decide to fake the calls, making-up results. Some believe, incorrectly, that alternating responses in an attempt to create an illusion of randomness conceals fakery. Unfortunately (for the students and the project), when the class ran a regression analysis, they found the residuals autocorrelated. When the first residual went up, the second in sequence went down. In an attempt to simulate randomness, the errant students created negatively correlated residuals. Table 16.1 shows a recording sequence and the hypothetical negative autocorrelation.

Table 16.1 Negatively Correlated Residuals

Sequence	Residuals	Lag 1 Residuals
1	6	
1	-1	6
3	0	-1
4	-5	0
5	1	-5
3	-3	1
7	4	-3
7	-1	4
9	1	-1
10	0	1
11	1	0
12	-2	1

Figure 16.1 shows a plot of negative serial correlation

Figure 16.1 Negative Serial Correlation

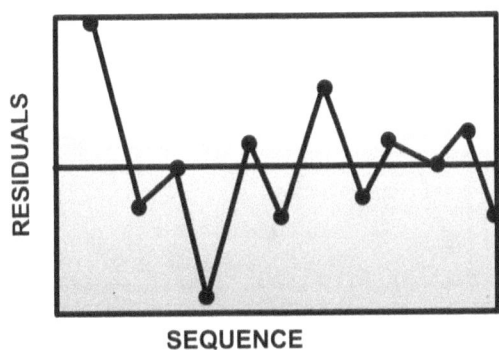

The residuals follow a pattern: As one goes down the next goes up, and so forth. Autocorrelations also follow positive patterns. Table 16.2 shows the residuals from a sample regression.

Table 16.2 Positive Autocorrelation

Sequence	Residuals	Lagged 1 Residuals
1	-0.04	
2	0.08	-0.04
3	0.12	0.08
4	0.17	0.12
5	0.05	0.17
6	0.01	0.05
7	0.03	0.01
8	0.03	0.03
9	0.09	0.03
10	0.15	0.09
11	0.16	0.15
12	0.21	0.16
13	0.18	0.21
14	0.11	0.18
15	-0.01	0.11

Figure 16.2 shows positively correlated residuals.

Figure 16.2 Positively Correlated Residuals

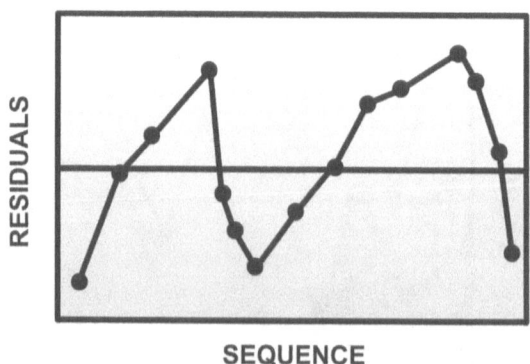

SEQUENCE

Here the pattern shows one going up and the next going up, then changing to down, down, and changing back again to up, up.

Detecting Autocorrelation

In addition to positive and negative, we also define autocorrelations as first-order, second-order, third-order, or higher. First-order autocorrelation relates the movement between the first and second observation, the second and third, the third and fourth, etc. Second-order autocorrelation relates the movement between the first observation and the third, the second and the fourth, the third and the fifth, etc. Third-order autocorrelation looks at first and fourth, etc. Figure 16.1 portrays a first-order autocorrelation.

Table 16.2 shows a first-order autocorrelation in a residual.

Table 16.2 First Order Autocorrelation

Recording Sequence	Residuals	Lagged 1 Residuals
1	-0.04	
2	0.08	-0.04
3	0.12	0.08
4	0.17	0.12
5	0.05	0.17
6	0.01	0.05
7	-0.03	0.01
8	0.03	-0.03
9	0.09	0.03
10	0.15	0.09
11	0.16	0.15
12	0.21	0.16
13	0.18	0.21
14	0.11	0.18
15	-0.01	0.11

The third column, lagged 1 residuals, matches the first residual in the sequence to the second, etc. The correlation coefficient between the residuals and the lagged-one residuals equals 0.399, the first-order autocorrelation coefficient. Statistical programs usually provide at least one diagnostic measure of autocorrelation, for example the Durbin-Watson statistic, a calculation that tests for first-order autocorrelation. The durbin-watson statistic tests the null hypothesis of zero first-order autocorrelation, i.e., that the autocorrelation in the sample probably occurred by chance. We give the equation for the durbin-watson statistic below.

$$D_w = \frac{\Sigma(Res - Res_{-1})^2}{\Sigma(Res^2)}$$

(Equation 16.1)

All the statistical programs we recommend calculate and test the statistical significance of an observed durbin-watson statistic. Some programs e.g., ncss and sx, also provide the autocorrelation coefficients, for all lags at sequences up to 48 and show the statistical significance of the coefficients. Table 16.1 provides an example.

Table 16.3 Autocorrelation of Residuals Out to 24 Lags, N = 200

Lag	Autocorrelation Coefficients	Lag	Autocorrelation Coefficients	Lag	Autocorrelation Coefficients
1	0.1312	9	0.2648	17	-0.0197
2	-0.1073	10	0.185	18	-0.0125
3	-0.0589	11	-0.0608	19	-0.0359
4	-0.2193	12	-0.2027	20	0.0389
5	-0.1939	13	-0.0362	21	0.0000
6	-0.0389	14	-0.0708	22	0.0000
7	0.1548	15	-0.0977	23	0.0000
8	-0.0002	16	0.1156	24	0.0000

If the absolute values of the autocorrelation coefficients are greater than 0.40, they are statistically significant.

Using our data with NCSS provided this table. From this sample, for all the lags shown we cannot reject the null hypothesis of no auto autocorrelation at any lag out to 24.

Correcting for Autocorrelation

Note: Autocorrelation in time series can be friend or foe. If our goal is forecasting, autocorrelation becomes friend. If our goal is estimating regression coefficients, autocorrelation becomes our foe.

We can correct for first-order autocorrelation using simple transformations of the dependent variable. We show the transformation below.

$$Y_{TRANSFORNED} = Y_1 - RY_{-1} \qquad \text{(Equation 16.2)}$$

Y_1 represents the observed value of Y at sequence 1, Y_{-1} the observed value of Y at sequence Y- 1. R signifies the lagged 1 autocorrelation. The equation instructs us to multiply the first order autocorrelation coefficient by the lagged 1 values of Y and subtracts those products from the unlagged values of Y. The transformation removes from Y the first-order autocorrelation.

An alternative correction runs a bivariate regression of the order $Y_T = B_0 + B_{T-1} + \epsilon$, and uses ϵ_i, the error term as a transformed Y with the effect of first order autocorrelation partialled. We can partial higher order autocorrelations by using multiple regressions with the requisite number of lags.

What is the point of the transformation? Imagine running regressions in an attempt to estimate the effects of certain X variables on Y. If autocorrelation infects Y and the independent variables, it is perfectly possible that an observed relationship among the X variables and the Y variables simply reflects their autocorrelations.

We will have much more to say about correcting for or using autocorrelation to forecast in part 5, time series analysis.

Heteroscedasticity

OLS regression assumes constant variance of the residuals around the regression line. We use the term heteroscedastic to label violation of the constant variance assumption and use the term homoscedastic to describe constant variance, conformity with the assumption. Heteroscedasticity does not bias the regression coefficients, but affects standard errors, significance tests, and confidence intervals. We note the relative strength of OLS in the face of heteroscedasticity. It takes a large violation of the assumption to create a measurable effect. Figures 16.3 a, b, and c show different patterns of heteroscedasticity: positive, negative, and bowed.[79]

Figure 16.3 Patterns of Heteroscedasticity

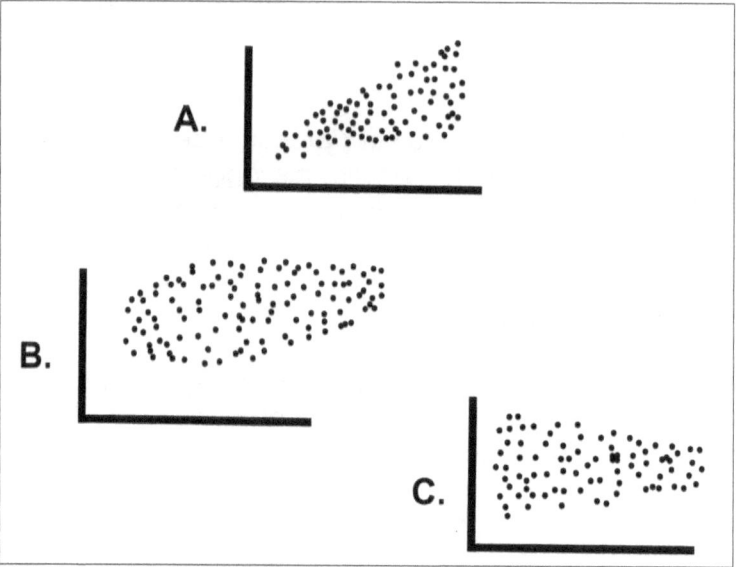

Detecting Heteroscedasticity

We outline the first procedure for detecting heteroscedasticity as follows:

1. Run the OLS regression and realize the predicted values of Y, Y_c and the residuals, ϵ. The predicted values of Y, Y_{cs} represent a linear combination of all the independent variables.
2. Square the residuals, ϵ^2.
3. Run an OLS regression of Y_c on ϵ^2.
4. With homoscedastic residuals, the R^2 from the regression in 3) will not show statistical significance. A statistically significant R^2 from the regression in 3) shows residuals growing larger or maller (depending on the sign of the regression coefficient) as Y_c increases and, therefore, signals heteroscedasticity.
5. A large R^2 bespeaks hetero severity requiring corrective action.

---

[79].   We note that outliers and curvilinearity can also cause heteroscedasticity.

6. Note: If you suspect curvilinear heteroscedasticity, i.e., smaller at both ends, middle bulging, run the regression in 3) as a quadratic, e.g., $Y_C = B_0 + B_1 \epsilon^2 + B_2 (\epsilon^2)^2$.

A second procedure for detection divides the residuals into three or four groups, calculates the residual variance for the highest and lowest groups, and runs an F test on the variances to test their statistical significance. Where F is

$$F = \frac{Varinace_{L \arg estGroup}}{Variance_{SmallestGroup}}$$

A statistically significant F test and/or a ratio of largest to smallest equal to or greater than 10, indicates severe heteroscedasticity and needs correction.

Correcting for Heteroscedasticity

In the case of increasing variance as Y_C increases, a log transformation of Y will often correct. If the log transformation does not work some other tail-pulling transformation, e.g., square root, may also work.

A second procedure uses weighted least squares regression analogous to the robust regression discussed in chapter 15. To use weighted regression, we have to estimate the weights as described below.

1. Run an OLS regression of Y on X's realizing predicted values, Y_C and residuals, ϵ.
2. Square the residuals, ϵ as realized in step 1.
3. Regress the squared residuals, ϵ^2, on the predicted values, Y_C (realized in step 1, i.e., $\epsilon_C{}^2 = B_0 + B_1 Y_C$.
4. The weights become the reciprocals of the predicted values realized in 3) or the weights $= \dfrac{1}{\epsilon_C^2}$.

Because all the statistical programs we recommend run a weighted regression, you simply save the weights created in step 4 and tell the program where you stored the weights. For some experts, weighted regression is not recommended because it compromises, MR^2, PTR^2, and FPR^2.

Note: Before taking action to correct for heteroscedasticity, nonconstant residuals might indicate unmodeled nonlinearity and/or interactions, a point discussed in chapter 18.

Multicollinearity

Multicollinearity refers to high correlation among the independent variables. In fully partialled regression models with a statistically significant MR^2, multicollinearity can cause statistical insignificance for all the independent variables, a condition that defies good sense. The process of partialling correlated independent variables dumps all the common variance into B_0, the intercept.

Very high multicollinearity also makes the regression coefficients highly unstable. Figure 16.4 helps you see why.

Figure 16.4 Severe Collinearity

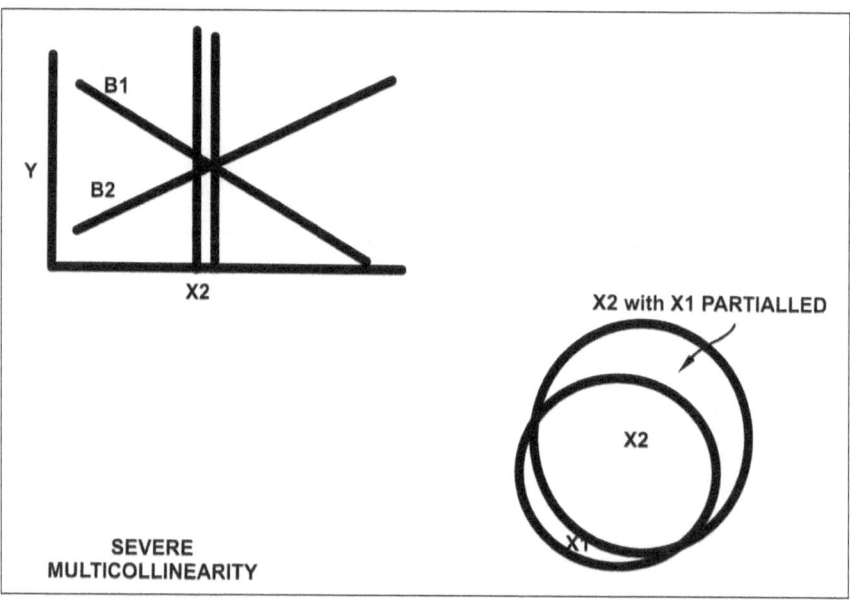

As the variance in the X variable becomes smaller, a large number of OLS regression coefficients will fit. Thus, reduced variance in the independent variable will make an estimate of its regression coefficient unreliable.

Quadratic or cubic transformations and interaction terms can also produce multicollinearity, an important reason for centering.

Detecting Multicollinearity

Tolerance

We calculate the value of tolerance for an independent variable by running a regression of the independent variable on all the other independent variables in the regression model and realizing MR^2. That regression tells us how much that variable, X_1 shares variance with the other independent variables, e.g., X_2, X_3 ... X_N. Subtracting MR^2 from one realizes Tolerance for X_1. So Tolerance for any independent variable equals

$$T = 1 - MR^2_{X1, X2, Xn}$$ (Equation 16.3)

Tolerance ranges from 0.00 to 1.00. If the MR^2 between X_1 and the other independent variables equals zero, the value of tolerance equals 1 or perfect tolerance. If the MR^2 equals one, tolerance equals zero, perfect intolerance. Some suggest, as a rule of thumb, that a tolerance less than 0.14 indicates a major collinearity problem. We take a more conservative position. Even modest MR^2_s of 0.20 with high tolerances in the range of 0.80 can result in drastic effects on estimated regression coefficients.

Variance Inflation Factor

We calculate the VIF by dividing one (1) by a variable's tolerance. Therefore

$$VIF = \frac{1}{Tolerance}$$ (Equation 16.4)

VIF is simply the reciprocal of Tolerance. If tolerance equals one, VIF equals 1. If Tolerance equals 0.0000, VIF equals infinity. The higher the value of tolerance, the lower is the value of VIF. Once again, even relatively low VIFs, e.g., 1.25, can signal trouble.

Remediating Multicollinearity

When research interests are limited to prediction or forecasting with no substantive interest in estimating the individual regression coefficients, then moderate multicollinearity does not pose a problem. If, however, our interest centers on estimating a variable's individual effect, the existence of multicollinearity demands remediation.

Centering Variables

In our discussion of curvilinearity, when analyzing a relationship characterized by a curve with a single bend, we discussed the use of a quadratic transformation. We created the quadratic by squaring the independent variable and running the multiple regression $Y_C = B_0 + B_1X_1 + B_2X_1^2$. By arithmetic necessity the squared second independent variable, X^2 must highly correlate with X_1. If we center X_1 before we square it, centering will reduce the correlation between the two variables thus reducing multicollinearity. Centering creates a variable whose mean is zero. What does centering do? Mathematically, two perfectly symmetrical centered variables with means of zero have zero correlation. And, as noted previously, centering does not change a variable's structure. The point: if scaling causes collineraity, centering can reduce it.

Respecifying the Model

In constructing regression models, we often use a series of X variables designed to measure some dimension of an underlying structure. For example, in measuring socio-economic class, we might include variables like familial income, education, father's occupation, etc., all highly correlated. One solution to collinearity might create a single variable representing the concept by summing the Z scores of the variables or combining the set of variables into index numbers. Another solution is to find uncorrelated variables that tap the same construct.

Principal Component Regression

We will discuss in detail the idea of Principal Component regression in chapter 25. Here, we give a common-sense introduction to the concept. The method of principal components is a technique that captures the common variance shared by a set of variables and creates new orthogonal variables consisting of that shared variance. In our example of measuring socio-economic class we might use income, education, and father's occupation as independent variables, all which are highly correlated. By running a principal component analysis we might produce one, perhaps not more than two, new orthogonal variables that capture the major proportion of the common variance in the set. These new variables can then be used in a regression that looks like the following:

$$Y_C = B_0 + B_1 \text{Component}_1 + B_2 \text{Component}_2 \text{ etc.}$$

NCSS, SPSS, SX, Systat, and OpenStat all have Principal Component modules.

Ridge Regression

Ridge regression represents an interesting alternative to Principal Component regression. Ridge regression works by adding constants to the variances of multicollinear variables. How does this process reduce multicollinearity? Assume two independent variables, X_1 and X_2. Remember, $R_{X1,X2}$ equals the $\dfrac{COVXY}{VarX_1 * VarX_2}$. That is, R_{X1X2}, the correlation coefficient between X_1 and X_2 equals the covariance of X_1 and X_2 divided by the variance of X_1 multiplied by the variance of X_2. Let us say the COVXY equals 240, the variance of X1, $S_{X1}{}^2 = 25$, and the variance of X_2, $S_X{}^2 = 10$, then $R_{X1,X2} = \dfrac{240}{250} = 0.96$, $R^2 = 0.92$, and VIF = 12.5. If we add a small constant to the variances of X_1 and X_2, say 2, it makes them 27 and 12 with their product equaling 324, then, dividing the COVXY, 240 by 324 reduces $R_{X1X2}{}^2$ to 0.55, increases tolerance to 0.45, and decreases VIF to 2.22. The addition of a small constant to the variances of X and Y lowers the VIF from an intolerable 12.5 to, we hope, an acceptably manageable 2.2.

Ridge regression biases the estimates of regression coefficients and reduces the standard errors. If we recognize ridge regression's limits, confronted by extreme multicollinearity, ridge regression sometimes provides an acceptable trade-off.

Correct Model Specification of Independent Variables

Ordinary Least Squares regression assumes the model includes only the correct independent variables. Failure to specify the model correctly carries dire consequences including biasing estimates of regression coefficients, MR^2, PTR^2s, PR^2s, the standard errors, and confidence intervals. Omitting critical variables that are related to the included independent variables causes correlations between the residuals. This means that the effects of the excluded independent variables are not reflected in the coefficients, PTR^2s, and PR^2s of the included independent variables.

Including wrong independent variables causes loss of degrees of freedom and, if the wrong variables are related to the X_s and irrelevantly related to Y, distorts the coefficient estimates.

In this arena, the notion of theory or good confederacy of hunches (soft theory) comes to the fore. Effective regression analyses are totally dependent on some conceptual scheme to guide the choice of independent variables. Running regression analyses using unreflected baskets of independent variables and retaining statistically significant effects constitutes bad practice. If analysis, willy-nilly, feeds enough independent variables into a regression, some, just by chance, will produce statistical significance. Avoid regressions without some theoretical guidance. Such an approach leads to ad hoc interpretations.

Therefore, your original choice of independent variables must be theory or good hunch based. If, however, new theory suggests a different model, then the model should face the new theory's test. We are saying here that regression facts are not just "out there"; we create them by theory, test them statistically, and validate them by action.

In our examples of productivity, we included the variables intelligence, energy, and gender using a theory as a guide. Unlike Mandrake the Magician, we did not pull them from a hat.

Non-normal Residuals

Non-normal residuals do not bias regression estimates and in large samples, have little effect on standard errors and confidence intervals. However, violation of the assumption of normally distributed residuals does signal possible misspecification of the regression model, and requires careful rechecking of other assumptions, e.g., linearity, etc. Non-normal residuals represent a critical sign of model misspecification.

For example, badly skewed residuals suggest the possible need for linear transformations as well as a rethinking of the variables in the model.

Detecting Non-normal Residuals

Detecting non-normal residuals is relatively easy. NCSS, SPSS, Systat, SX, and OpenStat all provide, at a minimum, measures of skewness, kurtosis, and normal plots. NCSS also provides other extensive tests of the normality assumption. We show a normal plot in figure 16.5 is an example.

Figure 16.5 The Normal Plot and Wilk-Shapiro (W) Test

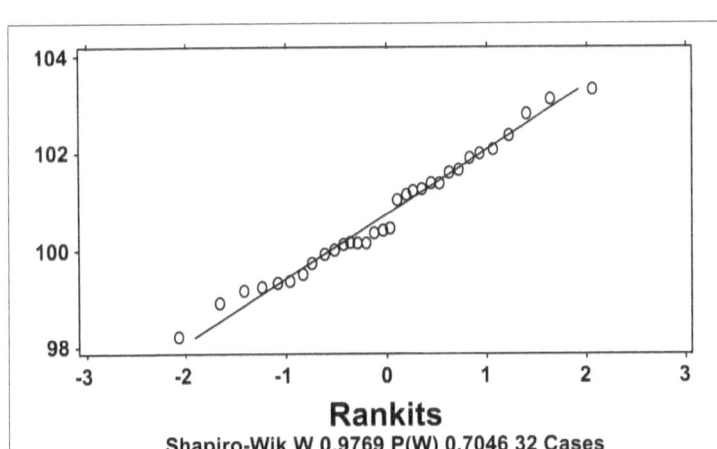

The normal plot depicts actual observations of a variable against values of the variable if it was normal. A straight line plot means a perfect correlation between the variable's actual observations and normality. Deviation from the straight line indicates a deviation from normality. Figure 16.5 also shows the Shapiro-Wilk (W) test, a correlation coefficient between the variable and normality. In figure 16.5, the Shapiro-Wilk value equals 0.9769. The null hypothesis tests the statistical significance of sample's deviation from normality. Or, what is the probability of finding a correlation coefficient as large as 0.9769 if the sample variable came from a non-normal population.

Of course we can also look at the statistical significance of coefficients of skewness and kurtosis or the tests that combine the two.

Summary

The critical concepts discussed in chapter 16 are the following:

1. Autocorrelation

    a. Definition

    OLS regression assumes uncorrelated residuals. Usually, violation of residual independence does not bias the regression coefficients, but affects probabilistic inference by making the standard errors smaller and increasing the probability of type II errors. However, in certain contexts, autocorrelation can dangerously affect our analytical interpretation. We have much more to say about that problem in our discussion of time series analysis.

    b. Detection

    The Durbin-Watson statistic tests the null hypothesis of zero autocorrelation, i.e., that the autocorrelation in the sample probably occurred by chance. Some programs show the autocorrelation coefficients for all lags at any sequence and identifying which are statistically significant.

    c. Remediation

    We enter an important note here: Autocorrelation in time series can be friend or foe. If our goal is forecasting, autocorrelation becomes friend. If our goal focuses on accurately estimating regression coefficients, autocorrelation becomes our foe.

    We can correct for first-order autocorrelation using a simple transformation of the dependent variable. We show the transformation below.

    $$Y_{transformed} = Y_1 - RY_{-1}$$

    An alternative correction runs a bivariate regression of the order $Y_T = B_0 + B_{T-1} + \epsilon$, using ϵ, the error term, as a transformed Y with the effect of first order autocorrelation partialled or removed.

We can partial higher order autocorrelations by using a multiple regression with the number of lags necessary.

2. Heteroscedasticity

    a. Definition

    OLS regression assumes constant variance of the residuals around the regression line. We use the term heteroscedastic to label violation of the constant variance assumption. We use the term homoscedastic to describe constant variance, conformity with the assumption

    b. Detection

    The simplest procedure for detection divides the residuals into groups, for example the or four, calculates the residual variance for each group, and runs an F test on the variances to test for statistical significance. Where F is

    $$F = \frac{Varinace_{l\,arg\,est}}{Variance_{smallest}}$$

    c. Remediation

    We can usually correct for heteroscedasticity by using one of the linear transformations discussed previously. For example, in the case of increasing variance as Y_c increases, a log transformation of Y will often correct. If the log does not work some other tail-pulling transformation, e.g., square root, may also work. Or we can use a weighted regression.

3. Multicollinearity

    a. Definition

    Multicollinearity involves a condition in which the independent variables are highly correlated

    b. Detection

Two simple procedures are effective in detecting multicollinearity: Tolerance and variance inflation.

c. Remediation

We can correct for multicollinearity by centering variables, respecifying the model, using principal components, or ridge regression.

4. Correct Model Specification

Ordinary Least Squares regression assumes that the model includes only the correct independent variables. The consequences of violating the correct model assumption are dire. We have already seen that the failure to include and partial critical variables affects the estimates of regression coefficients, MR^2, $PARTR^2$s, FPR^2s, the standard errors, and confidence intervals.

5. Non-Normal Residuals

Non-normal residuals map one the most critical signs of model misspecification. For example badly skewed residuals suggest the possible need for linear transformations as well as a rethinking of the variables in the model.

Chapter 16

Appendix

Your tasks are the following:

1. Define the dependent and independent variables.
2. Using the given data set, run a fully partialled multiple regression.
3. Check for violations of the OLS assumptions or other problems like multicollinearity.
4. Using your best judgment, deal with missing data, outliers, collinearity, and any other problems you see in the data set
5. And fully explain the rationale for your actions.

ID	Var1	Var2	Var3	Var4	Var5
1	74.73	158.96	37.94	147.75	0
2	79.89	228.7	64.25	153.13	M
3	105.62	223.77	61.4	147.65	0
4	74.59	208.61	62.61	157.59	0
5	M	245.14	53.38	158.05	0
6	93.28	103.73	56.62	157.72	0
7	84.56	186.53	63.65	140.42	0
8	130.13	272.75	76.85	154.63	1
9	81.84	142.74	60.26	179.8	1
10	72.97	207.52	74.76	141.8	0
11	80.14	204.35	61.42	153.7	1
12	82.03	127.36	41.88	144.39	0
13	82.98	201.27	M	155.23	0
14	102.7	272.32	M	117.71	1
15	97.75	126.61	57.75	146.34	0
16	94.78	222.65	67.64	157.12	0
17	135.21	241.03	50.84	146.59	1
18	110.1	236.28	51.86	159.35	1
19	91.21	M	55.52	175.93	1
20	87.92	196.77	60.84	193.57	1
21	109.98	230.94	61.51	157.35	0
22	136.36	239.83	70.94	161.49	1
23	101.29	230.42	75.9	169.46	1
24	98.61	226	71.23	174.39	1
25	104.58	186.01	59.16	161.72	1
26	95.31	208.69	67.34	152.64	1
27	93.01	117.72	40.17	146.4	0
28	106.19	256.29	87.98	170.18	1
29	77.84	M	59.39	M	1
30	130.06	224.59	77.72	164.67	1

PART IV

Special Issues in Multiple Regression

Introduction

Chapters 17 through 20 treats special issues in multiple regressions, all aimed at particular circumstances that, if left untreated, deter distinguishing signal from noise. Chapter 17, for example, focuses on nominal variable coding beyond two categories, i.e., coded as categories with more than two discrete categories of things. Chapter 18 explains the process of modeling and interpreting interaction terms (be patient, we'll explain!) among independent variables, once again showing how the form of measurement is no hindrance. This chapter is very much in the spirit of sharpening the search for and enhancement of signal. Chapter 19 addresses the special problems posed by regressing proportions, ranks, counts, and nominal variables as dependent variables.

Chapter 20 conveys a more recent (some would say more advanced) form of OLS regression devoted to making more definitive and reliable causal inferences, structural and reduced form equation models. We also show that what's called the fully partialled OLS model can be used for inferring causality between variables of varying types. Finally, we deal with mutual causation, i.e., when it is just as likely that the dependent variable chosen for a model is also an independent variable. Often times, both are, indeed, dependent and independent at the same time.

CHAPTER 17

NOMINAL VARIABLE CODING

In our earlier discussion of multiple regression, we introduced a model in which the dependent variable was cardinal and one of the independent variables was categorical or nominal. In the discussion, we showed the generality of multiple regressions: how multiple regressions can incorporate and interpret effects of nominal variables with two categories. The coding scheme we used, dummy variable coding, assigned a value of one (1) to females and zero (0) to males implying no hierarchy. We interpreted the regression coefficient, if statistically significant, as the mean difference between the two groups, female and male; the sign of the coefficient indicating the direction of the effect. If the sign was positive, females in our sample, coded 1, were, on average, more productive than males, i.e., the mean difference between female and male productivity favored females. If the sign was negative, the opposite would hold. In this chapter, we show how to code and interpret nominal variables with more than two categories. We also show an alternative to dummy variable coding, effects coding.

Dummy Variable Coding

Two Categories

In this section, we review a regression analysis with a cardinal dependent variable and a nominal independent variable. To illustrate, we take data from one of our previous examples ($N = 200$), and run a regression between productivity and gender, females coded 1; males coded 0. Table 17.1 summarizes the bivariate regression.

Table 17.1 Bivariate Regression of Productivity on Gender

Variables	Coefficient	Standard Error	T (Z)	P
Constant	46.61	0.79	58.86	0.0000
Gender	6.43	1.068	6.02	0.0000

$R^2 = 0.15$

$F = 36.25$

$P = 0.000$

In table 17.1, we show the result of running a nominal independent variable, gender. The regression constant, B_0 and coefficient, B_1, were statistically significant and positive. Thus, we conclude that females were, on average, approximately 6.42 points more productive than males; the R^2 was about 0.15 suggesting that gender explains about 15% of the total variance in productivity. We can estimate the mean productivity of the two categories, female and male, by substituting one and zero into the equation and solving. Female mean productivity equals the constant 46.61 plus the coefficient, 6.42, 53.03. The mean productivity of males equals the constant, 46.61.

If we add another independent variable, energy, to the regression, we see a different picture in table 17.2.

Table 17.2 Regression of Productivity on Gender and Energy

Variables	Coefficient	Std Error	T	P	VIF
Constant	10.0002	2.68635	3.72	0.0003	
GENDER	2.84965	0.80186	3.55	0.0005	1.1
ENERGY	0.16663	0.01195	13.94	0.0000	1.1

R Squared = 0.5748

F = 132.98   P = 0.0000

The R^2 increased from 0.15 to about 0.57. The partialled regression coefficient for gender decreased from 6.43 to 2.85. The VIFs for gender and energy, while relatively low, still indicate some collinearity that accounts for the sharp decline in the gender coefficient. We attribute this change to the addition to the equation and partialling of energy from gender. The effect of gender is only about three productivity points, the difference between females and males, holding energy constant. To estimate the mean productivity of females and males, we substituted the mean of gender into the equation and 1 and 0 for females and males. Substituting the mean of energy effectively holds energy constant in our estimation of female and male productivity. Energy's mean is about 232. Substituting 232 and 1 and 0 into the equation yields $0.17 \times 232 + 2.85 \times 1 = 39.44 + 2.85 = 42.29$, the productivity mean of females with energy held constant. The productivity mean of males with energy held constant is 39.44. If we had centered energy, its mean would have equaled zero and we could have solved the equation for 1 and 0.

Dummy Variable Coding: More Than Two Categories

If the nominal variable has more than two categories, say three, the coding scheme creates two variables. You may ask, "Why not three?" Remember that the two-category case required only one variable because one and zero coding captured all the information in the gender variable. The rule in coding nominal variables: the number of variables required equals $K - 1$ where K is the number of categories.

To illustrate, assume that we have a four-category variable measuring political preference: radical, liberal, moderate, and conservative. Here, the coding scheme creates three variables: radical, liberal, and moderate with conservative as the reference group. Note: It makes no difference as to which group is used as reference. Table 17.3 shows the coding.

Table 17.3 Political Preference with Conservative as the Reference Group

Individual	Radical	Liberal	Moderate
1	1	0	0
2	0	1	0
3	0	0	1
4	0	0	0
5	1	0	0
6	0	0	0

The first individual in our sample is a radical, therefore we code her 1 and, because she is neither liberal, moderate, nor conservative, we code her zero on liberal and moderate with conservative understood. The second individual, a liberal, is coded 1 as a liberal, zero as radical, zero as moderate, and again, conservative understood. For the third individual, a moderate, the coding follows the same scheme. The fourth individual, a conservative, is neither radical, liberal, nor moderate, and, therefore is coded as 0, 0, and 0. We call the group coded 0, 0, 0 (in this case Conservative), the reference group. We identify the fifth and sixth persons as radical and conservative respectively.

In the regression, we interpret significant coefficients for any of the three variables as the mean differences between the variables and the reference group. In our example, if Y equals income, and all three variables show statistical significance, the regression coefficients correspond to the mean differences in income between radicals, liberals, moderates and the reference group, conservatives. The sign of the coefficient indicates the direction of the difference. The equation: $Y_C = B_0 + B_1 \text{Radical} + B_2 \text{Liberal} + B_3 \text{Moderate}$ defines the regression.

If we wanted to change the reference group, for example to radical, we would reconstruct the matrix as illustrated in table 17.4.

Table 17.4 Political Preference with Radical as the Reference Group

Individual	Category	Conservative	Liberal	Moderate
1	Radical	0	0	0
2	Liberal	0	1	0
3	Moderate	0	0	1
4	Conservative	1	0	0
5	Radical	0	0	0
6	Conservative	1	0	0

We dropped the variable, Radical, and brought Conservative into the regression equation. The equation now looks like this: Y_C (Income) = $B_0 + B_1$ (Conservative) + B_2 (Liberal) + B_3 (Moderate). We have made radical the reference group and we see him (number 5) in tables 17.3 and 17.4 coded 1, 0, 0 and 0, 0, 0. It makes no difference which group we treat as reference, MR^2, F, and estimates of mean incomes will be the same.

Suppose we ran a regression of the data in table 17.3 extended to 20 individuals and their income added. The result of the regression is summarized as table 17.3B.

**Table 17.3B Regression of Income on Political Attitudes
with Radical as Reference Group**

Independent Variable	Regression Coefficient	Standard Error	T Value	Prob
Intercept	35.20	8.63	4.07	0.00
(Conservative=1)	4.36	4.71	0.93	0.37
(Liberal=1)	19.02	4.18	4.55	0.00
(Moderate=1)	0.86	5.18	0.17	0.87

$R^2 = 0.599$ F Ratio = 7.975

The equation $Y_C = B_0 + B_1 \text{Liberal} + B_2 \text{Moderate} + B_3 \text{Conservative}$ is statistically significant at $P < 0.01$ and the Liberal variable is significant at $P < 0.001$. We interpret the result to mean that liberal's income is about $19,000 greater than radicals and the differences in incomes among moderates and conservatives, and radicals are not statistically different. An estimate of the mean income of liberals substitutes 1 for liberal, 0 for conservatives and 0 for moderates and solves the equation. The estimated mean income for liberals is about $54,200 (the regression coefficient B_1 + the constant B_0). For conservatives, the estimate is about $39,600 ($B_3 + B_0$) , for moderates around $36,000 , and for radicals about $35,200 ($B_0$) . Note: We do not know whether the differences in incomes between liberals and moderates and conservatives are statistically significant.

Again, it makes no difference which group we designate as reference, the mean estimates will remain the same. If we wanted to test the significance of the difference between the means of all the groups, we could run four regressions with each of the groups as reference. This method of coding also returns all the appropriate tests of significance and allows for interval estimates.

For another example, let us assume that we have a random sample (N = 30) with productivity as the dependent variable and a three category nominal set, red, yellow, and green, as independent variables. The three colors represent a unit's compliance with company procedures. Red means not following procedures, yellow means partly following procedures, and green means full compliance with procedures. Table 17.6 shows an abbreviation of the data set with dummy coding.

Table 17.5 Dummy Coding of Compliance with Red as the Reference Group

Production Unit

	Category	Yellow	Green
1	yellow	1	0
2	green	0	1
3	red	0	0

We ran a regression with red as the reference group with table 17.6 summarizing the results.

Table 17.6 Regression of Productivity on Compliance with Red as the Reference Group

Variables	Coefficient	Std Error	T	P	VIF
Constant	51.9711	1.24847	41.63	0.0000	0.0
yellow	-6.08855	2.99374	-2.03	0.0519	1.0
green	-7.20677	2.40612	-3.00	0.0058	1.0

R Squared	0.2892
$F = 5.49$	$P < 0.0500$

Note: We have included red in the title of table 17.7, emphasizing that red was not omitted from the analysis, just specified as the reference group. The over-all model is statistically significant at alpha < 0.05, with an MR^2 of about 0.29. Thus, the overall set, compliance with procedures, explains about 29% of the variation in productivity. The regression coefficient for green, about - 7.21 says that groups coded green, on the average, are about 7 productivity points lower than groups coded red. Yellow, also statistically significant, indicates that groups coded yellow were, on average, about 6 productivity points lower than groups coded red. We interpret that regression to mean that, on average, groups not complying with company procedures are more productive. We suspect that the difference between yellow and green will not show significance, although we do not know that from the data in table 17.7; we estimate the means of the three categories by substitution into the equation. To calculate the mean for yellow, we substitute 1 for yellow and 0 for green; for green, we substitute 0 for yellow and 1 for green and for red, we substitute 0 and 0. The mean for yellow is about 51.97 - 6.088, 45.9, for green 51.97 - 7.21, 44.76, and for red, 51.97, the constant.

To demonstrate that our choice of reference groups does not affect MR^2, F, and P, we reran the regression with green as the reference group; Table 17.8 summarizes that regression

Table 17.8 Regression of Productivity on Compliance with Green as the Reference Group

Variables	Coefficient	Std Error	T	P	VIF
Constant	44.7643	7.2067	21.76	0.0000	0.0
yellow	1.1182	3.4109	0.33	0.7456	1.4
red	7.2067	2.4061	3.00	0.0058	1.4

R Squared   0.2892

F = 5.49 P<0.05

Notice that MR^2, F, and P for the overall model in these two tables are the same. By making green the reference group, we see that the difference between yellow and green is not statistically significant. The positive sign on the yellow and red coefficients suggests that, on average, the less a unit follows company procedures, the more productive the unit. The coefficient red, about 7.21, when added to the constant gives the same mean as in the previous regression. If we estimate the mean of the three categories, we will obtain the same result as in the equation with red as reference.

Unweighted Effects Coding[80]

Many research problems require comparisons, not between groups, but between the groups and the mean of the entire set (the Grand mean). Effects coding follows the same scheme as dummy coding with one exception: effects coding uses minus one (-1) for the reference group instead of zero. The consequence of this scheme makes the regression coefficients represent the mean difference between the specific group and the unweighted mean of all the groups. Here, the term unweighted means that this coding scheme ignores the differences in the number of units in the categories, treating them as if they all had the same size Ns. Table 17.9 shows this pattern with red as the reference group.

Table 17.8 Compliance with Procedures with Red as the Reference Group

Individual Production Unit	Category	Yellow	Green
1	yellow	1	0
2	green	0	1
3	red	-1	-1
4	red	-1	-1
5	yellow	1	0
6	green	0	1

Here we used -1 instead of 0 for the reference group. This coding will yield regression coefficients that, if significant, show the difference between the mean of a specific group and the unweighted grand mean for the three groups. The unweighted grand mean for the three categories equals:

$$\frac{Mean_{yellow} + Mean_{Green} + Mean_{Red}}{3} . \qquad \text{(Equation 17.1)}$$

---

[80]. JCC, pp. 320-328.

This scheme gives equal weight to the three categories ignoring any differences in the size of the category. It also gives significance tests for each of the variables in the analysis. Note: We cannot see the effect of red by simply substituting 0 and 0 into the equation as with dummy coding. To see the effect of red, we would construct another table with a different reference group and rerun the regression. Usually the group designated as reference, reflects the group in which we have the least interest.

Effects coding also works with a two group nominal variable. In that case, one group is coded 1, the reference group -1. The regression coefficient of the single variable would represent the difference between the mean of each group and the unweighted grand mean.

We note that it is not a good idea to calculate the FPR^2_s for unweighted effects coding because the means are not adequately represented. Table 17 9 shows data summarizing the regression using unweighted effects coding with red as reference.

Table 17.9 Regression of Compliance using Unweighted Effects Coding with Red as the Reference. Group

Variables	Coefficient	Std Error	T	P	VIF
Constant	47.53931	.21075	39.26	0.0000	0.0
yellow	-1.65678	1.98339	-0.84	0.4109	2.1
green	-2.77499	1.69592	-1.64	0.1134	2.1
R Squared	0.2892				
F = 5.49	P < 0.0500				

Compare this regression with the dummy coded regression. R^2, Model F, and probabilities are the same. The regression coefficients now represent the difference between the category mean and the unweighted mean of the three groups, 47.5393. Neither yellow nor green show statistical significance. We interpret that to mean that the difference between yellow and green and the unweighted Grand Mean is probably attributable to chance. So the mean of red is probably different from the unweighted Grand mean because, as we note below, the overall model is significant.

We can calculate the group means for yellow and green by substitution; of course, they will be the same as in our dummy coded regression. But we cannot calculate the mean of red by simply substituting -1 for yellow and -1 for green because that only yields the unweighted mean of productivity. To see red's effect, we must change the reference and rerun the regression. Therefore, when using effects coding the choice of reference group should be the variable of least interest.

Weighted Effects Coding[81]

When we take random samples of populations with nominal variables, we can reasonably assume that the nominal variables in the random sample, if large enough, would reflect the population proportion. For example, if a population contains 60% males and 40% females, a reasonable expectation is that a random sample will roughly reflect that proportion. In those cases, it's appropriate to use a weighted effects code. Table 17.11 illustrates how weighted coding works.

Table 17.10 Weighted Effects Coding

Individual Production Unit	N = Size of Category	Category	Yellow	Green
1	NYellow = 4	yellow	1	0
2	NGreen = 7	green	0	1
3	NRed = 19	red	-0.211	-0.368
4		yellow	1	0
5		red	-0.211	-0.368

We have added a column showing the size of each group. To construct the weighted effects coding, we do not use -1 as the reference group code. Instead, we create the reference group code as a value with a negative numerator, the number of observations in the group coded one and a positive denominator, the size of the reference group,. For example, from table 17.10, we have 30 observations in the sample. The category green has 7 units, yellow has 4, and red has 19. If we choose red as the reference group, the reference group code for

81.   OP Cit

green equals $\dfrac{-N_{K1}}{N_{reference}} = \dfrac{-7}{19} = -0.368$ and the reference group code

for yellow equals $\dfrac{-N_{K2}}{N_{reference}} = \dfrac{-4}{19} = -0.211.\cdot$

For weighted effects coding, instead of -1 as the reference group, we code the reference group as -0.368 for and -0.211 (see table 17.10 above). In the regression using weighted effects coding, the coefficients represent the difference between each of the groups in the equation and the weighted grand mean of the all three groups. The grand weighted mean of the three groups equals

$$\frac{Mean_{Red}(19) + Mean_{Green}(7) + Mean_{Yellow}(4)}{30} = 49.48. \quad \text{(Equation 17.2)}$$

You can see from Equation 17.2 that this procedure uses the weighted grand mean.

Table 17.11 summarizes a regression of Productivity on green, yellow, and red with red as the reference group.

Table 17.11 Regression with Weighted Coding

Variables	Coefficient	Std Error	T	P	VIF
Constant	49.4778	0.99356	49.80	0.0000	0.0
green	-4.71356	1.80163	-2.62	0.0144	1.0
yellow	-3.59535	2.53224	-1.42	0.1671	1.0

R Squared =	0.2892
F = 5.49	P < 0.05

Compare these results to table 17.9. Everything is the same except the size of the regression coefficients and the statistical significance of green. We attribute this difference to the weighting scheme because the coefficients in table 17.11 represent the differences between the means of the categories and the weighted grand mean, the actual mean of productivity. In the weighted regression the constant, B_0, is the weighted grand mean of productivity, 49.477. We interpret the statistical significance of green from the weighted grand mean as the green mean is lower by about 4.7. Solving the regression equation for green yields a productivity mean of 44.7642. We did not solve for yellow because it was not significantly different from the weighted grand mean. If we want to test for red, we need to respecify the reference group and rerun the regression.

In addition to understanding the mechanics of the coding schemes, it is important to grasp that the strategies will yield the same results for MR^2, F ratio, and P. Tests of significance of the constant and the independent variables will show differences.

Missing Data in Nominal Variables

In chapter 15, we discussed techniques for handling missing data in variables. In this brief section, we extend that discussion to nominal data with more than two categories. Let us, again, assume that we have a data set like the Compliance analysis in this chapter with missing data across the three categories. Our preferred solution to this condition assigns zeros to the missing data and creates another categorical variable coding missing as 1 and actuals as 0. The single new categorical variable accounts for all the missings in the set. If we have two sets of categorical variables representing two constructs, e.g., political preference and religion, we will need two missing data variables.

In the regression, we use the missings variable as a separate independent variable. Table 17.13 illustrates the idea.

Table 17.12 Missing data in a Data Set of Nominal Independent Variables Using Dummy Variable Coding with Missing Data

ID	Productivity	Green	Yellow	Red
1	90.16	0	0	0
2	88.21	1	0	0
3	82.32	1	0	0
4	101.33	0	0	1
5	83.72	0	1	0
6	109.44	0	0	1
7	89.07	0	0	1
8	132.19	M	M	M
9	137.56	0	1	0

Table 17.13 is an extract of a new data set showing only the first 9 observations out of 30. Observations 8, 13, and 21 have missing values in all three categories. We used weighted effects coding with red as the reference group. First, we ran a regression ignoring the missing data. Table 17.13 summarizes this regression.

Table 17.13 Regression of Productivity on Compliance with Red As Reference Using Dummy Coding Ignoring Missing Values

Variables	Coefficient	Std Error	T	P	VIF
Constant	4.57211	0.03151	145.09	0.0000	0.0
green	-0.14519	0.05577	-2.60	0.0156	1.0
yellow	0.08152	0.06407	1.27	0.2154	1.0

R^2	0.24
F	3.84
P	0.036

The difference between green and red was statistically significant. This analysis tells us nothing about the possible effect of missing data.

In table 17.14, we assigned zeros (0s) to the missing data and created another categorical variable coded missing equal one and nonmissing equal zero. In this example, we continued using unweighted effects coding. .

Table 17.14 Missing Data Coding with Dummy Variables

ID	Green	Yellow	Red	Missing	Productivity
1	0	0	0	0	90.16
2	1	0	0	0	88.21
3	1	0	0	0	82.32
4	0	0	1	0	101.33
5	0	1	0	0	83.72
6	0	0	1	0	109.44
7	0	0	1	0	89.07
8	0	0	0	1	132.19
9	0	1	0	0	137.56

We reran the regression using the missing data variable as the reference group. Table 17.15 summarizes that regression.

Table 17.15 Regression of Productivity on Compliance with Dummy Coding and the Missing Data Variable as the Reference Group

Variables	Coefficient	Std Error	T (Z)	P
Constant	4.43309	0.06198	71.52	0.0000
green	0.25666	0.09602	2.67	0.0128
yellow	0.17930	0.07506	2.39	0.0245
red	1.52320	0.54294	2.81	0.0094

R Squared = 0.2508
F   3.44
P   0.03134

The overall model along with all the variable coefficients was statistically significant from the missings. Our analysis tells us that we probably have a serious missing data problem. To compare the groups, we could change the analysis to make red the reference group and retain the missing data variable as an independent variable.

Summary

The ideas and concepts of most importance in chapter 17 are summarized as follows:

1. Nominal variables

   We call variables that measure qualities, e.g., gender, religion, and political preference, that have no natural hierarchy and cannot find expression as cardinal, interval, or ranked data.

2. Dummy variable coding of nominal variables with two groups

   An example of dummy variable coding of a nominal variable with two categories, we use a regression with a cardinal variable as dependent and a nominal variable as independent. In this case, the cardinal dependent variable equals productivity, the dependent variable, gender, with females coded 1 and males coded 0.

3. Dummy variable coding of nominal variables with more than two groups

If the nominal variable has more than two categories, say three, the coding scheme creates two variables. You may ask, "Why not three?" The two category case required only one variable, because 1 and 0 coding captured all the information. The rule in coding nominal variables: the number of variables required = K -1, where K is the number of categories.

4. Unweighted effects coding

Many research problems require comparisons, not between groups, but between the groups and the mean of the set (the mean of all). Effects coding, follows the same scheme as dummy coding with one exception: effects coding uses minus one (-1) for the reference group instead of zero. The consequence of this scheme makes the regression coefficients represent the mean difference between the specific group and the unweighted mean of all the groups.

5. Weighted effects coding

When we take random samples of populations with nominal variables, we can reasonably assume that the nominal variables in the random sample, if large enough, would reflect the population proportion. For example, if the population from which we take a sample contains 60% males and 40% females, we can reasonably expect that our sample will reflect that proportion. In those cases, then, it's appropriate to use a weighted effects code.

6. Coding nominal variables for missing data

Assume that we have a data set like the Compliance analysis in this chapter with missing data across the three categories. One effective solution to this condition assigns zeros to the missing data and creates one additional nominal variable coding missings as 1 and actuals as 0.

Appendix

Chapter 17

In the data set given below, the dependent variable represents attitudes toward management (high scores represent better attitudes). The independent variable is a nominal set, political orientation: liberal, moderate, conservative. Your tasks are the following:

1. The data have missing observations. You should make an effort to ameliorate the effect of missings.

2. Using dummy variable coding, run a multiple regression without the missing adjustment.

3. Using dummy coding, run a regression with a missing compensation. Compare your results with the findings in 2 above.

4. Using unweighted effects coding, run a multiple regression and compare your results with 3 above.

5. From the data set, drop the last 3 observations in C, drop the last 5 observations in M, and drop the last 7 observations in L. Then run a multiple regression using weighted effects coding and compare your result with 4 above.

6. In all the tasks, explain and interpret your output.

ID	Attitude	L	M	C
1	90.16	0	0	1
2	88.21	1	0	0
3	82.32	1	0	0
4	102.28	0	1	0
5	83.72	M	M	M
6	109.44	1	0	0
7	89.07	0	1	0
8	132.19	0	0	1
9	137.56	0	1	0
10	101.71	0	1	0
11	81.66	0	0	1
12	96.52	0	0	1

13	82.81	M	M	M
14	126.59	0	0	1
15	117.85	0	0	1
16	108.37	0	0	1
17	97.19	1	0	0
18	110.57	0	1	0
19	77.21	1	0	0
20	147.65	0	0	1
21	97.78	M	M	M
22	96.51	1	0	0
23	110.76	0	1	0
24	81.02	1	0	0
25	100.31	0	0	1
26	85.22	1	0	0
27	104.88	0	1	0
28	90.64	0	1	0
29	95.15	1	0	0
30	89.99	M	M	M

CHAPTER 18

INTERACTIONS

In our example of the study of productivity using intelligence, energy level, and gender as independent variables, we treated the effects of intelligence, energy, and gender as *additive* where $Y_C = B_0 + B_1X_1 + B_2X_2 + B_3X_3$. We call these effects additive because, as the equation instructs, we sum the effects. Addition, however, is not the only way to conceive an effect. Based on theory, we can envision the effect of intelligence and energy or gender and intelligence on productivity as multiplicative or conditional. Suppose that when a worker combines high intelligence and high energy, there may be a separate, non-additive effect on productivity. Alternatively, when a worker combines low energy and high intelligence, the productivity effect may be negative. Interaction means that the whole may be greater or less than the sum of its parts.

If we add gender to the mix, the effect of gender interacting with energy may show that high-energy females are even much more productive than the additive effects of high energy and gender would suggest.

In other words, if two independent variables have an interactive effect on Y, the effect of one variable depends on (is conditioned by) the value of the other. In this chapter, we show how to create, test, and interpret interaction effects using combinations of continuous and nominal variables.

Before we begin, please consider another caveat: Because modern computer programs make interactions easy to create and test, persons doing regressions face great temptations to create and test interactions indiscriminately. Interactions should not be added without theoretical guidance. Unless you can hypothesize and explain the effect of an interaction effect *before* you create it and run the regression, you should avoid the temptation. Post hoc explanations of interactions are taboo. In addition, many times analysts confuse interaction effects with nonlinearity or outliers. Examining data for nonlinearity and outliers takes precedence over running interactions. This reason alone illustrates why we should not introduce interactions without at least a good confederacy of hunches.

Interactions among Cardinal Independent Variables

Using our previous productivity example, we first explain how to create and test for interaction effects; then explain how to interpret them. Assume the following model:

$$Y_{C\ Productivity} = B_0 + B_1 \times \text{Intelligence} + B_2 \times \text{Energy Level}$$

The equation assumes that energy and intelligence only have additive effects. After some thought, we revised the theory to include the notion that energy and intelligence have a positive conditional effect. As the revised theory goes, when nature and hard work combine at high levels of energy and intelligence, the effect on productivity transcends the additive, main effect of the two variables.

To test this interaction hypothesis, we create a third interactive independent variable by multiplying the two variables X_1 and X_2 together producing $X3$. Our new model becomes:

$$Y_{C\ Productivity} = B_0 + B_1 \times \text{Intelligence} + B_2 \times \text{Energy Level} + B_3 \times \text{Interaction (I} \times \text{E).}$$

It should be obvious that multiplying intelligence times energy produces a highly collinear new independent variable because the new interaction variable is their product. So before you create the interaction, center the two variables X_1 and X_2 to reduce collinearity. Remember, centering a variable, means subtracting its mean. This reexpression is not a transformation because it does not change a variable's structure.

After creating a new interactive (multiplicative) variable, we rerun the analysis. In the new model, the regression analysis partials the two main effect variables from the interactive variable, leaving a residual that represents only the interaction. If statistically significant, the interaction adds something to our understanding beyond the main additive effects. A statistically significant interaction coefficient increases MR^2. We note here that, most of the time, the interaction R^2 added to MR^2, will be relatively small, no greater than 0.05. Therefore, when contemplating running an interaction the sample size should be large enough (have enough power) to find a statistically significant interaction effect.[82]

To interpret the effect of an interaction, we use the following example. Table 18.1 is an excerpt from a hypothetical sample ($N = 50$).

Table 18.1 Excerpt of a sample, N = 50

id	Productivity (Y)	Intelligence (X_1)	Energy (X_2)
1	55.69	9.65984	5.5482
2	47.139	-16.41516	-11.5518
3	73.028	40.63184	15.2982
4	59.841	29.83984	-14.6818
5	48.597	-29.35616	-11.3718

Now. we run a regression of productivity on intelligence and energy. Table 18.2 is a summary of the additive multiple regression model: $Y_C = B_0 + B_1$Intelligence $+ B_2$Energy.

Table 18.2 Regression of Productivity on Energy and Intelligence

Variables	Coefficient	Std Error	T	P	VIF
Constant	51.2296	0.72886	70.29	0.0000	
Energy	0.37580	0.03969	9.47	0.0000	1.0
Intelligence	0.05673	0.04048	1.40	0.1677	1.0

R^2 = 0.6603
F = 45.68
P = 0.0000

The overall model shows statistical significance ($P < 0.0000$), model MR^2 equals 0.660, and the energy independent variable's positive coefficient also shows statistical significance. The positive coefficient for the independent variable, intelligence, does not show statistical

---

[82]　For an elaboration of this Idea, see JCC, chapters 7, 8, and 9.

significance. *VIFs* of one suggest no problem of multicollinearity. In this example, assume that the analyst included intelligence, because she thought that it would condition the effect of energy on productivity, not because she thought it would necessarily have a main effect on productivity. The fully partialled PR^2 for energy is about 0.65. Table 18.3 shows an excerpt of our sample with a newly created interaction variable (intelligence and energy multiplied together).

Table 18.3 Excerpt of Data with the Interaction Term

Id	Productivity	Intelligence	Energy	Interaction (I × E)
1	55.69	9.6598	5.5482	53.5947
2	47.139	-16.4151	-11.5518	189.6246
3	73.028	40.6318	15.2982	621.5940
4	59.841	29.8398	-14.6818	-438.1025
5	48.597	-29.3561	-11.3718	333.8323

We created the interaction variable by multiplying the energy by intelligence. Our new model is $Y_C = B_0 + B_1 \text{Intelligence} + B_2 \text{Energy} + B_3 \text{Interaction } (X_1 \times X_2)$. Now, we run the regression using all three variables. Table 18.4 is a summary of that regression.

Table 18.4 Regression with Interaction

Variable	Coefficient	Std Error	T ratio	Prob
Intercept	51.2445	0.7060	72.581	0.0000
Cenergy	0.3421	0.0419	8.164	0.0000
Cintellig	0.0423	0.0399	1.062	0.2937
Inter	0.0048	0.0024	2.023	0.0489

$MR^2 =$ 0.688
F = 33.819
P = 0.000

The overall model, statistically significant at $P < 0.0000$, shows a MR^2 of 0.688, an increase over the model without the interaction of slightly less than 0.03. The letter *C* preceding energy and intelligence shows centering the variables before running the regression. The statistical significance of the Interaction variable ($P < 0.05$) signals that the relationship is multiplicative (conditional). In addition, interaction effects always trump main effects; interpretation of the main effects making up the interaction must defer to the statistically significant interaction.

How do we interpret the interaction? Interactions can form three distinct patterns: (1) *intensifying*, in which both independent variables have the same sign (predictors affect the dependent variable in the same direction); (2) dampening, in which the two independent variables have opposite signs with one predictor weakening the effect of the other; and (3) conflicting, in which both predictors have the same sign, but the interaction variable has the opposite sign. In this example, all the coefficients have positive signs signifying that, as intelligence increases, the effect of high energy on productivity produces an effect even stronger than its main (additive) effect.[83]

Beyond that general statement, we can glean more information by looking at the size of the coefficients. To extend our interpretation, we locate the values of the centered main effect independent variables at +/- one standard deviations from the mean. The means of zero for energy and intelligence represent middle intelligence and energy. Then we can solve the regression equation for productivity for the nine combinations of low intelligence × low energy, middle intelligence × low energy, high intelligence × low energy, and the rest. We show the regression below:

$$Y_c = 51.2445 + 0.3421 \times Energy + 0.0423 \times Intelligence + 0.0048 \times Interaction.$$

The high, middle, and low values for centered energy and intelligence, +/- 1 standard deviation from the means, equal -19, 0, 19 and 18, 0, 18 for centered intelligence. Now, we solve this equation for nine values of productivity: (1) low intelligence, low energy; (2) middle intelligence, low energy; (3) high intelligence, low energy; (4) low intelligence, middle energy; (5) middle intelligence, middle energy; (6) high intelligence, middle energy; (7) low intelligence, high energy; (8) middle intelligence, high energy; and (9) high intelligence, high energy.

---

83.   JCC, chapters 7, 8, and 9

Solving for the first combination (low intelligence, low energy) by substitution into the equation yields:

Productivity = 51.2445 + (0.3421× -19) + (0.0423 × -18) + (0.0048 × -18 × -19) = 44.39.

Table 18.5 summarizes the other eight calculations.

Table 18.5 Interaction of Intelligence and Energy

	LI	MI	HI	Slopes
HE	55.3414	57.7444	60.1474	2.403
ME	50.4831	51.2445	52.0059	0.761
LE	45.6248	44.7446	43.8644	-0.880

The values in the cells of table 18.5 represent the mean productivity for the various combinations of the interaction effect, for example, the mean productivity of high intelligence × high energy is about 60.15 and for low intelligence × low energy, about 45.62. The slopes in the far right column of the table reflect the difference between low, middle and high intelligence at the three levels of energy. To clarify the issue, we plot these numbers as figure 18.1.

Figure 18.1 Interaction Plots of Productivity on Intelligence and Energy

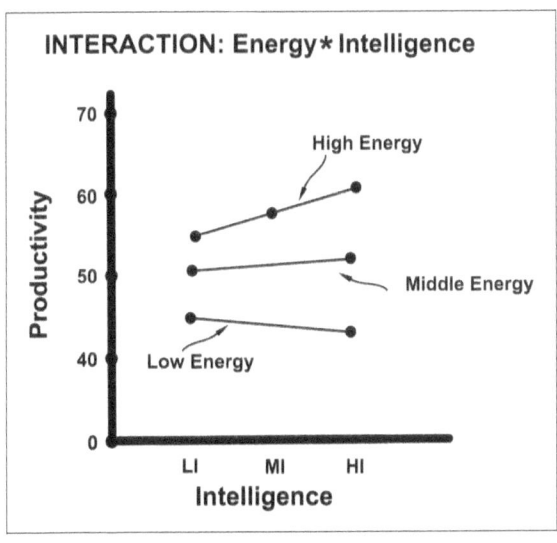

With no interaction, these lines would plot as parallel, simply reflecting the additive main effects. In this graph, we made the lines equal energy with the levels of intelligence on the x-axis. By plotting the graph this way, intelligence conditions energy's effect on productivity, consistent with our earlier hunch, the justification for testing the interaction. The graph confirms our interpretation of the positive signs on the main and interaction coefficients. At every level of intelligence, high energy people are on the average more productive than middle and low energy prospective employees. Unexpectedly, it also tells us that low energy interacting with intelligence results in declining productivity as intelligence increases. Intelligence's conditioning effect increases the slope of high-energy people over middle energy and low energy. We placed the slopes of high, middle, and low energy on the right hand side of table 18.5. The positive slope of middle energy suggests that as intelligence increases, middle energy persons increase productivity. Note: these lines are all linear because of linear main effects.

Because all interactions are reciprocal, if it made sense, we could check the interaction effect of energy on intelligence, by putting energy on the x-axis and plotting intelligence lines. In figure 18.2, we changed the look of the graph, now showing energy conditioning intelligence's effect on productivity.

Figure 18.2 Interaction Plots of productivity on Energy × Intelligence

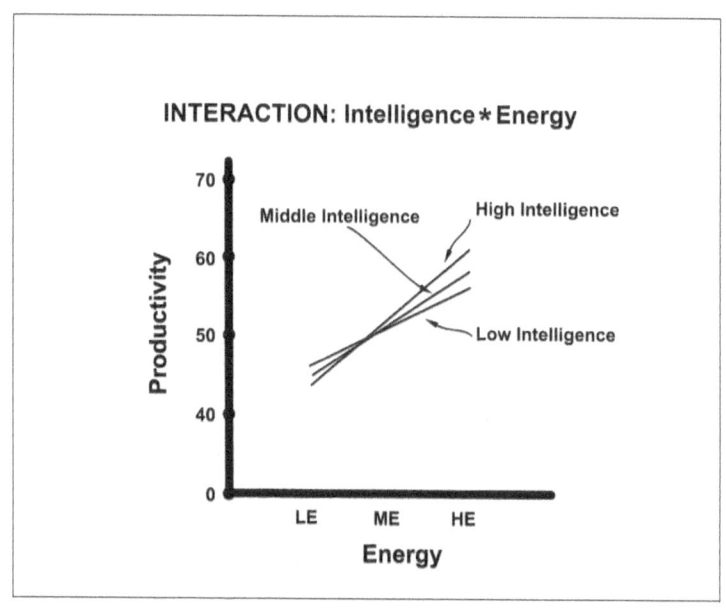

In this graph, the conditioning effect of energy reinforces what we saw in figure 18.1. High intelligence, low energy people have the lowest productivity, while the combination of high intelligence and high energy display the highest productivity. Intelligence has a positive effect on productivity throughout all levels of energy because all levels of intelligence have positive slopes, but low energy and high intelligence have the lowest mean productivity.

Graphing both conditioning effects, does not change the values in the table, but shows the always reciprocal effect of interactive conditioning. The question arises, what conditions what? The answer: both condition each other. Our preference in interpreting interaction effects must show the effect given by theory. The only reason we made this picture was to show the reciprocal nature of all interactions. Whether it makes sense to say that energy conditions the effect of intelligence on productivity is a matter defined by theory.

We note that we can extend the idea to triple interactions involving more than two independent variables and to polynomials modeling curvilinear effects as well. Such interactions are justified only when theory points the way.

Interactions with Continuous and Nominal Variables

Assume that our theoretical productivity model strongly suggests the addition of another variable, gender. In response, we take another random sample ($N = 50$), and reproduce a small part of that sample (the first 5 observations) as table 18.6.

Table 18.6 Excerpt of a Random Sample of 50

ID	Productivity	Centered Intelligence	Centered Energy	Gender	Interaction Gender × Energy
1	55.690	9.65984	5.5482	0	0
2	47.139	-16.41516	-11.5518	0	0
3	73.028	40.63184	15.2982	1	15.2982
4	9.841	29.83984	-14.6818	1	-14.6818
5	48.597	-29.35615	-11.3718	0	0

Notice in table 18.6, when the nominal variable is zero (0), the interaction nominal × continuous variable also equals zero (0). This is not a mistake.

Now, we run the model without the interaction. Table 18.7 summarizes this regression: $Y_C = B_0 + B_1$Intelligence $+ B_2$Energy $+ B_3$Gender.

Table 18.7 Summary of Additive Main Effect Regression

Regression Equation: $Y_C = B_0 + B_1$Intelligence $+ B_2$Energy $+ B_3$Gender.

Independent Variable	Coefficient	Standard Error	T-Value	Prob
Intercept	48.9593	0.7339	66.711	0.0000
Cenergy	0.2498	0.0402	6.208	0.0000
Cintellig	0.0314	0.0330	0.954	0.3450
Gender	8.7322	1.6955	5.150	0.0000
R^2	0.7845			
F	55.829			
P	0.0000			

The overall model, statistically significant at $P < 0.0000$, shows the model MR^2 as about 0.78. The energy and gender independent variables also show statistical significance at $P < 0.0000$; The fully partialled FPR^2s for energy and gender respectively equal 0.8100 and 0.7740, both strong effects.

Table 18.8 Summary of Interaction Effect

Regression Equation: Y_C productivity $= B_0 + B_1$Energy $+ B_2$Intelligence$+ B_3$Gender $+ B_4$Interaction (gender × energy).

Variable	Regression Coefficient	Standard Error	T(Z) ratio	Prob
Intercept	49.0124	0.7011	69.903	0.0000
Cenergy	0.2211	0.0403	5.481	.00000
Cintellig	0.0187	0.0319	0.585	0.5614
Interaction				
Cenergy*	0.0044	0.0019	2.335	0.0241
Gender	8.5803	1.6203	5.295	0.0000

$R^2 = 0.810$

$F = 47.286$

$P = 0.000$

The overall model, still strongly significant, produced a model MR^2 almost 3% higher than the additive analysis; the interaction effect is statistically significant and positive.

In this example, our interest lies in the possible conditioning effect of gender on energy with respect to productivity. Interpreting that effect, again requires a close look at the coefficient signs of the additive main effects and their interaction. The positive coefficients tell us that females (coded 1) and high energy increase productivity beyond their additive main effects. To understand more, we again build an interaction table by solving the equation for high, middle, and low centered energy and gender. We use +/- one standard deviations from the mean of energy, -19, 0, +19 as low middle and high levels of energy and solve the equation for six conditions: (1) male × low energy, (2) female × low energy, (3) male × middle energy, (4) female × middle energy, (5) male × high energy, and (6) female × high energy. Because we used centering, we can ignore intelligence; its mean of zero automatically controls its effect. Table 18.9 is the interaction table.

Table 18.9 Interactive Effects of Energy × Gender

	Male	Female	Slopes
LE	45.6959	54.2102	8.5413
ME	49.0124	57.5927	8.5803
HE	52.3289	60.9752	8.6463

By looking at the slopes in table 18.9, we can see very small differences and this is reflected in the picture portrayed in figure 18.3. Figure 18.3 portrays the interaction showing how gender conditions energy. The slope[84] of the high-energy line is slightly larger than the slopes of middle and low energy suggesting the synergistic effect of females and energy. Therefore, the interactive effect, while statistically significant, seems to be trivial.

[84.] These slopes represent the differences between male and female for low, middle, and high energy.

Figure 18.3 Interactions of Productivity on Gender × Energy

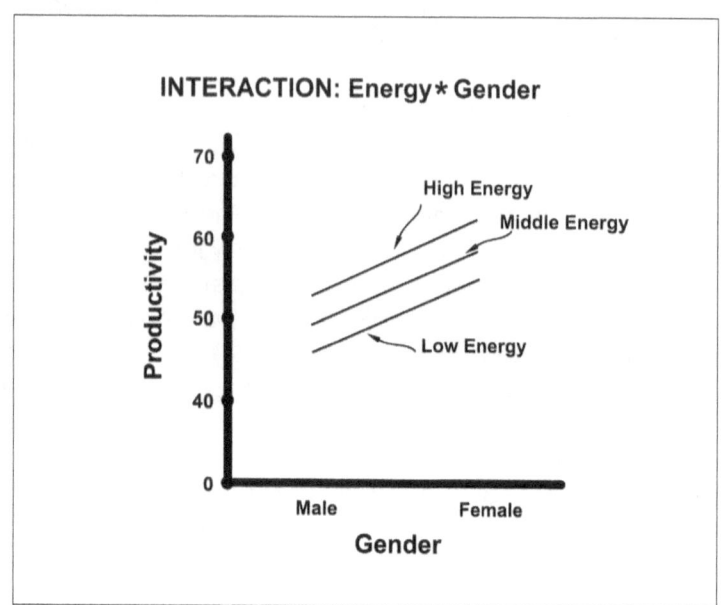

Reversing the interaction in figure 18.4 shows the conditioning effect of energy on gender.

Figure 18.4 Conditioning Effect of Energy × Gender

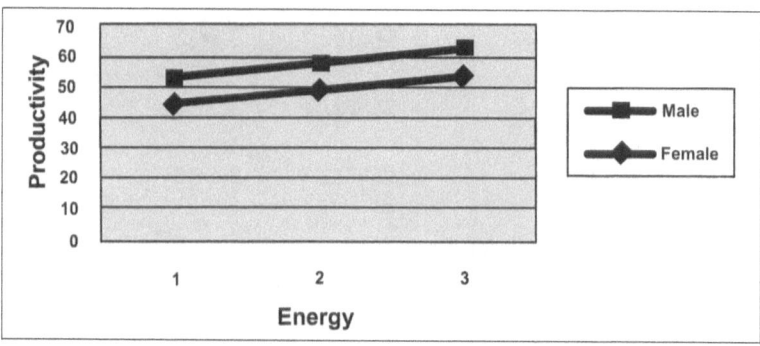

Again, the lines plot almost parallel; while the interaction shows statistical significance, its effect does not change the additive main effects in any important way.

Note: By centering the main effect variables, we made their mean equal zero. Therefore, in solving the interaction equation, we effectively controlled or held constant intelligence by substituting its mean, zero. Centering, saved us the trouble of making the additional calculation of multiplying the mean of intelligence by its coefficient.

Statistical Significance of Interaction Slopes

The statistically significant interaction variable coefficient tells us that the interaction was probably not attributable to chance. Aside from that test, tests of significance between the slopes of various combinations of interaction effects, while available, are probably not worth the effort. Making the calculations shown above will tell you which of the interactions represent statistical significance.

Interaction of Nominal, Nominal Variables

If, we added to the examples given above, another nominal variable, the skin color of the individual, e.g., white = zero, black = one, we could run another interaction, gender × race with the hypothesis that gender conditions the effect of race on productivity. Table 18.10 shows an excerpt of a random sample of 50 with a race variable and an interaction between gender and race.

Table 18.10 Interaction Two Nominal Variables

ID	Productivity	Centered Intelligence	Centered Energy	Gender	Race	Race × Gender Interaction
1	55.69	9.65984	5.5482	0	1	0
2	47.139	-16.41516	-11.5518	0	0	0
3	73.028	40.63184	15.2982	1	0	0
4	59.841	29.83984	-14.6818	1	1	1
5	48.597	-29.35616	-11.3718	0	1	0

First, we run the regression without the interaction. Table 18.11 is a summary of that regression.

Table 18.11 Regression without Interaction

Regression of Productivity without Interaction

Variables	Coefficient	STD Error	T	P	VIF
Constant	45.9168	1.03084	44.54	0.0000	0.0
Race	6.77352	1.42634	4.75	0.0000	1.7
Cinergy	0.25234	0.03697	6.83	0.0000	1.6
Cintel	0.05923	0.03249	1.82	0.0749	1.2
Gender	4.45951	1.28522	3.47	0.0012	1.1

This statistically significant model at $P < 0.0000$ produced MR^2 of 0.82. The independent variables race, energy, and gender show statistical significance at $P < 0.01$ or lower.

Now, we run the regression with the interaction of race and gender: $Y_C = B_0 + B_1E + B_2I + B_3G + B_4R + B_5G \times R$. Table 18.12 summarizes that regression.

Table 18.12 Regression of Productivity with Interaction of Race and Gender

Variables	Coefficient	Std Error	T	P	VIF
Constant	46.1615	0.98508	46.86	0.0000	0.0
Race	6.32943	1.36823	4.63	0.0000	1.7
Cenergy	0.22289	0.03720	5.99	0.0000	1.8
Cintellig	0.07423	0.03150	2.36	0.0230	1.2
Gender	-0.71972	2.47217	-0.29	0.7723	4.7
INTGR	6.69095	2.77657	2.41	0.0202	5.0

R^2	=	0.8418
F	=	46.84
P	=	0.0000

The overall model is still statistically significant at $P < 0.0000$; the interaction of gender and race, statistically significant at $P < 0.05$, has opposite main effect signs suggesting that gender has a dampening effect on Race. To see how the interaction changes our main effect conclusions, we make another interpretative matrix, in this case, two by two. We solved the four equations that illuminate the interaction effects, with results shown in table 18.13.

1. White/Male = B_0 (46.1615) + B_1X_1 (6.32943) × 0 + B_4X_4 (-0.71972) × 0 + B_5X_5 (6.69095) × 0 = 46.1615.

2. White/Female = B_0 (46.1615) + B_1X_1 (6.3294) × 0 + B_4X_4 (-0.7197) × 1 + B_5X_5 (6.6910) × 0 = 45.4418

3. Black/Male = B_0 (46.1615) + B_1X_1 (6.3294) × 1 + B_4X_4 (-0.7197) × 0 + B_5X_5 (6.6910) × 0 = 52.4909

4. Black/Female = B_0 (46.1615) + B_1X_1 (6.3294) × 1 + B_4X_4 (-0.7197) × 1 + B_5X_5 (6.6910) × 1 = 58.4612

Table 18.13 Interaction Effects of Gender × Race on Productivity

	Male = 0	Female = 1	Slope
White = 0	46.1615	45.4417	-0.7197
Black = 1	52.4909	58.4612	5.9715

Figure 18.5 is a plot of the interaction.

Figure 18.5 Interaction of Productivity on Gender × Race

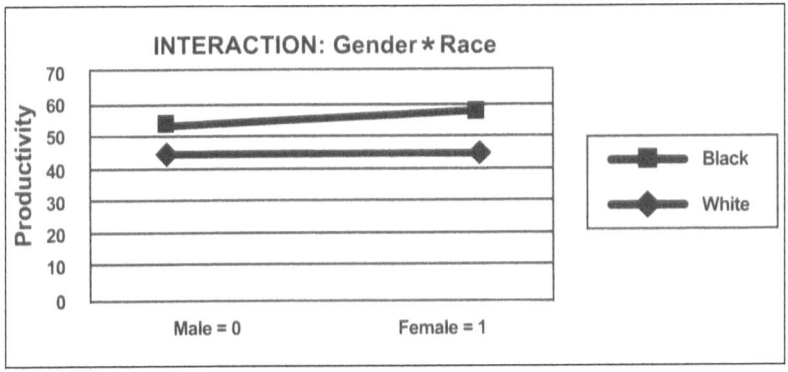

In the model without the interaction, race and gender are both positive and statistically significant. Without the interaction, we would conclude that females are more productive than males and blacks are more productive than whites. We can see from table 18.13 and figure 18.5 that black Females are much more productive than the additive main effects would suggest. White females are less productive than white males.

Our discussion has shown the power of interaction effects and how easily we can create interactions. But, once again, we emphasize that we should not run interactions unless we have strong justification. In addition, when an interaction shows one variable conditioning another we mean that the variable is conditioning another variable's effect on the dependent variable. A final point: even when main effects do not show statistical significance, we can still run interactions and, if the interaction shows significance, its effect overrides the statistically insignificant main effect variables.

Chapter 18

Summary

The critical ideas discussed in this chapter are outlined as follows:

1. The concept of interaction

   So far, we have restricted our regression discussions to additive or main effects. We can envision the effect of intelligence and energy or gender and intelligence on productivity as multiplicative or conditional. Suppose that a worker combines high intelligence and high energy, there may be a separate, non-additive effect on productivity. Alternatively, when a worker combines low energy and high intelligence, the productivity effect may be negative. Interaction means that the whole may be greater or less than the sum of its parts.

2. The construction of interaction variables

   We can create interaction variables from two or more independent variables simply by multiplying the independent variables.

3.   The theory driven constraint on interactions

Because modern computer programs make interactions easy to create and test, persons doing regressions face great temptations to indiscriminately run and test interactions. Interactions should not be added without theoretical guidance. Unless you can hypothesize and explain an interaction effect before you run the regression, you should avoid the temptation. Post hoc explanations of interactions are taboo. In addition, many times analysts confuse interaction effects with nonlinearity or outliers. Examining data for nonlinearity and outliers takes precedence over running interactions. This reason illustrates why interactions should not be introduced without at least a good confederacy of hunches

4.   Cardinal × cardinal interaction

After creating a new interactive (multiplicative) variable, we rerun the model. In the new model, the regression analysis partials the two main effects from the interactive variable, leaving a residual that represents only the interaction. If statistically significant, the interaction adds something to our understanding beyond the main additive effects. A statistically significant interaction coefficient increases MR^2. We note here that, most of the time, the interaction R^2 added to MR^2, will be no greater than 0.05. Therefore, when contemplating running an interaction the sample size should be large enough (have enough power) to find a statistically significant interaction effect.

5.   Cardinal × nominal and nominal × nominal interactions

Run, testing, and interpreting interactions between cardinal and nominal variables and between two or more nominal variables is perfectly possible and acceptable. The key to meaty interpretations rests on a close examination of the signs on the main effect and interaction variables. These signs give a summary overall assessment of the interaction effects. For a more detailed interpretation, interaction plots and tables tell the story.

Appendix

Chapter 18

The following data set ($N = 80$) has an interaction. The dependent variable, Attitude, is an index of employee attitudes toward management. The independent variables are gender and time with the firm. Your tasks are the following: (1) create the interaction variable, (2) Run a regression of attitudes on gender and time without the interaction, (3) run a regression of attitudes on gender and time with the interaction, and (4) interpret your findings.

ID	Attitudes	G	Time
1	0.6482561	1	10
2	0.911808	1	19
3	0.5928161	1	10
4	0.6243637	1	18
5	0.9629504	1	18
6	0.7098677	1	20
7	0.9089067	1	33
8	0.4732104	1	10
9	0.8168971	1	20
10	0.8015126	1	17
11	0.7849105	1	12
12	0.6533302	1	13
13	0.6927937	1	14
14	0.8278656	1	15
15	0.663597	1	30
16	0.5872789	1	10
17	0.5680993	1	16
18	0.7648623	1	6
19	0.6035823	1	7
20	0.4956369	1	3
21	0.4941238	1	10
22	0.6291418	1	1
23	0.5962087	1	3
24	0.669385	1	14
25	0.5521937	1	8
26	0.6875961	1	18
27	0.7650282	1	22

28	0.5221901	1	4
29	0.6086801	1	2
30	0.6147784	1	7
31	0.5201276	1	0
32	0.8270476	1	15
33	0.6729276	1	16
34	0.6996269	1	6
35	0.7444049	1	14
36	0.7980752	1	16
37	0.6589021	1	10
38	0.8591181	1	23
39	0.6369232	1	2
40	0.7404907	1	15
41	0.5261319	1	1
42	0.5719522	1	13
43	0.6406714	1	13
44	0.8157865	1	9
45	0.7717141	1	20
46	0.6277689	1	27
47	0.5705192	1	4
48	0.9696599	1	29
49	0.6099989	1	2
50	0.9934724	1	17
51	0.4891145	1	17
52	0.7972263	1	18
53	0.6458896	1	11
54	0.8801563	1	26
55	0.7448187	1	20
56	0.7515732	1	23
57	0.4583914	1	3
58	0.6317773	1	7
59	0.6289477	1	18
60	0.7153702	1	13
61	0.8291868	0	0
62	0.7487506	0	11
63	0.6727731	0	9
64	0.8321876	0	4
65	0.8769433	0	6
66	0.7155682	0	1
67	0.6581318	0	6
68	0.6637319	0	6

69	0.7259826	0	3
70	0.8274871	0	11
71	0.8674096	0	1
72	0.7707885	0	9
73	0.8710758	0	18
74	0.6025174	0	16
75	0.9889952	0	7
76	0.5951415	0	11
77	0.5453484	0	18
78	0.8963854	0	14
79	0.7055226	0	2
80	0.7291549	0	14

CHAPTER 19

PROPORTIONS, COUNTS, RANKS,
AND DICHOTOMOUS DEPENDENT VARIABLES

This chapter expands our discussion of measurement modes (chapter 1) and Transformations (chapter 14). In those discussions, we said that proportions, rank ordered variables, counts, and nominal data, while statistically analyzable, often require special analytical treatment, i.e., variations of OLS, than those used with cardinal and interval measures. These problems occur especially when using these data types as dependent variables.

Proportions (Continuous Logit OLS)[85]

When we use proportions as dependent variables, we must pay attention to their zero (0) and one (1) boundaries. Any analysis using dependent variables with boundaries zero and one runs the real risk of estimates falling outside those boundaries, a conceptual impossibility. Running an OLS regression using a proportion as a dependent variable can easily produce an equation yielding predictions that are negative or exceeding one hundred percent. This condition requires a nonlinear transformation. The nonlinear relationship is shown in figure 19.1.

---

[85] Our examples use proportions, counts, ranks and dichotomous data as dependent variables. Used as independent variables, these variables may require similar transformation to achieve linearity.

Figure 19.1 Proportions as a Dependent Variable

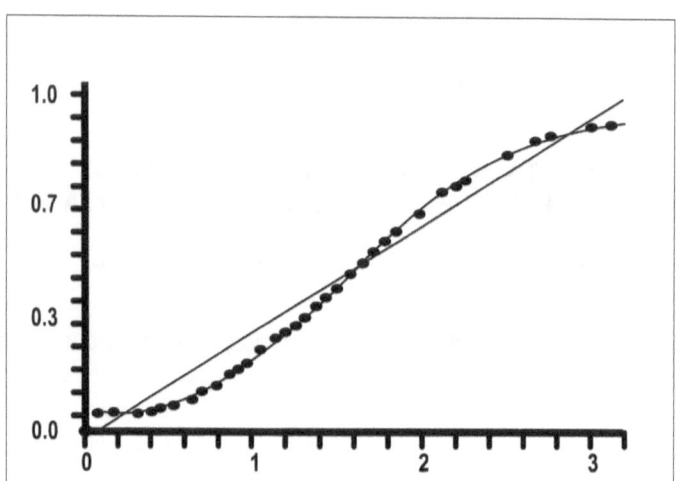

This nonlinear relationship is an S shaped curve, constrained so any estimate of a proportion must be greater than zero or less than one. One transformation (our preference), a Logit, simultaneously makes the relationship linear and constrains estimates to greater than zero and less than one. The logit transformation proceeds in three steps:

1. Express the data as a decimal fraction;

2. Transform the decimal fraction into an ODDS ratio, treating the proportion as a probability of winning. The equation is shown below

$$\text{ODDS} = \frac{P(win)}{1 - P(win)} \qquad\qquad \text{Equation 19.15}$$

To obtain ODDS from a probability, divide the proportion, P(win), by 1 - P(win). For example, if P(win) = 0.50, the odds ratio is $\dfrac{0.50}{1-0.50} = 1$. This means that, if the probability (proportion) of winning is 0.50, a bet of one (1) to get one (1) represents a fair bet. i.e., bet 1, win 1. If the probability (proportion) equals 0.25, the odds (ratio) of winning equals

$\dfrac{0.25}{1-0.25}$ equals 0.33, interpreted to mean a fair bet is three (3) to one

(1), i.e., bet one (1), win three (3). If we run a regression against the

ODDS ratio, any estimate of the odds ratio, by mathematical necessity, when converted back to a proportion, constrains the estimate to greater than zero (0) and less than one (1). Why? Because, conversion from an ODDS ratio back to a proportion is

$$P = \frac{ODDS}{ODDS + 1}.$$ (Equation 19.2)

So if ODDS equal 0.10, $P = \dfrac{0.10}{1 + 0.10}$ equal 0.091. Whatever the ODDS ratio, it cannot convert back to a proportion less than 0 or greater than 1.

3. A regression on ODDS ratio is still nonlinear. Thus, we can make the regression on an ODDS ratio linear by a natural log transformation. The Logit, then, equals LnODDS.

    $Logit = LnODDS$ (Equation 19.3)

Table 19.1 represents an abbreviation of a larger sample ($N = 100$) in which the dependent variable is a proportion expressed as a decimal fraction. We took the sample from a survey of consumer attitudes toward a company's customer service division. The dependent variable equals attitudes toward customer service for our hypothetical firm, StatSys. This attitude scale has a range of 0.05 (terrible) to 0.95 (excellent). The independent variables equal income and intelligence.

Table 19.1 Proportions as a Dependent Variable

id	Attitudes	Income	Intelligence
1	0.26	63.49	99.14
2	0.11	49.52	124.04
3	0.55	45.52	93.54
4	0.66	46.38	95.75
5	0.05	44.28	138.02

To run this problem as an OLS regression (called a continuous Logit), we transform attitudes to a Logit, center income and intelligence, and run an OLS regression. Table 19.2 summarizes this regression.

Table 19.2 OLS Logit Regression

Independent Variable	Regression Coefficient	Standard Error	T Value	Prob
Intercept	-0.3088	0.1930	-1.600	0.1128
cincome	0.0669	0.0276	2.424	0.0172
cintellige	0.0420	0.0119	3.520	0.0007

$R^2 = 0.17$

$F = 9.01$ $P = 0.0010$

The statistically significant overall model ($p < 0.01$) produced a mr^2 of about 0.17. Both of the independent variables, statistically significant at $p < 0.05$, have positive signs suggesting that the higher the income and intelligence of customers, the higher their satisfaction with customer service.

Remember, our regression coefficients are scaled in logits because we scaled the dependent variable in logits. How do we make the interpretation of the nonlinear effect of the independent variables on satisfaction more intelligible? We can solve the equation for predicted values of attitudes for five values of intelligence and income: 10th, 25th,, 50th, 75th, and 90th percentiles and call these values low-low, low-middle, middle, high-low, and high. Then we can convert the logits to satisfaction scores to odds by talking anti-log of lnodds and converting the odds to satisfaction scores with equation 19.2. Table 19.3 shows the results of these calculations.

Table 19.3 Predicted Satisfaction Scores for Five Levels of Income and Intelligence

	Low-Low 10th Percentile	Low-Middle 25th Percentile	Middle 50th Percentile	High-Middle 75th Percentile	High 90th Percentile
Predicted satisfaction Income	0.28535	0.350001	0.40532	0.523133	0.586848
Predicted satisfaction Intelligence	0.32124	0.391048	0.513502	0.63269	0.708423

From table 19.3, shows us how satisfaction increases as income and intelligence increase, but, as we would expect, we also see the curvilinearity of the increase as the *rate* of increase in satisfaction declines as intelligence and income approach the highest level. The slopes of the two curves are clearly not linear.

Counts

Counts present a problem when used as a dependent variable. Count data often follow a Poisson probability distribution, a distribution that shows the probability of several discrete occurrences taking place during a given time-interval. The number of useful creative suggestions offered by employees in a particular work environment over a given period represents one example of count data. Characterized by many zeros and a strong right skew, counts as a dependent variable make an OLS regression nonlinear. Also, OLS does not work well here, in part, because OLS often produces impossible negative count estimates and partly because it cannot handle the resulting heteroscedasticity.

OLS with A Square Root Transformation

A simple square root transformation of count dependent variables often provides a satisfactory solution to this problem. Table 19.4, is an excerpt of a hypothetical random sample (n=200) of the number of useful creative suggestions made by employees in a certain work context over a given time period, say one year. The dependent variable equals Counts. The independent variables are gender, intelligence, loyalty, and time with the firm.

Table 19.4 Count Data

ID	Counts	Gender	Intelligen	Loyalty	Time	Square Root Counts
1	1	1	80	38.86	10	1.0000
2	2	1	94	50.03	10	1.4140
3	3	0	113	67.23	12	1.7320
4	2	0	115	65.61	10	1.4140
5	0	0	118	55.63	11	0

Table 19.5 summarizes an OLS regression of the untransformed counts on gender, intelligence, loyalty, and participation.

Table 19.5 Regression on Untransformed Counts

Independent Variable	Regression Coefficient	Standard Error	T Value	Prob
Intercept	4.4935	1.1307	3.974	0.0001
Gender	-0.0701	0.1599	-0.438	0.6617
Intelligence	-0.0285	0.0055	-5.204	0.0000
Loyalty	-0.0395	0.0119	-3.329	0.0010
Time	0.2284	0.1198	1.906	0.0581

$MR^2 = 0.2280$

$F = 14.396$

$P < 0.0000$

The MR^2 for this regression model, significant at $P < 0.0000$ equals 0.23. Intelligence and loyalty, both statistically significant at $P < 0.01$, surprisingly, show negative signs. Also, time, shows significance as a one-tail test at one-half 0.0581, or about 0.03

In table 19.6, we summarize an OLS regression of the *square root of counts* on gender, intelligence, loyalty, and time.

Table 19.6 Regression of the Square Root of Counts

Variables	Regression Coefficient	Standard Error	T ratio	Prob
Intercept	3.8269	0.6395	5.984	0.0000
(Gender=1)	-0.0463	0.0905	-0.512	0.6096
Intelligence	-0.0229	0.0031	-7.384	0.0000
Loyalty	-0.0232	0.0067	-3.447	0.0007
Time	0.0946	0.0678	1.396	0.1644

$R^2 = 0.3345$

$P < 0.0000$

Comparing the results of the two analyses, both models, while statistically significant at $P < 0.0000$, have a difference in MR^2, about 0.23 for untransformed counts and about 0.33 for the square root of counts. Thus, we see the square root of counts as a better fit. In both models, however, the independent variables, intelligence and loyalty, significant at $P < 0.05$, still have negative signs, but we cannot claim even one-tail significance for time. Because the output of the analysis is scaled in square roots, any interpretation must return to the original metric.

Poisson Regression

The Prussian army used the distribution first to analyze deaths by horse kicks. As compared to OLS, which assumes a normal distribution of errors, Poisson distributions are skewed, non-negative, and heteroscedastic. Poisson regression is perfectly suited for count analysis particularly when the count occurrences are rare because it adjusts for skewness using a log transformation, prevents negative predicted values, and models heteroscedasticity.

The programs NCSS, SPSS, and SX have Poisson statistical routines; these routines use Maximum Likelihood Estimators, the interpretation of which parallels OLS

OLS, as we have seen, uses an algorithm minimizing the sum of squared errors. MLE uses a nonlinear estimation based on trial and error by trying and retrying (iterating) a fit until it finds a solution having the highest probability. The principle of maximum likelihood estimation (MLE), originally developed by R. A. Fisher in the 1920s, identifies the desired probability distribution that makes the observed data "most likely," which means that the procedure seeks the value of the observations that maximize the likelihood function.

A Poisson Regression with Counts

Table 19.7 is a summary of a Poisson regression using count data (the same data used in the previous OLS square root transformation).

Table 19.7 Poisson Regression of Counts

Variables	Coefficient	Std Error	SE	Prob
Constant	2.29128	0.86383	2.65	0.0080
Gender	-0.08059	0.12443	-0.65	0.5172
Intelligence	-0.01868	0.00412	-4.53	0.0000
Loyalty	-0.02770	0.00923	-3.00	0.0027
Time	0.16076	0.09306	1.73	0.0841

Deviance   237.98
P Value   0.0193 (for type I error)
Degrees of Freedom   195
Pseudo R^2 = 0.15

Compare the Poisson regression with the square root transformed OLS. The Poisson regression uses deviance and a Chi-square test where deviance is roughly comparable to squared error in OLS and the model Chi-square as an alternative to the F test. The statistically significant overall model still shows the independent variables, intelligence and loyalty, statistically significant with negative signs and brings back time as significant with a one-tail test. The Poisson regression calculates a pseudo R^2, about 0.15 based on reducing the deviance from $\approx$ 282 to $\approx$ 238, i.e., $\dfrac{238}{282} \approx 0.15$.

In this case, the Poisson and square root transformations work about equally well.

Ranked data

Like proportions, ranked data as a dependent variable are curvilinear. Data collected using a Likert scale, a measuring device named after its creator, Rensis Likert, represents a prime example of ranked data. One variation of this scale uses seven points with the respondent asked to rank a service from one to seven where one is very poor and seven is excellent. This form of measurement produces a dependent variable on which an OLS can produce estimates that fall outside the one, seven range. Like proportions, ranks as a dependent variable, require transformation to overcome nonlinearity.

For this purpose, we return to the *logit transformation*; here is the way we do it with ranked data:

First, reexpress the ranks as percentiles. Programs like NCSS and Statistix have subroutines that do this for you. This transformation does not change the rank order.

Second, reexpress the percentile ranks as decimal fractions.

Third, transform the decimal fractions to Logits as we did with proportions.

Fourth, run an OLS on the Logits. We interpret the results just as we interpreted proportions. Table 19.8 is an excerpt of a random sample of customers who have responded to a questionnaire about StatSys's customer service. The dependent variable equals the ranks produced by a seven point Likert scale. The independent variables are gender, income, and age of the respondents. The table's sixth column is a percentile rank; the seventh, eighth, and ninth columns are the decimal fractions of percentile ranks, the odds ratios, and Logits.

Table 19.8

ID	Ranks	Gender	Age	Income	Percentile ranks	Decimal	Odds	Logit
1	4	1	30	30555	62.6866	0.62687	1.68	0.51879
2	3	0	32	34440	31.3433	0.31343	0.45	-0.78410
3	4	1	21	28996	62.6866	0.62687	1.68	0.51879
4	4	1	41	37183	62.6866	0.62687	1.68	0.51879

Table 19.9 summarizes an OLS regression of the untransformed ranks on the three independent variables, gender, age, and income.

Table 19.9 OLS Regression on Untransformed Ranks

Variables	Coefficient	Std Error	T	P
Constant	-3.02707	0.91302	-3.32	0.0011
Cage	-0.06988	0.51598	-0.14	0.8924
Cincome	2.59264	0.74133	3.50	0.0006
Gender	0.81100	0.21842	3.71	0.0003

$F = 9.43$
$P = 0.0000$

The overall model shows statistical significance at $P < 0.0000$, with centered income and gender also statistically significant at $P < 0.001$. Table 19.10 summarizes an OLS regression on the ranks transformed to Logits.

Table 19.10 OLS Regression on Ranks Transformed to Logits.

Independent Variable	Coefficient	Standard Error	T	P
Intercept	-3.0271	0.9130	-3.315	0.0011
Age	-0.0020	0.0145	-0.135	0.8924
(Gender=1)	0.8110	0.2184	3.713	0.0003
Income	0.0001	0.0000	3.497	0.0006

$F = 9.431$
$P < 0.0000$

Both models fit the data about equally well, neither of which have large MR^2s. We prefer the logit model because the residuals seem to adhere more closely to the OLS assumptions.

To interpret the output, create high, middle, and low values for income and solve the equation for income and gender. When solving for income, substitute the mean of gender to hold it constant and, if we want to solve for gender holding age constant, because we centered age, substitute zero as its mean. Because our dependent variable is in logit form, our estimates also require conversion back to the original metric.

We note that the Logit transformation works well in this case, because the measured data permit each rank to appear more than once. Other situations permit a specific rank to appear only once in a data set. For example, assume your supervisor asks for an evaluation of employees under your direct supervision by ranking each employee from first to last, or, let's say, from 1 to 35. Here, some prefer the Probit (normal) transformation. The procedure follows the steps outlined above, converting the data to decimal percentile ranks, but instead of Logits, transforms to Probits. These transformed values represent positions in a normal distribution. Some statistical programs have routines that automate this transformation. You may want to ignore the Probit altogether and just use the Logit because the differences between Probit and Logit transformations are usually small.

Nominal (Dichotomous) Dependent Variables

We refer here to a s condition in which we coded the qualitative (nominal) dependent variable of interest, one (1) and zero (0). Here are some examples: pass/fail, win/lose, promoted/not promoted, gender, religious preference. We can use two methods for analyzing nominal data as dependent variables: (1) the linear probability model (OLS) and (2) Logit (Maximum Likelihood).

The Linear Probability Model (OLS)

To illustrate this method, we use a hypothetical random sample of 200 employees working at a certain level in our company with the dependent variable defined as one equals promotion and zero equals passed over for promotion. The independent variables are age, education, time with the company in months, gender (0 = Female, 1 = male) race/ethnic (three categories: white, Hispanic, and black; white, equals the reference group). Table 19.11 summarizes an OLS regression of promoted on age, education, time, gender, and race/ethnic.

Table 19.11 OLS Regression of Promoted

Independent Variables	Coefficient	Std Error	T	P
Constant	-2.71649	0.49056	-5.540	0.0000
Age	0.03739	0.01292	2.890	0.0043
Black	-0.12090	0.06996	-1.730	0.0856
Education	0.12953	0.01909	6.790	0.0000
Gender	0.11005	0.06357	1.730	0.0850
Hispanic	-0.00306	0.06915	-0.040	0.9648
Time	0.00109	0.00192	0.570	0.5690

MR2 = 0.1995

F = 9.27

P = 0.0000

Once again, our interpretation of these data focuses on the regression coefficients. With nominal dependent variables, we interpret the regression coefficients as probabilities. In our example the overall model, statistically significant at $P < 0.0000$, shows age and education also statistically significant at $P < 0.01$ and < 0.0000. The positive sign on age suggests that an increase in age of one year increases the probability of promotion by about 0.037 or 4%. Older persons in this firm, on average, have higher probability of promotion than younger persons. An increase in education by one year, on average, increases the probability of promotion by about 0.13 or 13%. The better educated, the greater the chance of promotion. The variable black, with white as the reference group, shows statistical significance as a one tail test; if the negative sign had been hypothesized ($P = 0.0856$ divided by 2 = 0.0428), this finding requires additional study. As it is, black persons, as compared to whites, have lower probability of promotion, on average, by about 0.12 or 12%. Gender, as a one tail test, also shows statistical significance, with males having about 12% greater chance of being promoted than females. If our model has included all the critical variables affecting promotion, i.e., we left out no plausible variables, the model suggests a systematic bias in promotion in favor of whites and males.

In this example, we dummy coded the race/ethnic groups. If we had used a system of effects coding, we would see different probabilities, because our analysis would focus on the differences between the race/ethnic variables, gender and the weighted grand mean of the two groups.

You can guess the weakness of OLS. An OLS model is not the best fit, because probability estimates can easily fall outside the bounds of 1 and 0. Therefore, the preferred method is a Logit analysis using Maximum Likelihood.

Logit (MLE)

Using the data in table 19.11, we now run the same analysis of promotion, this time with Logit MLE.[86] Table 19.12 sums-up the output of the Logit MLE regression.

---

[86] Running the Logit analysis with a nominal dependent variable coded one and zero requires special software. All of the programs we recommend have this routine.

Table 19.11 Logistic Regression of Promoted

Independent Variables	Coefficient	Std Error	T	P
Constant	-22.1218	4.19394	-5.27	0.0000
Age	0.26255	0.09344	2.81	0.0050
Black	-0.89337	0.51611	-1.73	0.0835
Education	0.89506	0.16488	5.43	0.0000
Gender	0.83179	0.45815	1.82	0.0694
Hispanic	0.06127	0.48722	0.13	0.8999
Time	0.00739	0.01325	0.56	0.5773
Deviance	167.35			
Model P	< 0.0000			
Pseudo R^2	0.265			

We compare this MLE model to the previous OLS regression. For the overall MLE model, our test of significance uses a Chi-square statistic with degrees of freedom equal to the number of independent variables; the difference between the Deviance without the analysis, about 216, and the model deviance, about 167 represents the value of chi-square, 48. Putting it another way, the logit analysis reduced the deviance by 48, from 216 to 167. If we divide the number 48 by 216, we produce the pseudo R^2 of about 23%. So with six independent variables (df equals six) we reject the null at $P < 0.0000$. The over-all model significance and the significance of the independent variables parallel the output of the OLS analysis.

Interpretation of this analysis involves substitution into the model of low, middle, and high values (or as many groups as you prefer) for each of the continuous variables, e.g., age and plugging the means of the other independent variables to hold them constant (unless they were centered). The solved equation for low, middle, and high age takes the form of Logits, demanding a conversion of the Logits to their decimal fraction equivalences and observing the effect of age on the probability of promotion. The two variables, age and education are statistically significant as is black and gender as one tail tests. Converting the black Logit coefficient to an odds ratio we take the ratio to exponent of the natural log, 0.409274 and transform the odds a probability ($P = \dfrac{Odds}{Odds+1}$) = 0.409274/1.409274 = 0.290145).

Thus, black as compared to white reduces the probability of promotion by about 29%. Comparing the two models, the Logit gives a much stronger effect of race than the OLS.

Summary

The critical concepts discussed in this chapter are as follows.

1. Proportions as dependent variables

    When we use proportions as a dependent variable, we must pay attention to their zero (0) and one (1) boundaries. It is impossible to have parts of a whole falling outside the boundary of zero and one. Running an ordinary least squares regression using a proportion as a dependent variable can easily produce an equation that yields predictions falling below zero and above one. All this implies, correctly, that the relationship between a set of independent variables and a proportional dependent variable requires a nonlinear transformation.

2. Counts as a dependent variable

    Counts present a problem when used as a dependent variable. Count data often follow a Poisson probability distribution, a distribution that shows the probability of several discrete occurrences taking place during a given time-interval. One example of count data shows the number of useful creative suggestions offered by employees in a particular work context over a given period. Characterized by many zeros and a strong right skew, counts as a dependent variable make a regression nonlinear. Also, OLS does not work well here, in part, because OLS often produces impossible negative count estimates and partly because it cannot handle a resulting heteroscedasticity.

3. Ranks as a dependent variable

    Like proportions, ranked data as a dependent variable are curvilinear. Data collected using a Likert scale, a measuring device named after its creator, Rensis Likert, represents a prime example of ranked data. One variation of this scale uses seven points with the respondent asked to rank a service from one to seven where one is very poor and seven is excellent. Using OLS with ranks as a dependent variable can produce estimates that fall outside the range of the ranks. In our example of the Likert scale, the ranks range from one to seven. Like a proportion, ranks, as a dependent variable require a transformation to overcome nonlinearity.

4. Nominal variables coded one and zero as a dependent variable

We refer here to conditions in which we coded the qualitative (nominal) dependent variable of interest, one (1) and zero (0). Examples abound: pass/fail, win/lose, promoted/not promoted, gender, religious preference. We can use two methods for analyzing these kinds of data: (1) the linear probability model (OLS) and (2) Logit (Maximum Likelihood).

Chapter 19

Appendix

We scaled $X1$ as a percentages using decimal fractions, $X2$ as ranks, $X3$ as counts, and $X4$ as nominal.

Your task are the following: (1) name each of the four variables in such a way as to make regressions rational, (2) run four multiple regressions using each of the four variables as the dependent variable and the other three as independent, and (3) explain and interpret your findings.

ID	X1	X2	X3	X4
1	0.205	3	0	0
2	0.983	6	2	1
3	0.435	4	0	0
4	0.024	6.75	0	1
5	0.604	9	0	1
6	0.024	7	0	1
7	0.649	5	0	1
8	0.319	5	0	0
9	0.529	8	0	1

10	0.433	3	0	1
11	0.024	5	0	1
12	0.983	8	0	1
13	0.699	7	4	1
14	0.911	5	0	1
15	0.024	7	2	1
16	0.519	6	0	1
17	0.195	4	0	0
18	0.522	4	0	0
19	0.024	6	0	1
20	0.926	7	3	1
21	0.024	8.5	1	1
22	0.207	6	1	1
23	0.024	10	0	1
24	0.421	6.25	1	1
25	0.847	5.5	0	0
26	0.024	1	0	1
27	0.794	8	1	1
28	0.245	4	0	1
29	0.285	2	0	0
30	0.482	1	0	1

Chapter 20

CAUSAL INFERENCE WITH STRUCTURAL AND REDUCED FORM EQUATIONS

In chapter 9, our discussion focused on the difference between causation, relationship, and the importance of this distinction to decision making. While relationships are necessary to causal inference, they are not sufficient. The preferred sufficient condition, controlled experimental manipulation, is often foreclosed by ethical or resource limitations. An alternative method for making non-experimental causal inferences lies in the arena of passive regression. Thus, we argued that, under certain conditions, we could make causal inferences from passive non-experimental data using regression analysis.

Reprise: Causal Inference Using a Fully Partialled OLS Multiple Regression

In all our previous analytical illustrations using multiple regressions, OLS, Logit, MLE, we used fully partialled models. That is, we used models in which we entered all the independent variables at the same time so that the regression coefficient of any single independent variable reflected its effect on the dependent variable with all the other independent variables removed or held constant. As we noted, given even modest degrees of collinearity, fully partialled models have the major defect of dumping all the common variance of the independent variables into the regression constant. For some purposes, prediction, forecasting, or concern for the causal effect of a single variable, this defect is of no concern. But, if the purpose of an analysis aims at estimating the causal effects of two or more variables, then, in the face of collinearity, the fully partialled model loses efficacy.

In the following example, we reprise the use of a fully partialled model in providing evidence of causation. Assume that we have a case of an enterprise delivering computer on-site services to businesses and homes. This hypothetical firm has 200 employees consisting of a well educated mix of age, gender, and race (white and black), As a result of some conflict over of salary decisions, a group of employees file a complaint accusing management and the firm's ownership of racial wage discrimination.

In order to answer these charges, the firm conducted a statistical study designed to show that neither the firm's ownership nor its management uses racial considerations in determining wages. To make the study credible, the persons conducting the analysis used input from management and employees including those bringing the complaint and, using this input, developed a model of the plausible variables affecting salary decisions. We show the final model agreed to by all the participants as follows:

Y_C Salaries = X_1 Formal Education. X_2 Experience in the Field, Age X_3, Gender, X_4, white/black X_5.

Note: Developing the model as a fully partialled regression, suggests that the principal interest in the variables education, experience, gender, and Age relates to their role as covariates. Here, the term covariate means that we are not interested in estimating their effects, but only in partialling their effects from the variable of interest, Race. If, after partialling the effects of these covariates and, given a consensus among qualified people, that no other critical variables have been omitted, a negative substantial effect of race produces reasonable evidence of racial bias. Our model specification must include all the plausible explanations for variation in salaries. Once again, the term plausible does not refer to all possible effect variables, but only those that are substantial (more than nominally small). If substantial effects result from regressing salary on race and partialling the covariates, we have credible evidence that salary decisions are racially biased.

Using the firm's records of current employees, a group appointed by management and complainants specified the model and collected the data. Table 20.1 shows an excerpt of these numbers.

Table 20.1 Salary, Age, Education, Gender, and Race

ID	Age	Education	Experience	Gender	Race	Salary
1	28	11	6	0	1	27348
2	26	11	6	0	1	28131
3	32	11	7	0	1	32889
4	24	12	6	0	0	33208
5	31	11	7	1	1	37435
6	29	12	7	0	1	28622
7	32	10	7	0	1	34234
8	29	11	7	1	1	34799
9	26	11	7	0	0	28109
10	31	11	8	0	1	38877

Using these data, the firm ran a fully partialled OLS analysis. Table 20.2 summarizes the results of that analysis.

Table 20.2 Fully Partialled OLS Regression of Salaries on Age, Education, Experience, Gender, and Race

Variables	Coefficient	Std Error	T	P
Constant	11070.3	3723.27	2.97	0.0033
Age	-7.1082	108.728	-0.71	0.4791
Education	-669.354	299.678	-2.23	0.0267
Experience	4626.36	384.405	12.04	0.0000
Female	650.351	542.247	1.20	0.2318
White	1110.35	559.89	1.98	0.0488
R square	0.5837			
P	< 0.0001			

The overall model is statistically significant ($P < 0.0001$). The gender variable was not significant. This finding undermines the claim that gender bias (caused) managerial decisions on wages. The race variable, (black as the reference group) was positive and statistically significant confirming the claim of racial bias in wage decisions. By holding constant the plausible variables affecting wage differentials:

age, gender, education, and experience, we found that white's annual Salary was on average greater than blacks by about $1,100.00.

It is critical to note once again that this finding is only as good as the model specification. If the analysis excludes one or more plausible explanations for wage differential from the model specification, the issue remains inconclusive. Also note that we used no interaction terms, for example, gender × race that might have confirmed a gender bias.

OLS Structural Equation Modeling

If interests and needs transcend the testing of a single causal effect extending to building and testing a causal theory, then we must turn to structural equation modeling, a statistical method that takes brings into focus all the joint variance between the independent variables that fully partialled analyses ignore. This method requires a better understanding of the causal forces than the simpler model discussed above.

The first step in building a structural model is identifying the plausible causal variables that affect the dependent variable. This identification also demands specification of any nonlinear effects and plausible interactions. Our illustration assumes all the variables are linear with no interactions.

The second step develops a series of equations that specify the hierarchical causal effects of the independent variables on the dependent variable. This means that we must rank the variables in the model in terms of their theoretical contribution to explaining the variance in the dependent variable. This obviously is more than just a guess, requiring high-level insight into the causal forces affecting Y.

Step 3 runs a series of regressions reflecting the specified equations in step two. We call these equations, structural.

To illustrate these steps assume the following scenario. Analysts in the Bureau of Labor Statistics build a theoretical model purporting to explain wage differentials of nonsalaried employees in a particular industry. To test the theory, the analysts take a random sample (n =

200). The measured variables in the sample are wages, intelligence, time with the firm, formal training, and experience in the field. Table 20.4 is an abbreviation of that sample.

Table 20.4 Data Set of Wages, Experience, Intelligence, Time, and Training

ID	Exp	Int	Time	Training	Wage
1	36.67	90	11.93	29.46	42800
2	48.01	103	24.24	60	50237
3	23.88	91	4.24	18	38637
4	18	90	4.24	18	29219
5	50	115	24.24	57.32	52974
6	33.81	107	4.24	18	41276
7	32.62	102	8.97	37.57	43346
8	26.8	90	4.24	18	35379
9	50	112	20.65	60	50154
10	34.44	98	10.71	45.24	44255

The dependent variable is annual wages. The independent variables are experience, intelligence, time, and training. Following the outline, the analysis constructed a series of structural equations defining the relationships between the variables. We must have as many equations as independent variables.

Structural Equation 1: $Y_{Wages} = B_0 + B_1 Int$
Structural Equation 2: $Y_{Wages} = B_0 + B_1 Int + B_2 Training$
Structural Equation 3: $Y_{Wages} = B_0 + B_1 Int + B_2 Training + B_3 Experience$
Structural Equation 4: $Y_{Wages} = B_0 + B_1 Int + B_2 Training + B_3 Experience + B_4 Time$

The hierarchy of hypothesized causation defines the order of equation entry, beginning with intelligence. The first equation, Y_{Wages} on intelligence, was statistically significant. Table 20.5 summarizes the output of the equation 1 regression.

Table 20.5 Regression of Wages on Intelligence

Variables	Coefficient	Std Error	T	P
Constant	-43073.5	3625.53	-11.88	0.0000
Int	836.571	35.9012	23.30	0.0000

B_1 for intelligence is \$836.57. As intelligence (on the Luker Abbreviated Test of Intelligence) increases by one point, the annual wage increases, on average by about \$837. This (\$837) is the *total structural equation effect* of intelligence on wages.

The second equation, Y_{Wages} on intelligence and training, was also statistically significant. Table 20.6 summarizes the regression output of equation 2.

Table 20.6 Regression of Wages on Intelligence and Training

Variable	Coeff.	SE	T	P
Constant	-6451.96	4399.18	-1.47	0.1441
Int	347.734	52.8301	6.58	0.0000
Train	351.850	32.0817	10.97	0.0000

B_1 for intelligence declined to \$348 indicating the redundant relationship between intelligence and training. B_2 for training is \$352 connoting that an increase of one period of training increases the annual wage, on average, by about \$352. This (\$352) is the total effect of training on wages holding intelligence constant. Please note: the joint or shared variance between intelligence and training has been assigned to intelligence, the first variable in the hierarchy. Contrast this result with a fully partialled model that assigns the joint variance to the regression constant.

The third equation, Y_{Wages} on intelligence, training, and experience, was statistically significant. Table 20.7 summarizes the regression output of Equation 3.

Table 20.7 Regression of Wages on Intelligence, Training, and Experience

Variables	Coefficient	Std Error	T	P
Constant	5943.52	4112.47	1.45	0.1500
IQ	148.576	52.0169	2.86	0.0047
Train	170.909	35.6895	4.79	0.0000
Exp	**422.895**	**52.1978**	**8.10**	**0.0000**

When experience entered the model, the coefficient B_1 for intelligence declined to $149 expressing the redundancy between intelligence and experience and the coefficient for training, B_2 declined to $171 reflecting the redundancy between training and experience. B_3 for experience was $423; for an increase in experience of one, wages increased, on average, by $423, the total effect of experience on wages.

The fourth equation, Y_{wages} on intelligence, training, experience, and time was statistically significant. Table 20.8 summarizes the output of Equation 4.

Table 20.8 Regression of Wages on Intelligence, Training, Experience, and Time

Variables	Coefficient	Std	Error	T	P
Constant	7101.17	4305.62	1.65	0.1007	
IQ	140.519	52.7840	2.66	0.0084	
Train	136.733	51.7618	2.64	0.0089	
Exp	415.117	52.9122	7.85	0.0000	
Time	92.3429	46.2600	1.99	0.0479	

The coefficients on all the variables preceding time declined, showing their redundancy with time. B_4 for time was $92, the total effect of time on wages.

Table 20.9 summarizes the Total effects of the four structural equations on wages.

Variable	Total Effect on Wages
IQ	837
Training	352
Experience	423
Time	92

Now compare these effects with the coefficients in table 20.8, the equivalent of a fully partialled model. Because of the collinearity between the independent variables, the shared variance in the fully partialled model is thrown into the regression constant, losing its interpretative effect. In table 20.9, we assigned the shared variance to the higher variable in the hierarchy by using the value of the coefficient when it entered. In our example, all the independent variables remained statistically significant throughout the four equations. However, if, for example, in the second equation, intelligence became statistically insignificant, its significance in the first equation would take precedence.

Obviously, the degree to which these conclusions are useful depends on our theory; the conclusions depend on the variables used and the order entered. Another plausible order of entry confounds any inferences drawn from this causal model. So clearly, we cannot use this approach without a convincing perceptual scheme.

Reduced Form Equation Modeling with Limited Theory

Reduced Form Equation Modeling, an extension of structural equations, divides total effects into *direct and indirect* effects. The first step is identical: to structural equation modeling; we identify and rank the causal variables.

Step 2 is more complex, demanding the construction of a path diagram that not only portrays the hierarchical causal effect of the independent variables on the dependent variable, but also requires a specification of their relationship to each other. This means what it says. Theory or conceptual scheme must be sufficiently sophisticated to allow specifying how the causal independent variables relate to each other as well as how they relate to the dependent variable.

Step 3 runs a series of regressions reflecting the path diagram and interprets their effect with respect to the theory. As above, we still call each of these equations, structural.

To illustrate these steps, we will assume the same scenario as used in the discussion of structural equations. Wages represents the dependent variable. The independent variables are intelligence, time, training, and experience. Following the outline, we construct a path diagram defining the relationships. Figure 20.1 illustrates the path diagram.

Figure 20.1 Path Diagram Relating the Independent Variables to Each Other as well as to the Dependent Variable

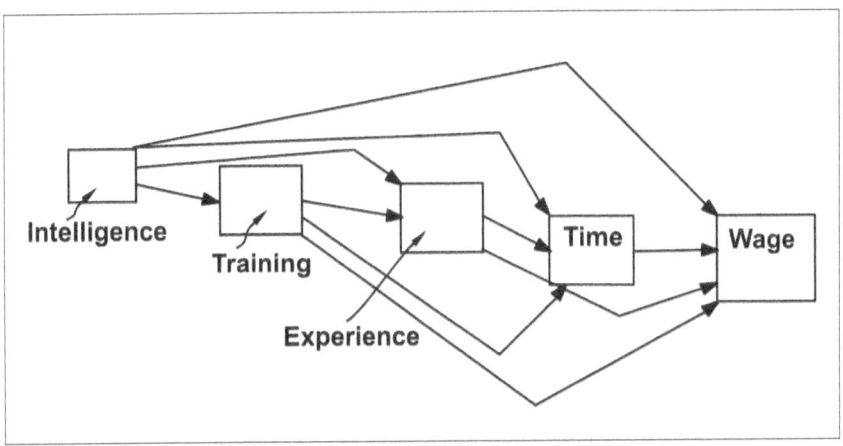

The path diagram may look a bit confusing, but it straightforwardly defines the theoretical causal relationships between the independent variables and the dependent variable as well as the relationship among the independent variables. Intelligence, on the far left, is called exogenous because it is not defined or influenced by any of the other variables in the model. The other variables are endogenous because they are influenced by other variables in the model. The arrow between intelligence and training indicates that intelligence partially explains (causes) training and, as a result, assigns part of the joint variance between intelligence and training affecting wages to intelligence. To put it another way, intelligence has an *indirect* effect on wages through its effect on training. The arrow running from intelligence to experience argues that intelligence partly explains (causes) experience and, therefore, indirectly affects wages through experience. The same logic applies to the arrow from intelligence to time signifying an indirect effect of intelligence on wages through its effect on time with the firm. The arrow from intelligence to wages signals a *direct* effect of intelligence on wages different from its indirect effects through the other variables.

Using the same logic, the arrow from training to wages shows the direct effect of training on wages. The arrows from training to experience and time show its indirect effects on wages through its effect on experience and time. Moreover, the arrows from experience to wages and time show the direct and indirect effects of experience on wages. The final arrow from time to wages shows the direct effect of time on wages. Time has no indirect effect because it is the last variable in the model and therefore affects none of the other variables in the equation.

The path diagram also defines the theory's structural equations, the first of which regresses wages on intelligence. Table 20.5 summarizes that first structural regression.

Table 20.13 Regression of Wages on Intelligence

Variables	Coefficient	Std Error	T	P
Constant	-43073	3625.53	-11.88	0.0000
Intelligence	836	35.9012	23.30	0.0000

B_1 for intelligence equals $836.57 showing that, as intelligence increases by one point, on the average, annual wages increase by about $837, the *total effect* of intelligence on wages. Table 20.14 summarizes the second structural equation, the regression of wages on intelligence and training.

Table 20.14 Wages on Intelligence and Training

Variables	Coeff.	Std Error	T	P
Constant	-6451	4399.18	-1.47	0.1441
Intelligence	348	52.8301	6.58	0.0000
Training	352	32.0817	10.97	0.0000

With intelligence held constant, the coefficient for training, B_2 equals $352. Thus, the total effect of training on annual wages equals $352 per unit increase in training. At this point we introduce the reduced-form idea. The coefficient for intelligence, B_1 fell to $348 because of the relationship between intelligence and training. We assign the difference between B_1 in the first equation and B_1 in the second equation, $490, the shared variance with training, to intelligence as the *indirect effect* of intelligence on wages through training. When we make that assignment, the equation becomes a *reduced-form* equation. Table

20.15 summarizes the third structural equation, the regression of wages on intelligence, training, and experience.

Table 20.15 Regression of Wages on Intelligence, Training, and Experience

Variables	Coefficient	Std Error	T	P
Constant	5943	4112.47	1.45	0.1500
Intelligence	149	52.0169	2.86	0.0047
Training	170	35.6895	4.79	0.0000
Experience	422	52.1978	8.10	0.0000

The coefficient for experience, B_3 equals $423 represents the total effect of experience with intelligence and training held constant. Note: B_1 and B_2 fell to $149 and $171 from $348 and $352 respectively. We assigned these two differences, the differences between equation 2 and equation 3, to intelligence and training as the *indirect effect* of intelligence through experience and the indirect effect of training through experience.

Table 20.16 summarizes the fourth structural equation, the regression of wages on intelligence, training, experience, and time.

Table 20.16 Regression of Wages on Intelligence, Training, Experience, and Time

Variables	Coeff.	Std Error	T	P
Constant	7101	4305.62	1.65	0.1007
Intelligence	141	52.7840	2.66	0.0084
Training	137	51.7618	2.64	0.0089
Experience	415	52.9122	7.85	0.0000
Time	92	46.2600	1.99	0.0479

B_4 for time, $92; with intelligence, training, and experience held constant, equals the total effect of time (with the firm) on wages for each month increase in time. Note again: B_1, B_2, and B_3 fell to $141, $137, and $415.12 respectively. These differences between equation 3 and equation 4 represent the indirect effects of intelligence, training, and experience through time. Table 20.17 summarizes all these effects.

Table 20.17 Summary of Reduced-Form Effects

Intelligence		Total	Direct	Indirect
		837	140.52	697
	Indirect Via Training			490
	Indirect Via Experience			199
	Indirect Via Time			8
	Total Indirect Effects			697
Training		Total	Direct	Indirect
		352	137	215
	Indirect Via Experience			182
	Indirect Via Time			33
	Total Indirect Effects			215
Experience		Total	Direct	Indirect
		422	415	7
	Indirect Via Time			7
	Total Indirect Effects			7

Time		Total	Direct	Indirect
		92	92	0
Total	Indirect			0
Effects				

Now compare this table to the path diagram in figure 20.1. The lines drawn from each independent variable to wages represent the direct effects. The lines drawn from one independent variable to another independent variable represent the indirect effects of those variables through the variable to which the line is drawn, Calculating the indirect effects involves subtracting the regression coefficient at step 2 from the regression coefficient at step 1, step 3 from step 2, etc. This assigns the redundancies to the previous independent variable in the hierarchical path diagram. For example, in equation 2, the B coefficient for training equals $352. In step 3 the B coefficient for training falls to $171. That is, entering experience in step 3 caused the regression coefficient for training to fall to $171. The difference between $352 and $171, $181, represents the indirect effect of training through experience. The B coefficient of each independent variable when it enters the equation represents the total effect. It is also the sum of direct and indirect effects.

We use the data in table 20.9 to test the theory defined by the path model by comparing our hypothesized paths in figure 20.1 with the actual paths. In this case, the data have not disconfirmed the theory. Does this mean that our theory is true? Not necessarily! Other reasonable (or even implausible) theories might also pass the same test. If theory fails to pass the structural equation empirical test, can we simply use the paths suggested by the analysis and assume that these paths confirm a new theory? The answer: emphatically no! If analysis disconfirms, we are required to reconstruct theory and test it with another independent sample.

Two-way Causation: Estimation of Reciprocally Causal Models

In our discussions of fully partialled OLS, structural equation modeling, and reduced-form modeling, one of our assumptions was one-way causal direction. That is, causal direction was not reciprocal. In many situations, this assumption does not hold.

Two examples illustrate the idea: Estimating the effect of attitudes toward economics and academic achievement in economics. In both of these cases, the dependency is reciprocal: poor attitudes toward the subject matter of economics reduce academic performance in economics and conversely, lower performance results in poorer attitudes.

To illustrate, consider the following models:

1. $X_3 = B_{03} + B_1 X_1 + B_2 X4 + \varepsilon_3$
2. $X_4 = B_{04} + B_1 X_2 + B_2 X_3 + \varepsilon_4$

In the two equations above, we regressed X_3 as a dependent variable on X_1 and X_4. Then we regressed X_4 as a dependent variable on X_2 and X_3. In other words, our equations are saying that X_3 is dependent on X_4 and X_4 is dependent on X_3. Figure 20.2 shows a path diagram of these relationships.

Figure 20.2 Path Diagram

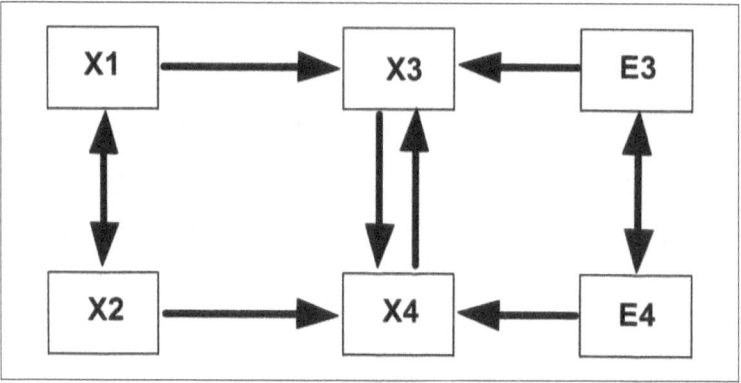

This path diagram depicts a mutual dependence between X_3 and X_4, and, if run as an OLS, because of that reciprocal dependence, makes the residuals ε_3 and ε_4 correlated. This means that in equation 1 above, the independent variable X_4 correlates with the residual ε_3 and in equation 2 the independent variable X_3 correlates with the residual ε_4. Thus, in both equations, we have violated one of the assumptions of the OLS model (see chapter 13), a violation that biases the estimates. Therefore, if two variables are mutually dependent, we *cannot* use *OLS* to estimate their effect on one another. Important note: we have constructed the path diagram in figure 20.2 to show a correlation (non causal) between X_1 and X_2, with X_1 correlated with X_4 only indirectly through X_3 and X_2 correlated with X_3 only indirectly through X_4.

There are multiple techniques for estimating this reciprocal model (e.g. Indirect least squares, instrumental variables, 2SLS). Our discussion focuses on two stage least squares (2SLS) with instrumental variables.

The following is an outline of the 2SLS procedure.

Run an OLS regression with each endogenous variable, X_3 and X_4 on *all* exogenous variables, X_1 and X_2. In this case, X_3 on X_1 and X_2, and regress X_4 on X_1 and X_2. Use the OLS estimates (Y_c) to construct instrumental variables, estimated X_{3I} and X_{4I}. We have appended the letter I to show that these variables are instruments.

$$X_{3I} = B_0 + B_1 X_1 + B_2 X_2$$
$$X_{4I} = B_0 + B_1 X_+ + B_2 X_2$$

These new instrumental variables are not correlated with the residuals, ε_3 and ε_4 because they are linear combinations of the exogenous variables X_1 and X_2.

In the second step, we replace any endogenous variable, e.g., X_3 serving as a independent explanatory variable in one of the structural equations by the corresponding instrumental variable. The estimated equations in this case are the following:

$$X_3 = B_0 + B_1 X_1 + B_2 X_{4I}$$
$$X_4 = B_0 + B_1 X_2 + B_2 X_{3I}$$

Given these substitutions, each explanatory variable in the modified structural equations are uncorrelated with the error terms, making OLS a satisfactory estimator.

The standard errors of 2SLS estimators are a partial function of the strength of the relationship between the instrumental variables and the variables they replace. That is, the stronger the relationship between the instrumental variables and the original variables they replace, the more efficient the parameters produced by 2SLS.

This is the reason why we use all (as opposed to some) of the exogenous variables as independent variables in the first stage.

To illustrate this idea, we created an abbreviated data set, in table 20.10 below.

Table 20.10 Creativity and Intelligence

ID	Creativity	Intelligence	X1	X2	Instrument 1 Creativity	Instrument 2 Intelligence
1	102	102	49	27	97.36	99.13
2	106	111	44	36	98.57	103.17
3	91	79	42	31	105.31	90.1
4	84	99	48	19	87.25	99.8
5	102	104	49	27	97.36	99.13
6	99	102	55	34	104.24	104.47
7	107	112	54	30	99.37	104.23
8	103	94	47	39	113.88	94.05
9	93	91	46	28	99.83	95.42
10	98	104	53	29	98.45	103.3

A path diagram, figure 20.3, shows the relationships among the variables in this equation.

Figure 20.3 Path Diagram

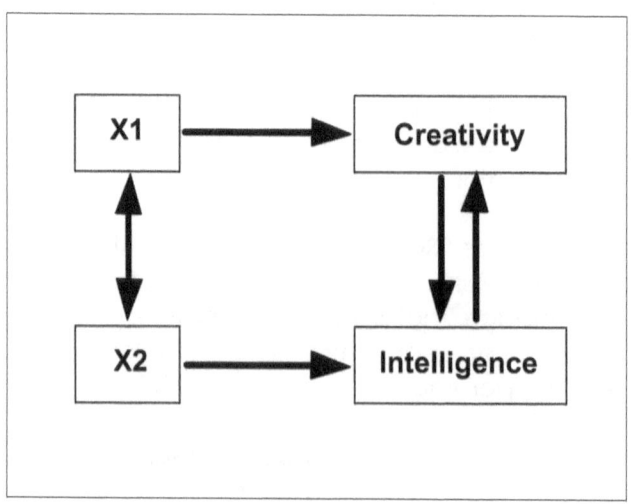

Our hypothetical firm wishes to see if employee Creativity is dependent on intelligence. To test this hypothesis they cannot run a simple OLS with Creativity as the dependent variable and intelligence as the independent variable because their relationship is reciprocal (mutually causal). Using our data set in table 20.18, we do two OLS analyses: one that ignores the reciprocal relationship and the other with 2SLS. Table 20.19 is a summary of the first OLS equation.

Table 20.19 Least Squares Linear Regression of Creativity on Intelligence and X2

Variables	Coefficient	Std Error	T	P	VIF
Constant	23.7320	10.5496	2.25	0.0270	0.0
Intel	0.47529	0.09173	5.18	0.0000	1.0
X2	1.02516	0.19594	5.23	0.0000	1.0

Table 20.20 is a summary of the OLS regression of Intelligence on Creativity and X1

Table 20.21 Regression of Intelligence on Creativityl and X1

Variables	Coefficient	Std	Error	T	P	VIF
Constant	-4.93465	8.62005	-0.57	0.5685	0.0
Creativityl	0.46754	0.05846	8.00	0.0000	1.0
X1	1.13327	0.11736	9.66	0.0000	1.0

Notice that in both of these OLS equations, the independent variables are all significant at $P < 0.0000$. Table 20.21 is a 2SLS Regression of Creativity on intelligence as an instrument.

Table 20.21 Regression of Creativity on Intelligencel As an Instrument

Variables	Coefficient	Std	Error	T	P	VIF
Constant	96.9248	16.9461	5.72	0.0000	0.0
Intelligencel	-0.33217	0.16554	-2.01	0.0479	1.0
X2	1.23574	0.21594	5.72	0.0000	1.0

Table 20.22 is a 2SLS Regression of Intelligence on the Creativity Instrument

Table 20.22 Least Squares Linear Regression of Intelligence on Creativityl as an Instrument

Variables	Coefficient	Std	Error	T	P	VIF
Constant	62.6909	16.7416	3.74	0.0003	0.0	
Creativityl	-0.17503	0.14093	-1.24	0.2175	1.0	
X1	1.09136	0.14969	7.29	0.0000	1.0	

Comparing OLS to 2SLS, intelligence as a dependent variable (table 20.21) is statistically significant at $P < 0.05$ forcing a conclusion that creativity is partially dependent on intelligence. In table 20.14, the instrument for Creativity is not significant at $P > 0.21$, suggesting that intelligence is not dependent on creativity. This is a different finding than given by OLS.

Two Important Notes Concerning 2SLS

1. Ideally, we could test these hypotheses by running an experiment in which we varied intelligence and observed its effect on creativity. Of course, such an experiment would not be ethical even if it were possible. The use of an instrument in 2SLS is another example of a passive regression analysis as an approximation of an experiment.

2. The success of 2SLS is highly dependent on finding effective variables for creating instruments. This is difficult because the variable X_1, as in this example, must be strongly related to Y_1 and only indirectly related to Y_2 through Y_1. Similarly, X_2 must be strongly related to Y_2 and only indirectly related to Y_1 through Y_2. We use the terms directly and indirectly exactly as we used them in discussing reduced-form equations. In our example, X_1 might be Alienation, a sociological concept strongly and directly related to creativity, but not to intelligence. X_2 might be income, strongly and directly related to intelligence, but not to creativity.

Summary

The critical ideas discussed in this chapter are outlined below.

1. Causal analysis with a fully partialled regression model

   Given modest degrees of collinearity, fully partialled models have the major defect of dumping all the common variance of the independent variables into the regression constant. For some purposes, prediction, forecasting, or concern for the causal effect of a single variable, this defect is of little concern. But, if the purpose of an analysis aims at estimating the causal effects of two or more variables, then, in the face of collinearity, the fully partialled model loses efficacy.

2. Causal analysis with a structural equation model

   If interests and needs transcend testing a singular causal effect extending to building and testing a causal theory, then you must turn to structural equation modeling, a statistical method that takes into consideration all the joint variance between the independent variables that fully partialled analyses ignore and assign to the regression constant. This method requires a better understanding of the causal forces than the fully partialled model.

3. Causal analysis with a reduced form model

   Reduced Form Equation Modeling, an extension of structural equations, divides total effects into direct and indirect effects.

4. Reciprocal causation

   Two examples illustrate the idea: Estimating the effect of intelligence on creativity or attitudes toward economics and academic achievement in economics. In both of these cases, the dependency is reciprocal: lower intelligence decreases creativity and lower creativity decreases intelligence; poor attitudes toward the subject matter of economics reduces academic performance in economics and conversely, lower performance results in poorer attitudes.

There are multiple techniques for estimating reciprocally causal models (e.g. indirect least squares, instrumental variables, 2SLS). Our discussion focused on two stage least squares (2SLS) with instrumental variables.

Chapter 20

Appendix

Your tasks are the following:

1. In the data set given below, treat X1 as a dependent variable with X2 through X5 as independent variables and run a fully partialled OLS multiple regression model.

2. Run an OLS structural equation model.

3. Run an OLS reduced form model.

4. In developing the structural equation and reduced form models, name the variables and provide a justification for your ordering decisions.

5. Compare the three models and explain your findings.

ID	X1	X2	X3	X4	X5
1	123.27	210.32	175.22	1	137.02
2	94.9	171.57	146.48	0	141.32
3	76.58	179.5	163.64	0	104.3
4	110.51	205.61	129.25	0	136.11
5	112.23	245.83	187.57	1	122.36
6	143.6	248.39	186.9	1	143.9
7	110.5	209.21	180.96	1	146.5
8	92.24	130.3	99.96	0	90.15
9	90.14	223.17	166.39	1	134.79
10	114.39	231.99	175.55	1	152.32
11	87.79	193.66	138.85	0	116.43
12	90.29	159.14	110.38	0	97.43
13	94.05	153.47	133.01	0	106.57
14	86.29	170.42	122.73	0	95.96

15	70.01	137.45	124.13	0	91.63
16	140.57	231.1	192.43	1	119.89
17	113.68	178.54	143.26	1	107.57
18	92.46	220.28	150.34	1	103.3
19	89.28	161.49	151.45	0	122.01
20	93.26	151.03	113.54	0	132.77
21	86.4	159.18	90.25	0	100.87
22	87.68	204.28	141.13	0	106.6
23	107.33	245.55	189.66	1	125.24
24	97.87	149.35	148.54	0	138.92
25	118.19	214.11	147.37	1	122.12
26	75.58	187.6	137.85	0	125.83
27	106.24	183.77	152.41	0	138.85
28	110.89	182.26	147.96	0	110.55
29	132.88	255.74	173.82	1	121.51
30	73.94	209.2	148.17	0	113.07
31	96.46	197.65	151.39	0	117.87
32	101.18	227.56	131.1	0	106.36
33	71.39	185.98	124.54	0	87.84
34	107.51	190.05	158.93	1	145.59
35	120.25	184.83	172.04	1	145.97
36	94.17	144.86	133.77	0	98.21
37	91.67	170.86	125.72	0	116.99
38	97.26	195.18	143.21	1	110.28
39	119.97	224.66	163	1	137.36
40	74.59	147.82	127.09	0	116.53
41	80.22	181.78	109.79	0	85.77
42	69.79	205.54	144	0	102.05
43	112.25	268.1	141.26	1	133.99
44	101.15	174.23	124.8	0	127.06
45	87.1	239.75	139.31	0	123.45
46	107.65	225.79	177.17	1	144.07
47	82.82	168.16	124.94	0	121.51
48	88.07	198.31	128.16	0	94.36
49	109.8	186.92	165.4	1	136.15
50	102.65	204.11	159.52	1	128.91

Time Series Analysis

Introduction

Both managers and managerial experts prick up their ears when they hear the topic of chapters 21-24. These are devoted to time series data. We show how to make estimates of the strength of statistical relationships between variables measured and moving through time, and time-series forecasts. To do this, we double back and explain the difference between forecasting and prediction, and the difference between time-series data and cross sectional data.

Then, in chapter 21, we do univariate forecasting with the "classical" decomposition model, i.e., decomposing the variable into the dynamic forces of trend, cycles, seasonality, irregular events such as strikes and weather, and residuals—the (sometimes noisy, sometimes not) leftovers of total variation after all those other factors have been stripped away. Chapter 22 focuses on univariate forecasting with moving averages (smoothing).

We then move in chapter 23 to the more complex noise-reducing technique of univariate forecasting with the AutoRegressive Integrated Moving Average (ARIMA) model, the Box-Jenkins approach to time-series modeling, and in this discussion, we adumbrate the concepts of stationarity, differencing, lagging, and a new take on moving averages. Again, all these concepts are critical in distinguishing noise from signal, and producing information from the peculiar complexities of time series data.

Chapter 24 focuses on the enormous power of time series analysis in teasing out causal inferences using the quasi-experimental methods called intruded time series and cross correlations as well as reduced form testing.

CHAPTER 21

FORECASTING WITH CLASSICAL DECOMPOSITION[87]

In our previous discussions, we restricted the discourse to cross-sectional data, i.e., data taken at a single point in time. Time series represent data measured in adjacent sequence over several points in time. Dynamism defines the major difference between cross-sections and time series; most time series have internal forces that cause them to "go somewhere." The statistical methods used to analyze time series draw upon methods discussed previously.

We can measure time series in minutes, hours, days, weeks, months and years, and, while we limit our discussion to economic time series measured in months, quarters (three month periods), or years, the methods we discuss are useful for any time series; economic, psychological, sociological and political measured in any length.

One of the reasons for a separate study of time series is that they provide powerful tools for forecasting and causal analysis. Notice here that we have introduced a new term, *forecasting*. In our discussion of cross sections, we used the term *prediction*. While both prediction and forecasting involve efforts to foretell the future, cross-sectional prediction is constrained by the range of data in the sample. For example, if we successfully relate hand-finger dexterity, X to ability

---

[87]. The Advent of powerful microcomputers and software programs forced the classical decomposition method discussed here out off general application. Nevertheless, it is still a useful technique, particularly for working with seasonal and cyclical data. It is also easy to understand and apply. In addition, for the German Statistical Service, classical decomposition and seasonal adjustment of economic time series has a long tradition. Using this procedure the German Service also provides the general public with information on trends and seasonally adjusted major business-cycle indicators. Their program BV4.1 is also down loadable without fee and usable for noncommercial purposes.

to fly an airplane, Y, we limit predictions to the range of our dexterity measure.

Forecasting, a special kind of prediction, always projects (extrapolates) beyond the constraints of the data. For example, if we have quarterly sales data for our firm, we want to project (forecast) beyond our data to future quarters.

Dynamics of Time Series

In the social sciences we usually define the internal forces in time series as trend, cycle, seasonal, calendar, irregular, and residual variation. Trend represents a longer term tendency to rise or fall. Figure 21.1 shows a time series plot of trend in sales of a hypothetical firm over a 50-year period.

Figure 21.1 A Non-linear Trend

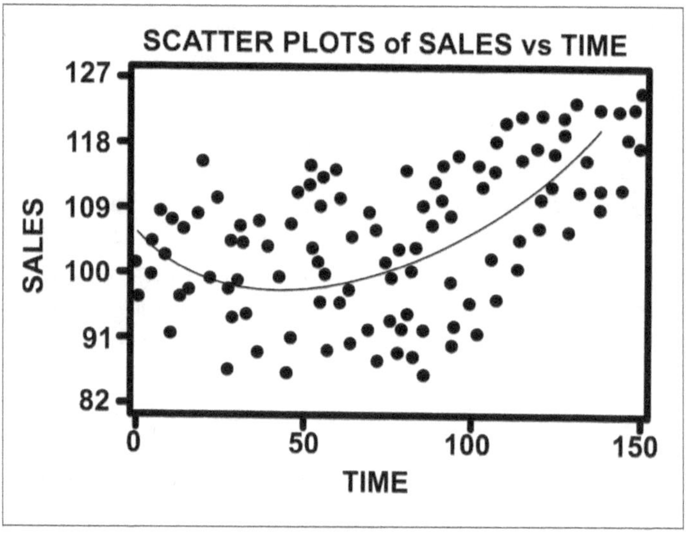

The curve running through the data depicts a nonlinear trend.

Cycles, the second dynamic, embody the shock of multiple forces having middle term recurring effects on a time series. Periodic economic recessions represent a familiar example of cycles. Figure 21.2 shows a time series plot of annual sales of another hypothetical firm depicting trend and cyclical variation.

Figure 21.2

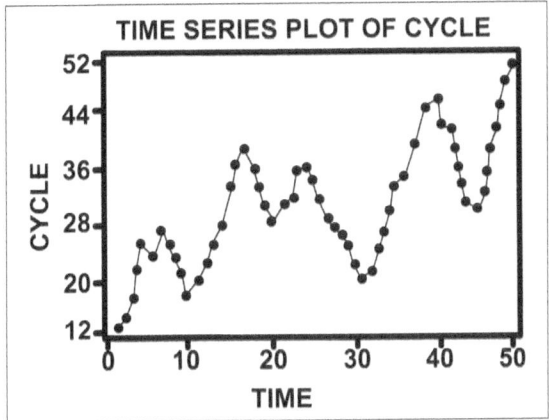

In figure 21.2, the plot captures the variation in sales attributable to trend and cycle.

Seasonal forces represent the effect of variables occurring at regular intervals within a year, e.g., daily, weekly, monthly or quarterly. Seasonal effects can be climatic or institutional, for example, the climatic variation in the output of corn or institutional variation in sales at Christmas, Thanksgiving, Easter, or Halloween.

Figure 21.3 depicts a time series plot showing seasonal variation of monthly sales of a third hypothetical firm.

Figure 21.3 Seasonal Variation

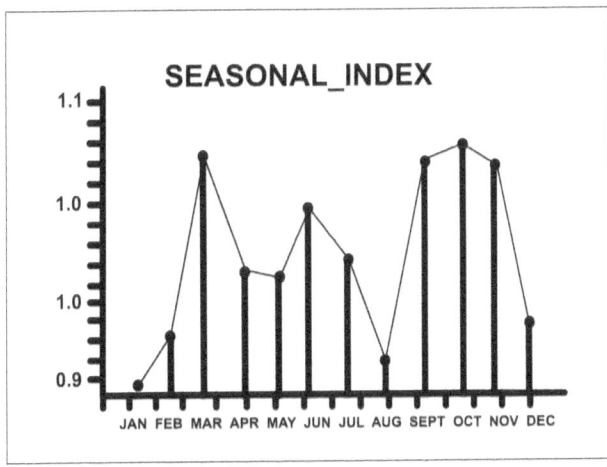

In this example, the seasonal variation is monthly. Look at the figure closely. Sales in months 4, 9, and 10, are relatively higher than other months. Sales in months 1, and 8, are relatively low, the worst sales months for the firm,

Irregular variation represents the effect of events like strikes, floods, hurricanes, and mass protests that intrude and disrupt the regular dynamic forces of a time series.

Calendar effects refer to different number of days in a month, holidays, or the length of workweeks. Calendar fluctuations are important to any cumulative data like sales. We already know the term residual; after regression of a dependent variable on a set of independent variables, the residual equals noise or risk, unexplained variance. Residuals represent unmeasured forces acting on the time series.

These internal forces represent dynamism, i.e., where the series is going without intervention. Depending on our research goals, these forces are friendly or hostile. If our goal is forecasting, the forces are friendly; we harness their effects for forecasting (extrapolation) into the future, treating the residuals as normally distributed risk. If causal inference defines our goal, these forces are hostile; we want to remove the confounding effects of where the series is going anyway. For example, if we want to test the effectiveness of an advertising campaign on sales, it is necessary to remove the confounding effects of trend, cycle, seasonal, irregular, and calendar variations. If, when we implemented an advertising campaign, sales were positively trending, we could mistakenly attribute the increased sales to the new advertising campaign when the sales were simply trending upward. We note that these forces all produce autocorrelation in a time series.[88] If we model a time series correctly, we remove their impact producing normal, uncorrelated, and homoscedastic residuals called "white noise."

Classical Decomposition

Classical decomposition, the focus of this chapter, is one method for modeling the internal dynamics of a time series. This method assumes some functional relationship between Trend, Seasonal, Cyclical, and

---

88.   Remember that autocorrelation occurs when sequential residual error terms from observations of the same variable move in the same or opposite direction.

Irregular variation. For example, the relationship may be additive $(T_{rend} + S_{easonal} + C_{yclical} + I_{rregular} + R_{andom})$ or multiplicative $(T \times S \times C \times I \times R)$.[89] In our examples, we assume multiplicative relationships.

Before we begin the discussion of the procedures used in decomposing a time series, a brief digression is necessary.

The Moving Average

At the heart of modeling the components of a time series, is the notion of a moving average. To help understand how this concept might be useful in analyzing a time series, consider the example in table 21.1.

Table 21.1 Three Period Moving Average

Period	Series	Three Period Moving Total	Three Period Moving Average
(Column 1)	(Column 2)	(Column 3)	(Column 4)
1	12		
2	14	39	13
3	13	45	15
4	18	52	17.3
5	21	60	20
6	21	64	21.3
7	22	61	20.3
8	18	63	21
9	23	65	21.7
10	24	69	23
11	22	72	24
12	26		

Table 21.1 shows a monthly time series (Column 2) on which we calculated a three period moving average. In the first calculation, we add the first three observations in the series and place that total, 39 in Column 3, a three period moving total, centered opposite the second period. The second calculation adds the values in periods two, three, and four

---

89. These two examples do not exhaust the possibilities, but represent those that apply most frequently.

centering this sum, 45 in the cell opposite the third period, repeating these calculations for the remainder of the table down through period eleven. Obviously, we cannot calculate a three period moving total for the cell opposite period twelve. For column 4, we divide the values in the third column by three, giving a three period moving average.

What is the effect of taking a moving average? In figure 21.4, we plotted the series (column 2) vs. the period (column 1).

Figure 21.4 Series on Time

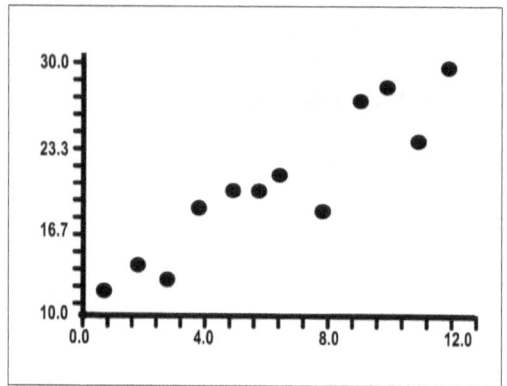

This plot shows all the variation in the time series.

Figure 21.5 plots the three period moving averages on time showing how the moving average smoothes the series, preserving the trend while dampening the seasonal or irregular variation.

Figure 21.5 Three Period Moving Average

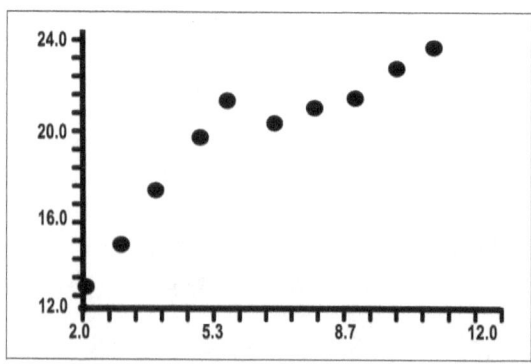

All the values are now closer to the regression line and the Loess plot.

STEPS IN MODELING A TIME SERIES USING CLASSICAL DECOMPOSITION[90]

The following is a systematic explanation of how to do a classical decomposition and develop a forecast.

Step 1: Adjust for Calendar Variation.

Quarterly and monthly data vary over a year. September, April, June, and November have 30 days. All the rest have 31 except February, 28 or 29. Any time series data affected by this variation, e.g., production, consumption, sales, must be adjusted for calendar variation. The adjustment may focus on calendar days or workdays. For example, if our series is sales, we adjust for workdays. Adjusting for work days involves counting all the Saturdays, Sundays, and holidays observed by the firm as non workdays, subtracting this number from the calendar days in the month and dividing this difference into the actual monthly sales to obtain sales per working day. If the adjustment involves consumption not affected by working days, then, we would, using the same logic, adjust for calendar day differences.

Step 2: Model the Seasonal Variation and
Construct a Seasonal Index

Table 21.2 is a hypothetical data set of monthly sales in thousands of dollars from January 1997 through December 2007 for the hypothetical statistical consulting firm, StatSys. Assume that we have made the Calendar adjustment as described above. For saving space, we omitted the middle seven years.

Modeling seasonal variation requires a moving average and, because we have monthly data, we use a 12-month moving average, with the same logic as the calculations in table 21.1. Given that the actual sales data in table 21.2 are shown at the middle of each month, the moving average must be centered on July 1997. Because our series uses 12 months, centering the data in the middle of the month requires an adjustment discussed below. Table 21.2 shows all the calculations needed to make that adjustment and construct a Seasonal index.

[90.] For a more detailed discussion of this concept, see CC, chapters 11 through 16.

TABLE 21.2 Construction of a Seasonal Index

Date Column 1	Sales Column 2	Twelve Month Moving Total Column 3	Twelve Month Moving Average Column 4	Centered Two Month Moving Total Column 5	Centered Twelve Month Moving Average Column 6	Seasonal and Irregular Column 7	Seasonal Index Column 8
Jan-97	101.40						0.90
2	96.20						0.93
3	105.20						1.08
4	108.50						0.99
5	99.90						1.00
6	107.90	1243.60	103.63				1.03
7	103.40	1240.90	103.41	207.04	103.52	1.00	0.99
8	92.90	1246.50	103.88	207.28	103.64	0.90	0.91
9	107.10	1252.90	104.41	208.28	104.14	1.03	1.07
10	114.00	1250.50	104.21	208.62	104.31	1.09	1.07
11	109.00	1251.00	104.25	208.46	104.23	1.05	1.07
12	98.10	1249.20	104.10	208.35	104.18	0.94	0.96
Jan-98	98.70	1247.00	103.92	208.02	104.01	0.95	0.90
2	101.80	1244.20	103.68	207.60	103.80	0.98	0.93
3	111.60	1247.30	103.94	207.63	103.81	1.08	1.08
4	106.10	1244.40	103.70	207.64	103.82	1.02	0.99
5	100.40	1240.70	103.39	207.09	103.55	0.97	1.00
6	106.10	1238.80	103.23	206.63	103.31	1.03	1.03
7	101.20	1224.30	102.03	205.26	102.63	0.99	0.99
8	90.10	1210.60	100.88	202.91	101.45	0.89	0.91
9	110.20	1205.00	100.42	201.30	100.65	1.09	1.07
10	111.10	1195.20	99.60	200.02	100.01	1.11	1.07
11	105.30	1187.00	98.92	198.52	99.26	1.06	1.07
12	96.20	1181.90	98.49	197.41	98.70	0.97	0.96
====	====	====	====	====	====	====	====
Jan-06	97.70	1340.30	111.69	223.13	111.57	0.88	0.90
2	108.90	1348.70	112.39	224.08	112.04	0.97	0.93
3	121.80	1351.40	112.62	225.01	112.50	1.08	1.08

4	106.00	1355.80	112.98	225.60	112.80	0.94	0.99
5	122.00	1361.40	113.45	226.43	113.22	1.08	1.00
6	111.50	1362.40	113.53	226.98	113.49	0.98	1.03
7	110.90	1372.60	114.38	227.92	113.96	0.97	0.99
8	109.00	1371.70	114.31	228.69	114.35	0.95	0.91
9	120.10	1372.50	114.38	228.68	114.34	1.05	1.07
10	119.90	1376.30	114.69	229.07	114.53	1.05	1.07
11	124.90	1371.70	114.31	229.00	114.50	1.09	1.07
12	109.70	1373.50	114.46	228.77	114.38	0.96	0.96
Jan-07	107.90	1374.80	114.57	229.03	114.51	0.94	0.90
2	108.00	1374.40	114.53	229.10	114.55	0.94	0.93
3	122.60	1368.40	114.03	228.57	114.28	1.07	1.08
4	109.80	1366.70	113.89	227.93	113.96	0.96	0.99
5	117.40	1364.10	113.68	227.57	113.78	1.03	1.00
6	113.30	1371.80	114.32	227.99	114.00	0.99	1.03
7	112.2						0.99
8	108.6						0.91
9	114.1						1.07
10	118.2						1.07
11	122.3						1.07
12	117.4						0.96

In column three of table 21.2, we calculated a 12-month moving total by summing the sales from january 1997 through december 1997 and placing that total, 1243.6 in the cell corresponding to june 1997. Then in column 4, we divided the 12-month moving total by 12 to achieve a 12-month moving average of sales. Now, because the 12 month moving average in column 4 is between july and june (we placed it in the june cell), and because the actual sales data were given in the middle of each month, for example the sales for july 1997, 103.4 were recorded in the middle of july, we must center the data in the middle of july and all the rest of the months. We center the data in column 5 by calculating a two-month moving total and centering it in July. That is, we calculate the first total in column 5 by summing the June and July averages (column 4), (103.63 and 103.41), and center that total, 207.04 in July, column 5. This operation centers the data in the middle of july. That process is repeated for each month through the rest of the series. In column 6, we simply divide column 5 by two, giving us a centered 12-month moving average. In column 7, we divide the actual sales (column 2) by the moving average sales in column 6.

Following the calculations, we explore their rationale. We achieved the 12-month moving total by calculating column 3 and the 12-month moving average by calculating column 4. But, because we needed to center our values in each month, we took a two-month moving total in column 5. and a centered 12-month moving average in column 6. In column 7, we divided the actual Sales in column 1 by the moving average in column 6. This gives the percentage of the actual sales against the moving average. What have we accomplished with these calculations? Assume, as we have throughout our discussion that the relationship is $T \times S \times C \times R \times I$. The twelve-month-moving average approximates T and C because it smoothes, eliminating most of the S, and I variation. If the 12 month moving average approximates T and C, then dividing the 12 month moving average into the original sales

series, $\dfrac{T * C * S * I}{T * C}$, gives an estimate of S, and I. And, if there are no

Irregular effects, dividing the 12 moving average into the sales data gives an estimate of Seasonal variation.

Now let's see if that's true. Figure 21.6 plots sales (Column 2) on time

Figure 21.6 Plot of Sales on Time

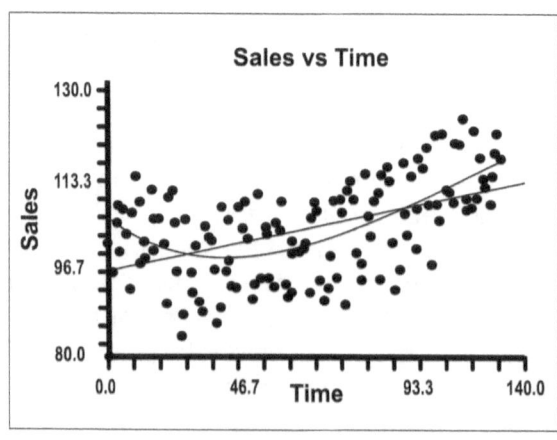

We see in this plot the nonlinear trend and all the rest of the variation in the entire series.

Figure 21.7 plots our estimate of Seasonal variation on time.

Figure 21.7 Plot of Seasonal and Irregular Variation on Time

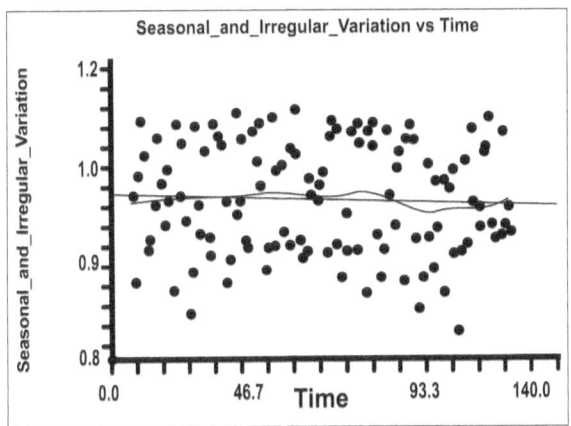

We see that, by dividing Sales by our estimate of Trend and Cyclical variation, we have eliminated the Trend and Cyclical components.

Our next step calculates the seasonal index for the 12 months, January through December. If our last column represents Seasonal and Residual variation, we can eliminate Residual variation by taking a trimmed average (ignoring the largest and smallest seasonal and residual values for each month) and averaging each month. These averages are an estimate of the seasonal variation in the series for each month. Doing that gives table 21.5 the seasonal index for our series.

Table 21.5 Seasonal Index

Jan	0.905
Feb	0.942
Mar	1.066
Apr	0.983
May	0.981
Jun	1.033
Jul	0.995
Aug	0.919
Sept	1.064
Oct	1.077
Nov	1.074
Dec	0.961

Notice that sales vary by month rising from January to March, falling in April and May, rising again in June, falling in July and August, rising in Oct and Nov and falling in December. The poorest month's sales are in January and August. The best month's sales are October and November. Now, we plot the Seasonal Index on Time in figure 21.8.

Figure 21.8 Plot of Seasonal Index

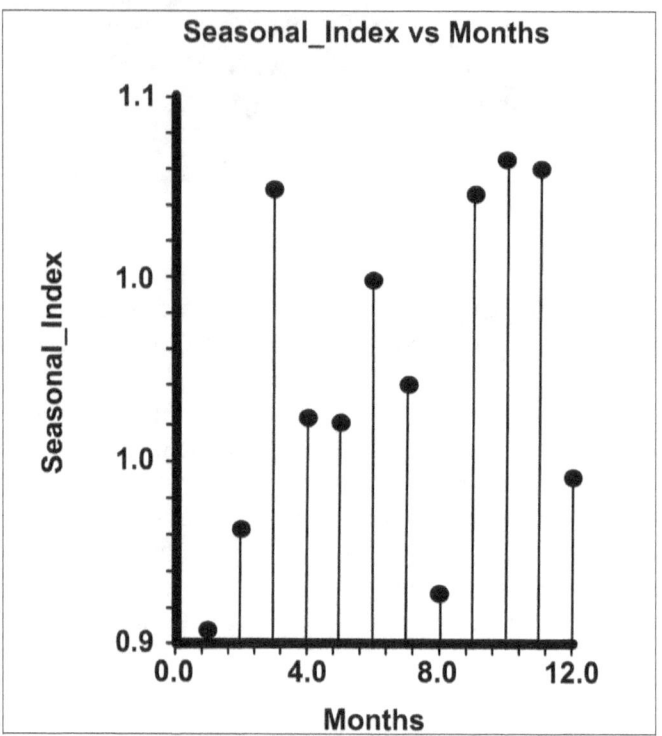

In figure 21.8, we see that the irregular and residual variation is gone; we have isolated the seasonal variation.

Step 3: Model for Trend

In figure 21.6, we plotted the actual sales against time and the Loess curve strongly suggested a nonlinear trend. So to calculate trend we ran a quadratic multiple regression of the form:

$$Y_c \text{ (Sales)} = B_0 + X_1 \text{Time} + X^2_2 \text{ Time Squared.}$$

Table 21.6 is a summary of that regression.

Table 21.6 Regression of Sales on Time and Time Squared

Independent Variable	Coefficient	Standard Error	T	Prob
Intercept	103.9562	1.8718	55.538	0.0000
Time	-0.2162	0.0650	-3.328	0.0011
Time²	0.0025	0.0005	5.299	0.0000

Estimated Model

$103.956176266482 - .216235218641027 \times Time + 2.50750910129789E\text{-}03 \times Time^2$

$R^2 = 0.3$

The overall model was statistically significant as were the two terms in the quadratic confirming our Trend Judgment. We saved the estimated Trend as the trend component.

Step 4: Model the Cyclical Component

Table 21.8 shows the Period, Actual Sales, Trend (Calculated in step 3), and Seasonal Index (calculated in step 2), cyclical and residual variation, and cyclical variation. For brevity, the table shows only the first two years.

Table 21.8 Cyclical Calculation

Period	Sales	Seasonal Index	Trend	Cycle and Residual	Cycle
Column 1	Column 2	Column 3	Column 4	Column 5	Column 6
1	101.40	0.90	103.74	1.09	
2	96.20	0.93	103.53	1.00	1.01
3	105.20	1.08	103.33	0.94	1.00
4	108.50	0.99	103.13	1.06	0.99
5	99.90	102.94	0.97	0.97	1.02
6	107.90	1.03	102.75	1.02	1.00
7	103.40	0.99	102.57	1.02	1.01
8	92.90	0.91	102.39	0.99	1.00
9	107.10	1.07	102.21	0.98	1.01
10	114.00	1.07	102.04	1.04	1.01

11	109.00	1.07	101.88	1.00	1.02
12	98.10	0.96	101.72	1.00	1.03
1	98.70	0.90	101.57	1.08	1.05
2	101.80	0.93	101.42	1.08	1.06
3	111.60	1.08	101.28	1.02	1.05
4	106.10	0.99	101.14	1.06	1.03
5	100.40	1.00	101.00	1.00	1.02
6	106.10	1.03	100.88	1.02	1.01
7	101.20	0.99	100.75	1.02	1.00
8	90.10	0.91	100.63	0.98	1.01
9	110.20	1.07	100.52	1.02	1.01
10	111.10	1.07	100.41	1.03	1.01
11	105.30	1.07	100.31	0.98	1.00
12	96.20	0.96	100.21	1.00	0.97

We calculated column 5 (cyclical and residual by dividing column 2, sales by the Trend and Seasonal estimates. Using that quotient, we eliminated the residual by taking a three-month moving average of column 5, creating in (column 6) our cyclical estimate. Figure 21.9 plots our Cyclical on Time.

Figure 21.9 Cyclical Index on Time

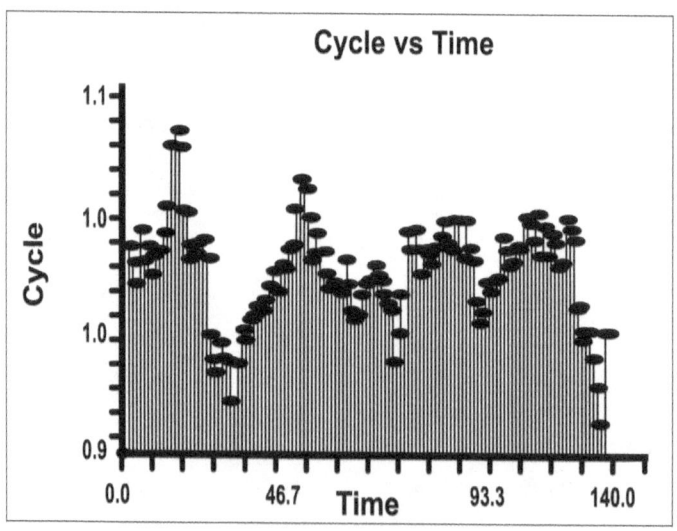

Here, we see the undulating cyclical movement of the series.

Step 5 Make a Forecast

To make the forecast, we enter the months we want to forecast into the quadratic trend equation. The last month for which we have data is 132, therefore we enter period 133 and 134 (if we want to make a two month forecast). The forecasting equation, a quadratic is as follows:

$$103.956176266482-.216235218641027 \times Time + 2.50750910129789E\text{-}03 \times Time^2.$$

Substituting 133 and 17689 for time and time squared and solving the equation yields the trend estimate for sales for month 133, 119.63 and month 134, 120.06. Now, we adjust the trend estimate by multiplying the trend estimate by the seasonal and cyclical indices. Using the seasonal indices for January and February (0.900 and 0.929) and multiplying these indices by our trend estimates, we obtain our deseasonalized estimates.

Projecting the cyclical values for January and February 2008 is now a matter of judgment. The tick marks of the plot in figure 21.6 equal about 5 months. Therefore, the length of a cycle for this firm is from trough to peak and peak to trough, on average, approximates 15 months with smaller fluctuations within that period. As we look at figure 21.6, our best judgment is that our sales are now recovering, so we use the cycle values of 0.978 and 0.980. Our forecast for sales adjusted for seasonal and cyclical variation in January 2008 (trend × seasonal × cyclical) or 119.63 × 0.900 × 0.978 is about 105.30. For February, the estimate is about 109.41.

Summary

The critical concepts discussed in this chapter are the following:

1. The notion of time series analysis

    Time series represent data measured, usually, but not necessarily, in adjacent sequence, over several points in time. The statistical regression methods used to analyze time series are based on methods used to analyze cross sectional data. The major difference between cross-sections and time series is dynamism; most time series have internal forces that cause time series to "go somewhere."

2. The moving average

    At the heart of modeling the components of a time series, is the notion of a moving average. To review this critical notion, look back at the example we developed in table 21.1.

3. Classical decomposition

    Classical decomposition, one method for modeling the internal dynamics of a time series, assumes some functional relationship between Trend, Seasonal, Cyclical, and Irregular variation. For example, the relationship may be additive ($T_{rend} + S_{easonal} + C_{yclical} + I_{rregular} + R_{andom}$) or multiplicative ($T \times S \times C \times I \times R$).

4. The concept of trend

    Trend is a time series' central tendency, a tendency to rise or fall driven by influences with long-term effects; examples are technological progress or decline, changes in scientific knowledge, and dynamic institutions.

5. The concept of seasonal variation

    Seasonal forces represent the effect of variables occurring at regular intervals within a year, e.g., monthly or quarterly. Seasonal effects can be climatic or institutional, for example, output of corn (climatic) or sales at Christmas, Thanksgiving, Easter, or Halloween (institutional).

6. The concept of cyclical variation

    Cycles embody the shock of multiple forces having middle term recurring effects on the time series. An example is economic recessions caused by a failure of aggregate demand

7. The concept of irregular variation

    Irregular variation represents the effect of events like strikes, floods, hurricanes, and mass protests that intrude and disrupt the regular dynamic forces of a time series

8. Classical decomposition procedures for modeling the time series components

Three procedures lie at the heart of classical decomposition: moving averages, indices, and regression.

9. Forecasting using classical time series analysis

Classical decomposition forecasts the trend of a series using a regression model with time as the independent variable. and, if necessary, dummy independent variables to control for irregular variation. This trend forecast is then corrected (adjusted) for seasonal and cyclical variation using seasonal and cyclical indices.

Chapter 21

Appendix

The data set given below is a time series of sales for the hypothetical firm StatSys. The data are quarterly over the period (2001-2009). The first observation in the series is the first quarter of 2001. Your tasks are the following: (1) Using classical decomposition, develop the trend equation and the seasonal and cyclical indices and (2) based on these analyses, forecast the series for the last three quarters of 2009. Explain and justify your analysis.

Sales	Time
39.94	1
39.86	2
38.65	3
42.23	4
39.13	5
39.94	6
36.86	7
38.16	8
39.06	9
39.64	10
43.62	11
44.58	12
44.78	13

43.43	14
44.84	15
45.38	16
44.43	17
42.99	18
45.10	19
42.73	20
46.80	21
49.24	22
47.02	23
46.89	24
48.76	25
48.30	26
45.93	27
47.44	28
48.01	29
48.01	30
52.84	31
50.84	32
51.64	33

CHAPTER 22

FORECASTING WITH MOVING AVERAGES

In this chapter, we extend our discussion of univariate forecasting to techniques called smoothing, built on the notion of moving average discussed in chapter 21. While we explain and illustrate using detailed hand calculations, the programs NCSS, SX, and OpenStat all provide routines accomplishing these techniques accurately and without tedium. Therefore, your task, as with all our previous discussions, is to understand how these methods work.

The Single (Simple) Moving Average

Smoothing is just another word for moving average. As we saw in chapter 21, taking a moving average makes a series less volatile by smoothing it out. The simple moving average provides a quick and easy forecast of data that contain neither trend nor seasonal components. Its major weakness is that it is not adept at signaling change. Here is an example:

To forecast using a single moving average we say that the forecast for all periods beyond T is just its moving average. This method only forecasts for one period ahead, updating the moving average as the actual observation for that period becomes available.

Table 22.1 shows quarterly sales for three years of some hypothetical enterprise. We calculated a three year moving average, placed it in the fourth year, and, repeating the calculation through the series, forecasted year 13. We also calculated and averaged the errors in our forecasts.

Table 22.1 Sales

Quarter	Sales ($1,000)	Three Period Moving Average	Sales minus Moving Average	Error Squared	Absolute Error	Percentage Error	Absolute Percentage Error
1	52						
2	51						
3	53						
4	48	52.00	-4.00	16.00	4.00	-0.08	0.08
5	45	50.67	-5.57	32.11	5.57	-0.13	0.13
6	48	48.67	-0.67	0.44	0.67	-0.01	0.01
7	50	47.00	3.00	9.00	3.00	0.06	0.06
8	52	47.67	4.33	18.78	4.33	0.08	0.08
9	54	50.00	4.00	16.00	4.00	0.07	0.07
10	52	52.00	0.00	0.00	0.00	0.01	0.01
11	53	52.67	0.33	0.11	0.33	-0.04	0.04
12	51	53.00	-2.00	4.00	2.00	-0.00	0.00
Forecast 1 for quarter 13		52.00		96.44	23.90		0.49
Sum				10.72	2.66		

Because the simple moving average does not work for trending or seasonal time series, and our example used annual data, we need a guide for determining the length of the moving average. A guide for deciding on the period length for a moving average is the absolute deviation (MAD).[91] In this example, the mean absolute deviation was about 2.66. Rounding 2.66 to 3 suggested a three period moving average. Taking a three period moving average, we put the first moving average in year four. By that logic we can use the three previous period's moving averages as forecasts for the next period and, in this example, forecast the thirteenth quarter.

Notice that we calculated a standard error using the squared difference between actual and forecasted sales (error). We also calculated the mean percentage error and the absolute percentage error as measures of goodness of fit. Our point forecast was $52,000.00

---

[91]   In this example, the mean average deviation (MAD) calculates the sum of the errors ignoring signs and divides by the number of forecasts.

and, with a 95% confidence interval, the forecast was $52,000 ± 1.96 × 3.27 or $58,400 to $45,600.

We used the program SX to check our work. Table 22.2 shows that output.

Table 22.2 Single (Simple) Moving Averages for annual Sales

Mean Squared Error (MSE)	**10.72**
Standard Error (SE)	3.273
Mean Absolute Deviation (MAD)	2.656
Mean Abs Percentage Error (MAPE)	0.054
Mean Percentage Error (MPE)	-0.43
Length of Series	12 Years

	95% CI		95% CI
Forecast	Lower	Forecast	Upper
1	45.5839	52.0000	58.4161

While our data set from chapter 21 contains trend and seasonal variation, just for the sake of comparison, we ran a single moving average on the sales data and compared the result with component forecasting. Table 22.3 shows the result.

Table 22.3 Single (Simple) Moving Averages Forecast for Sales (From Chapter 21)

Moving Average Length 12[92]	
Sum of Squared Errors (SSE)	5.935E+11
Mean Squared Error (MSE)	**2.603E+09**
Standard Error (Standard Error)	51019.4
Mean Absolute Deviation (MAD)	42106.4
Mean Absolute Percentage Error (MAPE)	5.59
Mean Percentage Error (MPE)	-1.73

---

[92]   We used the seasonal period, 12 as our guide for the moving average although the simple moving average was not designed for use with seasonal data.

Number of Cases 240

	95% CI		95% CI
Forecast	Lower	Forecast	Upper
1	534252	634250	734248

Comparing this estimate for January 2009, about $634,000 with our estimate in chapter 21 using component analysis, about $633,000, we see that it is not far off and certainly within the 95% confidence interval.

The Double Moving Average

With a trending series, we use a double moving average. This technique calculates a second moving average (M2) from the original (M1) using the same value for the moving average length. The principle behind this move is that this second Moving Average (M2) provides the basis for calculating a *series of changing regression equations*. We add the difference between the two moving averages, M1 and M2 to the first moving average, M1 to realize the constant B_0 for the regression equation in that period The difference between M1 and M2 yields the regression coefficient, B_1 for the regression equation for that period and, when added to B_0, the constant, produces the period forecast. Using the data in table 22.2, we constructed table 22.4 showing the second moving average, the calculations of B_0 and B_1 for all the periods, and forecasts starting with period 6.

Table 22.4 Double Moving Average (Trend)

Time Column1	Sales Column2	First Three Period Moving Average M1 Column3	Second Three Period Moving Average M2 Column4	Moving $B0 = (M1 - M2) + M1$ Column5	Moving $B_1, B_1 = M1 - M2$ Column6	Forecasts $= M1$ $= B0 + B_1$ Column7	Error = Sales - Forecast Column8	Error Squared Column9
1	52.00							
2	51.00							
3	53.00	52.00						
4	48.00	50.66						

5	45.00	48.66	50.44				
6	48.00	47.00	48.77	46.88		45.10	2.90	8.41
7	50.00	47.66	47.77	47.89	-1.77	43.46	6.54	42.7
8	52.00	50.00	48.22	47.55	-1.78	47.44	4.56	20.8
9	54.00	52.00	49.88	51.78	-0.11	53.56	0.44	0.19
10	52.00	52.66	51.55	54.12	1.78	56.24	-4.24	18
11	53.00	53.00	52.55	53.77	2.12	54.88	-1.88	3.53
12	51.00	52.00	52.55	53.45	1.11	53.90	-2.90	8.41
13				51.44	0.45	50.9		
					-0.55			

Here is a summary of a one period forecast using the double moving average method:

$(M_1 - M_2) + M_1 = B_0$
$(52 - 52.55) + 51 = 51.44 = B_0$
$B_1 = (M_1 - M2)$
$B_1 = 52 - 52.55 = -0.55$
$B_0 + B_1 = Forecast_1 = 50.9$

Notice that we lagged the calculations of B_0 and B_1 behind the original series. The theory undergirding this approach is that the first moving average, M1 lags behind the original series and the second moving average lags behind the first by the same amount. Also notice that this procedure produces moving regression equations to reflect the new actual sales. A 95% interval estimate is 50.9 ± 1.96 × 3.82 or 43.26 to 58.54.

In table 22.5 we show the double moving average forecast of the sales data from chapter 21. While this method does not generate an equation for handling the seasonal component, we could have deseasonalized the data and run our double moving average on that series. Or we could have calculated a seasonal index, run the data with its seasonal variation and corrected it with our calculated seasonal index.

Table 22.5 Double Moving Averages for Monthly Sales from chapter 21.

Moving Average Length 12

Sum of Squared Errors (SSE)	5.563E+11
Mean Squared Error (MSE)	2.563E+09
Standard Error (SE)	**50630.5**
Mean Absolute Deviation (MAD)	**38601.4**
Mean Abs Percentage Error (MAPE)	5.17
Mean Percentage Error (MPE)	5%
Length of Series	240 Months

	95% CI		95% CI
Lead	Lower	Forecast	Upper
1	532175	631410	730646

The forecast, about $631,000, is slightly lower than the decomposition forecast, about $633,000, but both our single and double moving average forecasts fall within 95% confidence intervals.

Single Exponential Weighted Average (Smoothing)

This technique, a weighted version of the single (simple) moving average, gives recent experience more or less influence and is useful for time series with neither trend nor seasonal variation. The computations are simple because only the estimate of the previous period and the current period determine the new estimate. The forecasting equation is as follows:

$$F_T = \forall Y_T + (1 - \forall) \times ExMa_{T-1}\text{[93]} \qquad \text{(Equation 22.1)}$$

---

[93]. "What," you may ask, "does this equation have to do with exponential smoothing?" The equation has no exponents in it. Without proof, the following equation is another form of equation 22.1:

$$F_T = \forall Y_T + \text{“} (1 - \forall)*Y_{T-1} + \forall (1 - \forall)2 *Y_{T-2} + \forall(1 - \forall)3 \times Y_{T-3} + \forall (1 - \forall)4 \times T - 4 + \ldots$$

This equation is a weighted average of the items in the series with the weights decreasing exponentially as the series moves back in time. A close look at this equation explains the name of the procedure: the weights decrease exponentially and, because moving averages smooth, the procedure is exponential smoothing.

where a equals the weight and $ExMa_{T-1}$ equals the exponential moving average for the previous period.

To demonstrate how this procedure works we once again use the data in table 22.1 and the forecast in table 22.7.

Table 22.7 Single Exponential Moving Average (Smoothing)

Time	Y_T	Exponential Moving Average	Forecast
0		50.75	
1	52	51.75	50.75
2	51	51.15	51.75
3	53	52.63	51.15
4	48	48.93	52.36
5	45	45.79	48.93
6	48	47.56	45.79
7	50	49.51	47.56
8	52	51.50	49.51
9	54	53.50	51.50
10	52	52.30	53.50
11	53	52.86	52.30
12	51	51.37	52.86
13			51.37

The first step in making a forecast using a single exponential moving average (smoothing) calculates a starting exponential moving average. That is, to make a forecast for period one we need the exponential average for the previous period Y_{T-1}, a value that we do not have (Equation 22.1). For that purpose, we used the mean of the series, 50.75. Another way to estimate the beginning value is to make a backcast by turning the series upside down, running a backward regression, and making a backward forecast. Notice from table 22.7 we put the starting value 50.75 in the column headed exponential average. The exponential average is a two-period weighted moving average. This two period weighted average then becomes the forecasted value for the next period.

Next we chose a weighting constant, some value greater than zero and less than one. The larger the value of ", the greater the emphasis on values closer to the present, the smaller, closer to the past. For our purpose, we chose 0.80. Some statistical programs, e.g., SX, either allow you to choose the constant or let the program choose to minimize some goodness of fit measure, e.g., MAD or ΣD^2.

Then we make the first forecast, F_{T1}, by simply using our estimated starting exponential moving average $F_{T1} = 50.75$.

Now, using equation 22.1 we calculate the exponential moving average for period one, a value that becomes the forecast for period two.

$F_{T2} = 0.80 \times 52 + (1 - 0.80) \times 50.75 = 51.75$

In this equation we substitute the actual value for period one, 52,

We repeat this process thru our forecast for period 13 using the weighted moving average for period 12 as F_{13}. This forecast is *51.37*.

Lead	95% CI Lower	Forecast	95% CI Upper
1	46.6	51.4	56.2

For comparison, once again, we apply this method to the Sales data from chapter 21.

Table 22.8 is a summary of our forecast of the data in chapter 21.

Forecast Summary Section

Variable	sales
Number of Rows	240
Mean	$769,000
Pseudo R Squared	0.977
Mean Square Error	480414940.273
Mean | Error |	17387.195
Mean | Percent Error |	2.294
Alpha	0.8
Forecast	$613,000

The estimate, using the NCSS simple exponential smooth routine, is about $613,000, considerably lower as compared with the component estimate of about $633,000. The mean percentage error is about 2.3%.

Holt's Exponential Moving Average (Smooth)

We use this procedure, also called Holt's linear trend, when trend is present in our data and we wish to weight either the early or more current observations. Holt's technique is an extension of simple exponential smoothing and the double moving average exponential smooth discussed previously. Holt's method estimates the trend component using continuously changing regression equations with different values of the regression constants, B_0, and regression coefficients, B_1, thus causing the forecasting equation to change from period to period. This method bases its forecasts on the following equations:

1. $B_{0T} = (\alpha Y_T + (1 - \alpha)) \times (B_{0T-1} + B_{1T-1})$ (Equation 22.2)
2. $B_{1T} = \beta \times (B_{0T} - B_{0T-1}) + (1 - \beta) \times B_{1T-1}$ (Equation 22.3)

We can see that Holt's technique extends the idea of weighting constants to the base and trend equations with the alpha (α) weights applied to the base and the beta (β) weights to the trend. These constants, selected by the forecaster and constrained by $0 < \alpha < 1$ and $0 < \beta < 1$ have an effect consistent with our earlier discussion. The statistical program, SX, allows the user or the program to select the weights. If you choose to let the program make the selection, its algorithm selects the weights that minimize some goodness of fit measure, e.g., MAD.

Now look at the two equations more closely. It is obvious that estimates of the linear constant, B_{0T} and B_{1T} require values for B_{0T-1} an B_{1T-1}, values that we do not have. Therefore, we must estimate them by running a backcasted least squares linear analysis using all the observations in the series. To illustrate this process, we once again use the same short data set, shown below, from table 22.1. We have added the columns B_0 and B_1 to show how this procedure makes the forecast(s).

Table 22.7

Time	Sales	B_0	B_1
1	52	49.55	0.1853
2	51		
3	53		
4	48		
5	45		
6	48		
7	50		
8	52		
9	54		
10	52		
11	53		
12	51		

Table 22.8 shows the result of that regression.

Table 22. 8 Backcasted Linear Regression

Predictor Variables	Coefficient	Std Error	T	P
Constant	49.5455	1.62064	30.57	0.0000
Time	0.18531	0.22020	0.84	0.4197

R Squared	0.0661
Resid. Mean Square (MSE)	6.93392
Adjusted R Squared	-0.0272
Standard Deviation	2.63323

F	P
.71	0.4197

The values for b_0 and b_1 are respectively 49.5488 and 0.18531. We placed those values in period one of our data set. Now using equations 22.2 and 22.3 we estimate the linear values for period 2 b_{0t2} and b_{1t2}.

$$B_{1T} = \$ \times (B_{0T} - B_{0T-1}) + (1 - \$) \times B_{1T-1}$$

We arbitrarily chose alpha and beta weights of 0.40.

$$B_{0T1} = \text{``}Y_T + ((1 - \text{``}) \times (B_{0T-1} + B_{1T-1}))$$

$$B_{0T2} = 0.40 \times 51 + (1 - 0.40) \times (49.5488 + 0.1853) = 50.24118$$

The equation says calculate the constant for the current period by multiplying the alpha weight, 0.40, times the current actual value, $Y_T = 51$, and, to that product, add 1 - the alpha weight (1-0.4), 0.60 multiplied by the sum of the linear constant T- 1 and the linear slope T - 1.

$$B_{1T} = \$ \times (B_{0T} - B_{0T-1}) + (1 - \$) \times B_{1T-1}$$

$$B_{1T2} = 0.40 \times (50.24118 - 49.5488) + ((1 - 0.40) \times 0.1853) = 0.350172$$

This equation says calculate the slope for the current period by multiplying the beta constant by the difference between the current constant and the constant T - 1 and adding that product to the result of multiplying 1 - the beta constant by the slope, T -1.

These two calculations give us the linear equation for period 2. and shows how the linear equation changes as time increases (Table 22.9). We repeat this calculation for period 3 with the result shown in table 22.9 below.

Table 22.9 Moving Linear Equations

Time	Sales	B_0	B_1
1	52	49.55	0.1853
2	51	50.21	0.3502
3	53	51.54	0.7432
4	48		
5	45		
6	48		
7	50		
8	52		
9	54		
10	52		
11	53		
12	51	52.81	0.4485
Forecast Period 13	53.2625		

For brevity, we omit the remainder of the calculations thru period 11 showing only the final equation for period 12.

We use this equation ($52.83 + 0.4485 \times T$) to make our forecasts. Thus for period 13, we substitute 1 for time and the forecast is $52.81 + (0.48 \times 1) = 53.2625$. For period 14, we substitute 2 for time the forecast is 53.7110 (the forecast for period 13 plus the slope, 0.4485.

Using a statistical program (SX), we made the same calculation, the result of which we summarize as follows:

Lead	Forecast
1	53.26

Using the same program, we forecasted the sales data from chapter 21, and show the result below.

Lead	Forecast
1	597740

The forecast for 1 period ahead is about $598,000, much lower as compared to about $633,000 using the decomposition method.

Seasonal Exponential Smooth[94]

To account for the seasonal variation, we use a technique that develops a seasonal index for the series (see chapter 21), divides the original data by the seasonal index, and makes, in the absence of trend, a forecast using a single exponential smooth. For trending and seasonal data, we deseasonalize the data, and use these deseasonalized values to make a double exponential smooth forecast. As an example, we again use the data from table 22.1.

---

[94]   We note that NCSS has a program that handles the seasonal component in a more sophisticated (and better) way than we discuss here by creating equations and a third set of weights for the seasonal component.

Table 22.10

Quarter	Sales	Seasonal Indices	Deseasonalized Sales
1	52	1.00	51.80
2	51	0.99	51.47
3	53	1.02	52.09
4	48	0.99	48.62
5	45	1.00	44.83
6	48	0.99	48.44
7	50	1.02	49.14
8	52	0.99	52.68
9	54	1.00	53.80
10	52	0.99	52.47
11	53	1.02	52.09
12	51	0.99	51.66
13	53		

Assuming quarterly seasonality, table 22.10 shows the quarterly seasonal indices, and the deseasonalized sales. Now, assuming a trend, we run Holt's exponential smooth on the deseasonalized sales with the following result.

The trend forecasting equation is $F_T = 52.4636 + 0.63575 \times$ T = 52.4636 + 0.63575 × 1 = 53.0994.

Lead	Forecast
1	53.0994

Now, using SX, we compare our decomposition forecast from chapter 21, about $633,000 to about $631,000 for our deseasonalized double exponential smooth. The result is very close and consistent with the other methods discussed in this chapter.

Summary

The critical concepts discussed in this chapter are the following:

1. Forecasting using the moving average

   The moving average provides a quick, easy, and accurate forecast of data. Forecasting using moving averages relies on the assumption that all periods beyond T is just its moving average M_T.

2. The single (simple) moving average

   This method only forecasts one period ahead, updating the moving average as the actual observation for that period becomes available. The method, single moving average, takes a moving average for two or more periods and uses that average as a forecast for the next period. We use this method for data that contain neither trend nor seasonal components.

3. The double moving average

   With a trending series, we use a double moving average. This technique calculates a second moving average (M2) from the original (M1) using the same value for the moving average length. The principle behind this move is that this second MA (M2) provides the basis for calculating a *series of changing regression equations*. We use these regression equations to make our forecasts.

4. The single exponential moving average (smooth)

   Exponential moving averages are characterized by weighting the observations in the time series in which the choice of weights gives greater or lesser emphasis on current or earlier observations. Except for the weights, the exponential weighting methods parallel their unweighted cousins.

5. The double exponential moving average (smooth)

6. The deseasonalized double exponential moving average (smooth)

Appendix

Chapter 22

The data set given below is the same time series used in the appendix to chapter 21. The data are quarterly over a nine-year period (2001-2009). The first observation in the series is the first quarter of 2001. Your tasks are the following:

1. Using single, double, and exponential moving averages, forecast the series for the first two quarters of 2010.

2. Using the decompositional method from chapter 21, forecast the first two quarters of 2010.

3. Compare your moving average forecasts to the decomposition forecasts and discuss why these methods differ.

Sales	Time
39.94	1
39.86	2
38.65	3
42.23	4
39.13	5
39.94	6
36.86	7
38.16	8
39.06	9
39.64	10
43.62	11
44.58	12
44.78	13
43.43	14
44.84	15
45.38	16
44.43	17
42.99	18
45.10	19
42.73	20
46.80	21
49.24	22
47.02	23

46.89	24
48.76	25
48.30	26
45.93	27
47.44	28
48.01	29
48.01	30
52.84	31
50.84	32
51.64	33
50.20	34
50.82	35
52.95	36

CHAPTER 23

ARIMA MODELS

In this chapter, we introduce a method called ARIMA, Autoregressive Integrated Moving Average. While more complex than decomposition and smoothing, ARIMA's success in producing effective univariate forecasts, justifies its complexity. However, before we begin a discussion of ARIMA forecasting, we must explain four critical notions related to this procedure: Stationarity, random walks, differencing, lagging, and a new take on moving averages.

Basic Notions

Stationarity

Stationarity, a precondition for using ARIMA, means that the arithmetic mean and variance of time series must not change over time. A series' seasonal component must also adhere to the same criteria.

One method for detecting nonstationarity uses a scatter plot with time on the x-axis and the series on the y-axis. Other, more precise, techniques for detecting nonstationarity exist, but, for this illustration, we begin with the scatter plot.

Figure 23.1, a time series plot of sales on time illustrating nonstationarity.

Figure 23.1 Time Series Plot of Sales of StatSys

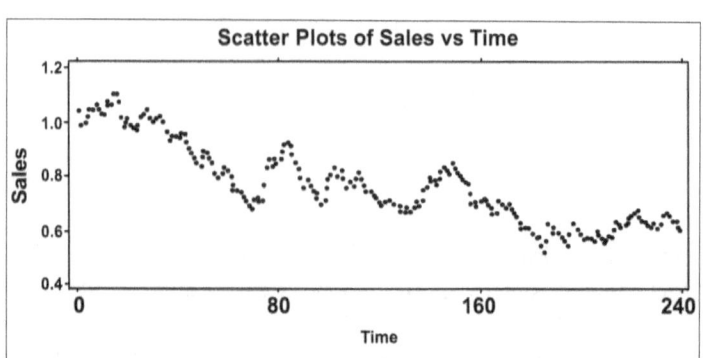

We can see from figure 23.1 the time series drifting downward, i.e., it is probably not stationary because its mean is changing over time.

The Random Walk

In our discussion of decomposition and smoothing forecasting the assumption was that if a series was moving up or down throughout the series' length it must be trending, requiring the fit of a linear or non linear equation. Because we are usually without adequate theoretical guidance, most of the time we don't definitively know that an upward or downward movement represents a trend. The movement may actually represent a *random walk*.

An example illustrates the concept of a random walk. If we roll an unbiased die 50 times, because of the independence of each role of the die from the previous roll, it is perfectly possible to roll 10 straight sixes followed by 10 straight ones. Unless we use a loaded die, an infinite number of rolls would yield one-sixth ones, one sixth twos, and so on. But, of course, we cannot roll the die an infinite number of times. The point is that the upward or downward movement of any observed time series (our sample slice out of time) might not reflect a trend, but random movement, i.e., a random walk.

Differencing

If we are not sure whether a series is trending or randomly walking, instead of fitting a trend, we can model the movement by differencing the series. To explain what we mean when we difference a series we use the data set from chapter 21. If the upward or downward movement is linear, first differencing (D1) will model (eliminate) the tendency. If the movement is curvilinear with one bend, a second difference (DD1) of the first difference will model the movement, if two bends, the third difference (DDD1) of the second difference, and so on. Using the data set from chapter 21, we plot the Sales of StatSys.

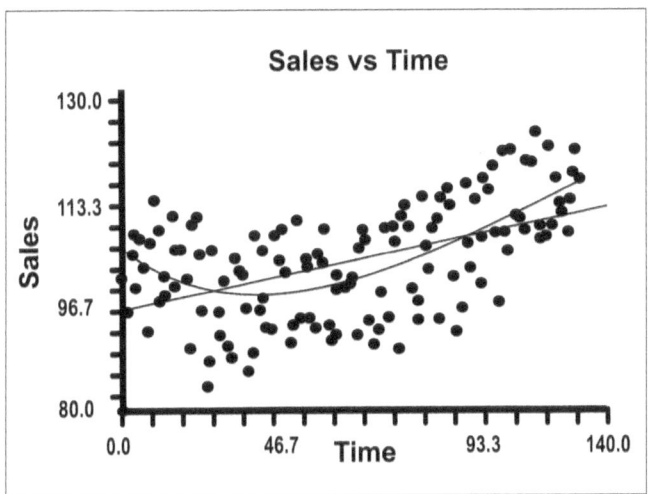

Based upon a visual inspection, we detect a movement down and then perhaps up. Consequently, we calculate the first and second differences. Table 23.1 excerpts the first nine months of the series differenced once and twice.

Table 23.1

Month	Sales	D1	DD1[97]
1	101.4		
2	96.2	-5.2	
3	105.2	9	14.2
4	108.5	3.3	-5.7
5	99.9	-8.6	-11.9
6	107.9	8	16.6
7	103.4	-4.5	-12.5
8	92.9	-10.5	-6
9	107.1	14.2	24.7

---

[95] DD1 refers to differencing the first difference.

To difference the variable once (D1) we subtract the first observation from the second, the second observation from the third, et cetera throughout the series. In table 23.1 we calculated the first difference, D1 by subtracting month one from month two, 96.2 - 101.4 equals -5.2, and placing that difference beside the second month. Then we subtracted month two from month three, 105.2 - 96.2 equals 9 and placed that difference beside the third month, and so on through the remainder of the series. We calculated a second difference, DD1 by taking the first difference of the first difference, 9 - (-5.2) equals 14.2, 3.3 - 9 equals -5.7, and so forth through the rest of the series. Figure 23.2 plots the differenced values.

Figure 23.2 Plots of First Differences (D1) of Sales on Time

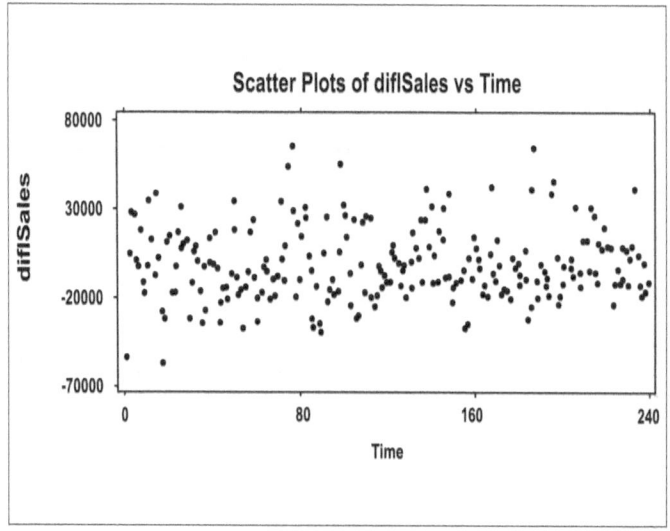

Differencing the sales series one time has probably eliminated the down movement, achieving stationarity. If taking one difference does not effectively eliminate the tendency, we difference again as we did in table 23.1.

Note: Differencing represents a transformation. Forecasting with a differenced series requires integration (transformation back to the original metric). Integration reverses the transformation back to the original metric, the undifferenced series. If we forecast a first difference, integration adds that difference to the previous undifferenced actual value.

Lagging and Autocorrelation

We introduced the idea of autocorrelation in chapter 13 when discussing the assumptions of the OLS model. We want to reintroduce the concept as it relates to time series where it is most applicable. To expand our understanding of autocorrelation, we need to grasp the concept of a lagged variable. Table 23.2 excerpts the first seven months of the sales data from chapter 21, lagged three times.

Table 23.2 Lagged Time Series

Sales	Lag1	Lag2	Lag3
101.4			
96.2	101.4		
105.2	96.2	101.4	
108.5	105.2	96.2	101.4
99.9	108.5	105.2	96.2
107.9	99.9	108.5	105.2
103.4	107.9	99.9	108.5

A first lag, Lag1 matches the first month against the second month, the second month against the third month, the third month against the fourth month and so on through the remainder of the series. Lag2 aligns the first month against the third month, the second month against the fourth, the third month against the fifth; Lag3 matches the first month against the fourth month, the second against the fifth, and the third against the sixth. Larger lags, e.g., Lag4, Lag5 . . . LagN follow the same logic.

Running a correlation between Sales and Lag1 produces the autocorrelation coefficient for Lag1. That correlation, if significant shows that period one correlates with period two, two with three, etc. Running the same analysis for Lag2, Lag3 . . . LagN produces correlation coefficients for these lags as well. Table 23.3 shows the autocorrelation coefficients for lags 1 through 10 for the sales series.

Table 23.3 Autocorrelations Lag 1 to Lag 10

Lag	Correlation
1	-0.29
2	-0.28
3	0.04
4	-0.06
5	-0.09
6	0.43
7	-0.24
8	0.10
9	-0.02
10	-0.37

We tested each of the correlation coefficients for statistical significance. The test of significance for autocorrelations is $T(Z) = \dfrac{R}{S \tan rdErrorofR}$ $= \dfrac{R}{\sqrt{\dfrac{1}{N}}}$. In this example, if the Lagged autocorrelation coefficient exceeds +/- 0.17677, we reject the null hypothesis of no autocorrelation at that lag.

We can use a plot (run on SX) of the lagged autocorrelation coefficients as another test for nonstationarity. Table 23.4 shows this plot.

Table 23.4 Autocorrelation Plot for Sales

```
        −1.0 −0.8 −0.6 −0.4 −0.2 0.0 0.2 0.4 0.6 0.8  1.0

Lag   Corr  +----+----+----+----+----+----+----+----+----+----+
 1   0.980              |  *** |********************
 2   0.958              |  *****|******************
 3   0.937              |  *******|***************
 4   0.915              |  ********|**************
```

```
 5  0.893                   |      ******** |************
 6  0.873                   |      ********* |***********
 7  0.854                   |      ***********|**********
 8  0.832                   |      ***********|**********
 9  0.808                   |      ***********|*******
10  0.783                   |      ************|*******
11  0.760                   |      *************|******
12  0.736                   |      *************|*****
13  0.713                   |     **************|****
14  0.694                   |     **************|***
15  0.676                   |     **************|***
16  0.661                   |    ***************|**
17  0.646                   |    ***************|*
18  0.634                   |    ***************|*
19  0.622                   |   ***************|
20  0.610                   |   ***************|
```

The vertical lines in table 23.4 show statistical significance, i.e., a coefficient falling outside the vertical lines means statistical significance. This pattern of autocorrelation coefficients is typical of a nonstationary time series. Table 23.5 shows the effect of differencing the series one time.

Table 23.5 Autocorrelation Plot for D1Sales

```
        -1.0 -0.8 -0.6 -0.4 -0.2 0.0 0.2 0.4 0.6 0.8 1.0

Lag   Corr +----+----+----+----+----+----+----+----+----+----+
  1  0.328                 |*** |*****
  2 -0.080                 |*** |
  3 -0.064                 |*** |
  4 -0.025                 | ** |
  5 -0.087                 |*** |
  6 -0.059                 | ** |
  7  0.060                 |*** |
  8  0.095                 |*** |
  9  0.073                 |*** |
 10 -0.045                 | ** |
 11 -0.078                 |*** |
 12 -0.145                 |****|
 13 -0.146                 |****|
```

14 -0.150	|****	|
15 -0.137	|****	|
16 0.044	| **	|
17 0.087	| ***	|
18 0.009	| *	|
19 -0.006	| *	|
20 0.018	| *	|

This pattern shows that the differencing has eliminated the overall or regular down movement of the series making the base series stationary. Because the series is monthly, there are still indications of seasonal non-stationarity. We determined that the series may be seasonally autocorrelated by looking at the pattern of autocorrelations January through December of the first year. The coefficients, falling, rising, and falling again, suggest the need for seasonal differencing, i.e., subtracting January of the first year from January of the second year, February of the first year from February of the second year, and so on. In our discussion of forecasting with ARIMA, we will show how ARIMA statistical packages do the seasonal differencing on command.

Note: We can model all the dynamic forces in any time series: trend, cyclical, seasonal by using differencing and a sufficient number of lags. Perhaps we can model a series with a small number of differences and lags, e.g., one or two differences and/or lags, or it might take many. The point: trend, cyclical, and seasonal variations cause autocorrelation. By using differencing and lagging, we can successfully model and partial trend, cyclical, and seasonal variation, making the residuals not only stationary but uncorrelated, homoscedastic, and approximately normal.

Moving Average: A Different Take

In chapter 21, we introduced the notion of simple and weighted moving averages as a univariate forecasting tool. In this section, we expand that concept in a way that, on the surface looks different, but is mathematically related.

Ordinary least squares regression, with a single lagged variable looks like the following: $Y_{TI} = B_0 + B_1 Y_{T-1} + \varepsilon_1$. For a reason explained later, we now change the symbol for the residual from ε to a so the equation becomes $Y_{TI} = B_0 + B_1 Y_{T-1} + a_i$. Now we run an OLS regression like the one specified above. We show the hypothetical data in table 23.6.

Table 23.6 New Take on Moving Averages

Y_T	Y_{T-1}	Y_{T-2}	Y_{T-3}	Y_{T-4}	a_{T-4}
988000					
999000	988000				
1013000	999000	988000			
995000	1013000	999000	988000		
977000	995000	1013000	999000	988000	-15028.51
974000	977000	995000	1013000	999000	-12424.74
990000	974000	977000	995000	1013000	1980.01
1020000	990000	974000	977000	995000	12213.66
1027000	1020000	990000	974000	977000	-6985.16
1037000	1027000	1020000	990000	974000	12259.03
1048000	1037000	1027000	1020000	990000	11428.71
1016000	1048000	1037000	1027000	1020000	-21054.93
1004000	1016000	1048000	1037000	1027000	11561.77
1010000	1004000	1016000	1048000	1037000	6050.145

The columns to the right of Y_T in table 23.6 are lagged variables of Y_T. If we ran a multiple regression of the form $Y_{TI} = B_0 + B_1 Y_{T-1} + B_2 Y_{T-2} + B_3 Y_{T-3} + B_4 Y_{T-4} + a_{T-4}$ we could specify the residual, a_{T-4} and, as we did, add it to table 23.6 as a new variable and call it a Moving Average. Obviously, the new independent variable, a_{T-4} is correlated with Y_T because it represents that part of Y_T not explained by the lagged AutoRegressive variables. So if we dropped two of the lagged variables, say Y_{T-4} and Y_{T-3}, we could run a new regression with a_i substituted as an independent variable. We define our new equation as $Y_{TI} = B_0 + B_1 Y_{T-1} + B2YT_{-2} + B_3 a_{T-4} + \varepsilon i$.

The equation with just the four lagged variables yields a MR^2 equal to 0.7275. The new equation with only two lagged AR variables and the moving average term yields a MR^2 equal to 0.9880, much higher than the equation with only the autoregressive terms. This is true because the moving average term, a residual, explains that part of Y_T not explained by the other variables in the model.

What is the point of this elaborate explanation? As we shall see, the notion of residuals as a moving average is a key element of the ARIMA approach.

ARIMA: A Heuristic Explanation

ARIMA models assume that a_t, the moving average, is the driving force of a time series Y_T. We can model a machine producing ice cream as an input-output process to illustrate the idea.

Ice Cream (raw material) $a_T \to Y_T$ (Ice Cream)[96]

The most important determinant of Y_T, the output of ice cream for the machine's production run is a_t, its input, how much ice cream raw material is used: the more material, the more ice cream output. To a lesser extent, a_{T-1}, the previous input, may also determine Y_T because a part of each input may remain inside the machine, delivered in the next output. Similarly, part of the previous run's output, Y_{T-1}, may remain in the machine, and determine Y_T. In a lesser way, further removed inputs and outputs, e.g., a_{T-N} and Y_{T-N}, may also determine Y_T. So output Y_T is a function of a_T, a_{T-1}, a_{T-N}, Y_{T-1}, Y_{T-N}.

From a regression perspective, the inputs, a_{T-1}, a_{T-2}, simply represent the residuals of the regression, $Y_T = f(Y_{T-1}, Y_{T-2})$, etc. This may not be intuitively obvious, so a brief explanation might help. If today's inputs (a_T) and previous inputs (a_{T-N}) and previous outputs, (Y_{T-N}) determine today's Y_T (output), partialling the previous outputs from the current output, $Y_T = Y_{T-1} + Y_{T-2} + Y_{T-N}$, leaves residuals as the previous inputs $(a_{T-1}, a_{T-2}, a_{T-n})$. The residual inputs, a_T define the Moving Averages in ARIMA. The Y_{T-N}, the previous outputs of Y, defines the AutoRegressive terms in ARIMA. The I term, Integrated, refers to the number of times a series must differenced to make a series stationary.

Therefore, the term ARIMA stands for AutoRegressive, Integrated, Moving Average. The technique also uses the symbolism PDQ in which

---

96.   McH, pp. 17-18.

P stands for the number of autoregressive lags used in the model, D stands for the number of differences necessary to make a series (base and seasonal) stationary and Q represents the number of moving average lags used in the model.

SARIMA is the model for time series with a seasonal component. For a seasonal model we use the lower case, pdq. A *base* or *regular* ARIMA PDQ model, for example 1, 1, 0, uses one autoregressive lag and one difference to achieve normal, uncorrelated, homoscedastic residuals (ARIMA calls these residuals *white noise*). A SARIMA model (PDQ = 1, 1, 0) and (pdq = 0, 1, 1) uses one AR lag and one difference in the regular (*base*) ARIMA and one difference and one moving average lag in the SARIMA model. It is obvious as to the origin of the AR terms of ARIMA. They are simply the lagged output of the series, Y_{T-N}. The meaning of the MA terms is not so obvious. As discussed in the previous section, we generate the MA terms by running a regressions of $Y_T = B_0 + B_1 Y_{T-1} + B_2 Y_{T-2} + B_N Y_{T-n} + a_{T-N}$. The MA terms, therefore, are the a_{T-N} residuals of an autocorrelated regression.

What is the point of generating the moving average variables, when we can obtain *white noise residuals* by using a large (sufficient) number of autoregressive lags to model a time series? The key word in answering this question is *sufficient*. With many time series it may take 20 or more autoregressive lags to obtain white noise residuals that, in effect loses 20 or more data points at the beginning of the series. By generating the residual moving average, it is possible, using a nonlinear routine like MLE, to backcast (backward forecast) the moving average residuals and use them either alone or in concert with fewer AR terms. In other words, if the series has constant variance, we can turn it upside down and estimate the residual MA terms backward. Therefore, the use of MA terms in the ARIMA model prevents the loss of data points in the analysis, making the model more efficient and parsimonious. The ARIMA software in the statistical packages we recommend do all the laborious calculations for you.

Univariate Forecasting with ARIMA

For our example, we use a data set of monthly sales (80 months) for a hypothetical firm. Figure 23.3 plots these data as a time series.

Figure 23.3 Time Series Plot of Sales

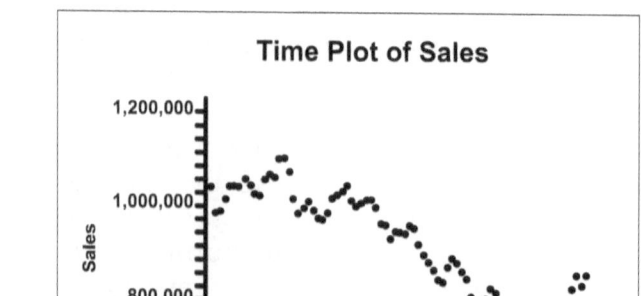

Step 1 in univariate ARIMA forecasting checks the series for stationarity. Remember that a linear or nonlinear trend or random walk can cause nonstationarity. While the time series plot in figure 23.3 points to a down and up movement, to buttress our visual impression of nonstationarity, we look at the autocorrelation plot in table 23.7.

Table 23.7 Autocorrelation Plot of Sales

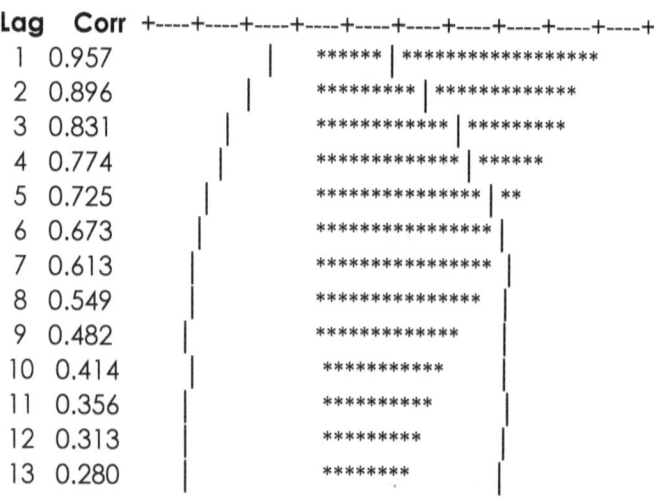

−1.0 −0.8 −0.6 −0.4 −0.2 0.0 0.2 0.4 0.6 0.8 1.0

Lag	Corr
1	0.957
2	0.896
3	0.831
4	0.774
5	0.725
6	0.673
7	0.613
8	0.549
9	0.482
10	0.414
11	0.356
12	0.313
13	0.280

We see again the pattern of slowly diminishing autocorrelation coefficients, characteristic of a nonstationary series. Clearly, our time series and autocorrelation plots indicate a nonstationary movement. To achieve stationarity, we difference the series one time and recheck. Here is a time series plot (figure 23.4) of the differenced series, D1Sales.

Figure 23.4 Time Series Plot of D1Sales

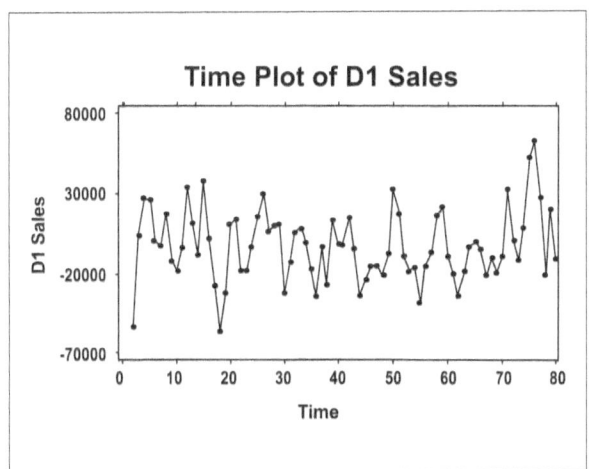

Figure 23.4 looks as if the first differencing of sales has made the base series stationary, but we remain unsure about nonconstant variance (heteroscedasticity)[97] and seasonal stationarity. Table 23.8 plots the autocorrelations of D1Sales.

Figure 23.5 Autocorrelation *Plot of D1Sales*

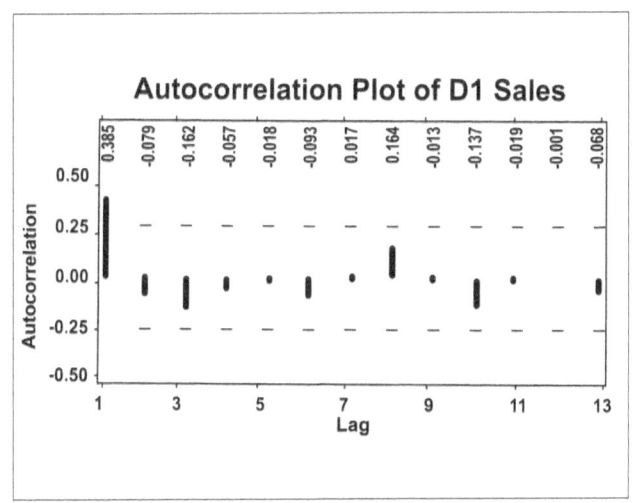

[97.]   if that variance is heteroscedastic, we try a natural log transformation.

While figures 23.4 and 23.5 show base D1Sales as stationary, we remain unsure about seasonal stationarity. To check on that possibility, we can difference the series at D12, i.e., from January of year one to January of year two, from February year one to February year two, etc, a differencing called seasonal first difference, D12. Quarterly data would require differencing at D4. We will return to the issue of seasonal stationarity in a moment. Just looking at figure 23.5, it does not look like the variable is seasonally nonstationary.

Step 2 in our ARIMA calculation determines the number of moving average and/or autocorrelation lags necessary to achieve white noise. This is a pivotal notion, because it addresses the issue of our starting points in the ARIMA analysis. To determine our starting point we use the autocorrelation and partial autocorrelation plots of D1Sales. Figure 23.6 plots the partial autocorrelations of the D1Sales series.

Figure 23.6 Partial Autocorrelation Plot for D1Sales

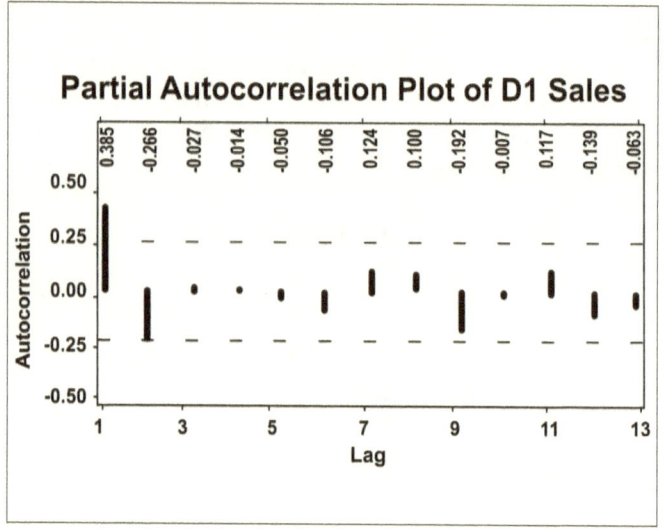

We use figures 23.5 and 23.6, the autocorrelation and partial autocorrelation plots to determine the need for one or more moving average lags and/or autocorrelation lags. The autocorrelation plot focuses on moving averages while the partial autocorrelation plot focuses on autocorrelation lags. We can see one significant autocorrelation at lag 1 and two significant partial autocorrelations at lags one and two.

If the autocorrelation plot(s) spike at early lags and quickly decline into insignificance and the partial autocorrelations all plot as statistically insignificant, the model is probably a base MA model. Depending upon the nature of the season, i.e., quarterly, monthly, biannually, if the autocorrelation plot(s) spikes significantly at 4, 8, or 12 (quarterly) or 12, 24, 36 (monthly) lags and the partial autocorrelations plot as insignificant, the model is probably a seasonal (pdq) moving average.

If the partial autocorrelation plot(s) spikes early, and the autocorrelations all plot as insignificant, the model is probably a regular AR model. If the partial autocorrelation plot(s) spike at seasonal lags and the autocorrelations plot at seasonal lags as statistically insignificant, the model is probably a seasonal auto regressive model.

If the autocorrelation plots and partial autocorrelation plots both spike at early lags, the model is probably a mixed regular AR and MA model. If the autocorrelation plot and partial autocorrelation plot both spike at seasonal lags, the model is probably a mixed seasonal $(pdq)_s$ AR and MA model. If the seasonal autocorrelation plot suggests seasonal nonstationarity, the model may have to use seasonal differencing as well.

Looking again at figure 23.5, the plot of autocorrelations, we see a statistically significant spike at Lag1, and no seasonal spikes at lag 12. Figure 23.6, the plot of partial autocorrelations, shows two significant spikes at Lags1 and 2 and no seasonal spike at lag12. This series is probably a mixed model with one base moving average lag and two base autoregressive lags with no seasonal component. Therefore, we can start the model with one regular moving average lag (MA1) and two auto regressive lags (AR^2), making our tentative ARIMA model $PDQ = 2, 1, 1; pdq = 0, 0, 0$.

Step 3 requires using special ARIMA software to test our model. These special ARIMA programs use curvilinear maximum likelihood estimation procedures instead of OLS and routinely calculate the lagged autocorrelation and moving average variables. In our example, we used the SARIMA routine in SX. The SX program offers a screen on which we enter our diagnosis. The screen asks for the variable, whether we want to difference it regularly and seasonally and for the specification of PQ and pq values. No matter which program used, you must give this minimal information. Table 23.8 is the output produced by the program SX.

Table 23.8 ARIMA Tests of Statistical Significance and Evaluation of Model Residuals

Nonseasonal Differencing of Order 1

Term	Coefficient	Standard Error	T ratios	P
AR1	0.555	0.325	1.71	0.087
AR2	-0.318	0.165	-1.93	0.054
MA1	-0.003	0.349	-0.01	0.993

DF	76
N Before Differencing	80
N After Differencing	79

Ljung-Box Portmanteau Lack-of-fit Diagnostics

LAG (DF) =	12(9)	24(21)	36 (33)	4 8 (4 5)
Chi-Sq (P) =	8.92(0.4451)	15.80(0.7808)	24.50(0.8573)	39.77(0.6924)

SX generated this output using a maximum likelihood regression of the form:

$$Y_{TI} = B_1 AR_{T-1} + B_2 AR_{T-2} + B_3 MA_{T-1} + \text{Residuals.}$$

In other words, our program created an equation that predicts Y_T while at the same time evaluating the model.

In table 23.8, we first observe the statistical significance of our coefficients, seeing immediately the insignificance of the AR1 and MA1 terms. This means that we must reject our model. For model acceptance, all the variables must be statistically significant. Thus, rejection of the model requires respecification. One obvious move drops the insignificant MA1 variable and reruns the model with AR^1, and AR^2 variables. Table 23.9 summarizes the test of the new model.

Table 23.9 ARIMA Summary of PDQ = 2, 1, 0; pdq = 0, 0, 0

Nonseasonal Differencing of Order 1

Term	Coefficient	Std Error	Coef/SE	P
AR 1	0.55729	0.10190	5.47	0.0000
AR 2	-0.31829	0.10239	-3.11	0.0019

DF	77
N Before Differencing	80
N After Differencing	79

Ljung-Box Portmanteau Lack-of-fit Diagnostics

LAG (DF) =	12(10)	24(22)	36(34)	48(46)
Chi-Sq (P) =	8.92(0.5399)	15.81(0.8251)	24.52(0.8841)	39.79(0.7287)

The coefficients in table 23.9 signal an acceptable model, both of which are statistically significant. While statistical significance of the coefficients is a necessary condition for model adequacy, it is not sufficient. Sufficient condition is that the residuals must be white noise: uncorrelated at all lags, normal and homoscedastic.

The Ljung-Box Portmanteau Lack-of-fit Chi-square (Q)[98] statistic tests the null hypothesis that the residuals are uncorrelated at all lags[99] Looking at table 23.9, we see that the test is insignificant at 12, 24, 36, and 48 lags. Therefore, we cannot reject the null hypothesis concluding;

---

[98]. McH, p. 99. We calculate the Q statistic as follows:

$$Q = (n(n+2)\sum_{}^{h} \frac{R_i^2}{n-i}$$ where n = the sample size, R_i is the autocorrelation

at lag i, and h is the number of lags. Degrees of freedom equal the number of lags minus the number of variables used in the estimate.

The equation says divide all of the squared autocorrelations for the number of lags tested by n - 1 (the sample size minus one), sum those quotients and multiply this sum by n × (n + 2) to yield Q, a X2 (a chi-square statistic).

[99]. In addition to Q statistic, the ARIMA coefficients must meet a final test of stationarity. Models with one AR term and two AR terms must be constrained to -1 <AR1 <1. Models with two AR terms (AR1 + AR2) must be < 1, AR2 - AR1 must be < 1, and the absolute value of |AR2| must be < 1. The stationarity bounds of MA terms, called invertibility, are the same as the AR constraints. In our example, the two AR terms meet the requirement. For a more extensive discussion of this point, see CD, pp. 239-240.

therefore, that there is no residual autocorrelation at all lags. We also tested (tests not shown) the residuals for normality and constant variance using the methods discussed in chapter 13. Our tests confirm normality, homoscedasticity, and stationarity.

White noise residuals entitle us to make a forecast using the equation produced in the analysis: $Y_{Cl} = B_1 AR_1 + B_2 AR_2 + \epsilon_i$. Table 23.10 shows a one month forecast.

Table 23.10 ARIMA One Month Forecast

Lead	95% CI Lower Bound	Forecast	Upper	95% CI Bound
1	800665	838743		876821

Summary

In chapters 21, 22, and 23 we looked at a series of univariate forecasting methods using decomposition, smoothing, and ARIMA. A wise forecaster will restrict forecasts to the short term because all these methods assume that the past is prologue and will also use more than one method to make forecasts. For example, using these same sales data with NCSS and SX, we compare methods. Table 23.11 compares our ARIMA forecasts with classical decomposition, and an array of moving averages.

Table 23.11 Comparison of ARIMA Forecasts Using NCSS and SX

	ARIMA	Single Moving Average	Classical Decomposition	Double Exponential Smoothing	Exponential Smoothing with Seasonal
NCSS					
One Period Forecast	6.1	6.04	4.98	6.11	6.07
Two Period Forecast	6.1	6.04	4.94	6.18	6.03
SX					
One Period Forecast	5.98	6.22	Not Available in SX	5.91	5.99
Two Period Forecast	5.40	6.18	Not Available in SX	5.89	5.88

NCSS and SX ARIMA statistical software include seasonal routines. The special software makes these tasks easy.

All the programs that we recommend, NCSS, SX, SPSS, and SYSTAT have good ARIMA routines

ARIMA uses backcasting to avoid losing observations. The essence of this idea is that the program literally turns the data set on its head and forecasts the predicted values backward. This works to forecast the lost residuals (MA values) backward. Homoscedasticity of the residuals makes this possible. It also works well because ARIMA uses the nonlinear MLE to backcast and forecast. ARIMA software also allows predictions as far out as is feasible

Finally, we temper any forecast using one or more of these methods by the judgment of those who know the territory.

Appendix

Chapter 23

Your tasks are the following:

1. Using the data set in the appendix to chapter 21, run an ARIMA model and forecast the four quarters of 2010.

2. Give a complete account of the method by which you arrived at the ARIMA model specification.

3. Compare your forecasts to those produced by decomposition and moving averages.

CHAPTER 24

CAUSAL INFERENCE WITH TIME SERIES

In previous chapters, we discussed causal inferences from cross sectional data. Here we discuss a powerful set of causal statistical techniques using time series. These methods defeat many, if not all, the confounders that plague other techniques used to infer causation.

Causal Inference using ARIMA: The Intruded Time Series

Arima Reviewed

Box and Jenkins (1976)[100] developed the ARIMA (autoregressive p, integrated d, moving average q) approach to time series analysis. The goal of this approach is to produce a stationary series (in terms of means and variances) from which we can forecast and make causal inferences. The *ar* (p) represents the autoregressive component of a time series, that part of previous observations carrying over to present observations/forecasts. The *i* (d) represents the degree of differencing needed (with or without ar/ma parameters). The *ma* (q) represents the moving average component, that part of previous errors/residuals that carry over to present observations/forecasts. Given a long enough series, ARIMA can incorporate seasonal (p, d, q) $_{models}$.

ARIMA and Causal Modeling

For causal inference the forces represented by ARIMA parameters are analytical confounders, sources of plausible alternative explanations for changes in the time series. Therefore, we must eliminate these confounders, "where the series is going anyway," before we can draw causal inferences. We must adequately estimate the ARIMA

[100.] Cited in McH, p. 13.

parameters, partial out their influence on the series and conduct the analysis on the resulting white noise residuals.

ARIMA and the Intruded Time Series

After achieving white noise residuals with ARIMA, we can use the intruded time series to assess the effects of an intrusion (X) on a white noise series. This quasi-experimental method treats political, social, economic, legal, business, advertising, and other purposeful, accidental, or natural intrusions as experimental variables. ARIMA methodology makes it possible to test the effects of many things that would be difficult or impossible to test in a classical experimental or quasi-experimental design. For example, the impact of a hurricane on social services, the effect of legislation on educational dropout rates, or increased advertising spending on sales. Figure 24.1 illustrates the classical intruded time series design.

Figure 24.1[101]

$$O_1 \ O_2 \ O_3 \ O_4 \ O_5 \ O_6 \ X \ O_7 \ O_8 \ O_9 \ O_{10} \ O_{11} \ O_{12}$$

The numbered O's (e.g., customer value, etc.) represent observational measurements taken before and after the intrusion, X (e.g., brand campaign). Except for the possible confounding effects of a simultaneous intrusion and, perhaps the decay of the measuring instrument, this design controls for all other threats to experimental internal validity i.e., the confounding effects of all plausible alternative explanations for the effect of X. Another virtue of this design extends to its atheoretical nature; it requires no complex model specification as, for example, with the path model discussed in chapter 20. All the caveats of measurement validity and reliability, of course, apply but the minimum amount of information needed is valid measurement of the Os, the intrusion's timing, its hypothesized effects, and the actuality of the intrusion. In addition, because the intrusion is a *natural*, not artificial event, its findings are highly generalizable.[102]

---

[101]. CS, p. 55.
[102]. It is possible to defeat all of the confounders by using a control group(s) that has strong similarities with the experimental group, e.g., cities.

Modeling Interventions

We can model interventions in at least three ways.

An Abrupt Constant Change[103]

We model an immediate (abrupt) constant change, assigning zero up to the moment of the intrusion and one to the moment of intervention and to all the later observations assuming the effect holds into the future. The intervention equation looks like this:

$$Y_T = \omega I_t + \text{ARIMA fit} \qquad \text{Equation 24.1}$$

Where I equals the intervention variable coded 0 before the intervention variable and 1 thereafter, ω equals the effect of the intervention, and the ARIMA fit equals the modeled time series.

We can rearrange 24.1 to:

$$Y_T - \text{ARIMA fit} = Y_T^* = \omega I_t \qquad \text{Equation 24.2}$$

Where Y_T^* equals the residuals of the ARIMA fit, the predicted values of Y_T minus the actual values of Y_T. Therefore, Y_T^* is the white noise time series and ω represents the effect of the intervention.

Abrupt constant change, the simplest intervention effect, assumes that the time series changes at the point of the intervention and remains in effect indefinitely. We scale the intrusion variable as 0 0 0 0 1 1 1 1 1. Figure 24.2 illustrates this intervention.

Figure 24.2 Abrupt Constant Change[104]

```
        1 1 1 1 1 1 1
        1
0 0 0 0 0 0 0 0 1
```

---

103. CC, p. 262.
104. CC, p. 262

A Gradual Constant Change[105]

We can also model the intervention as a gradual, constant change. The coding for this model assigns zeros before the intervention, and, if we know how long it will take to achieve the full effect of the intervention, e.g., three months, one third (0.33) to the first period of the intervention, two thirds (0.67) to the second, and one (1) to the third with ones thereafter. Or, if the analyst does not know how long it will take to achieve the full effect, zeros up to the intervention and ones thereafter. This model is often called a step function. The model's equation is generated by adding a second variable to the intrusion:

$$Y_T^* = \delta Y_{T-1} + \omega I_t \qquad\qquad \text{Equation 24.3}$$

The parameter δ, estimated from the data, controls how the series will change its level. In this context then, a significant δ measures the rate at which the intervention achieves its full effect. When δ is large, it takes a longer period for the series to reach its level. When δ is small, the change reaches its level almost immediately. Obviously, when δ is zero the intervention is not gradual, but abrupt and immediate as in equation 24.2. Figure 24.3 illustrates a gradual constant positive change

Figure 24.3 Gradual Constant Change

```
        1 1 1 1 1 1 1
        1
0 0 0 0 0 0 0 1
```

We scale the independent variable zero (0) before the intervention and one (1) thereafter, just as in the abrupt constant model. The difference between the two is in the intervention equation. In the abrupt, constant model, the equation contains only one variable, the intervention variable.[106] In this model, the equation contains Y_{t-1} plus the intervention variable.

---

[105]. CC, p. 262

[106]. Or we can scale the intrusion variable as fractions, depending on how long it takes to reach its maximum. For example, if the series is monthly and the analyst believes that it will take nine months to achieve its maximal level, the post intervention is scaled 1/9th, 2/9ths, 3/8ths . . . 1.

The time series change begins at point *I* with the final total change equal to:

$$\frac{\omega}{1-\delta}$$

Equation 24.4

From equation 24.4, we see the total value of the change. With a zero or small δ, the total effect of the change approximates ω. With a large δ, e.g., 0.67, the effect triples.

A Gradual Temporary Change[107]

The time series can be modeled as an abrupt temporary (pulse) change, assigning zeros to all observations before the intervention, one to the intervention, and zeros to all observations following the intervention. This model assumes the effect is abrupt, losing its power immediately [or quickly] after the intervention. The intervention equation takes the following form:

$$Y_T^* = \delta Y_{T-1} + \omega I_t$$

Equation 24.5

Where Y_T^* is the white noise time series, I_t is the intervention at point T, and ω is the effect of the intervention. Figure 24.4 shows a gradual temporary change.

Figure 24.4

```
    |
    ||
    |||
0 0 0 0 | | | | 0 0 0 0 0 0
```

We know that this is a temporary change if δ is small and the decline or increase pattern is quick. If δ is large the decline or increase pattern is slow.

[107.] CC, p. 264

Which model should we use? The answer to this question depends on our knowledge. If we have a good idea about what to expect from an intervention, we simply select the appropriate model. If, however, we do not know what to anticipate, we proceed as follows.

1. Test a general effect hypothesis without reference to its nature. To make this test, begin with an abrupt, constant model, i.e., 0 0 0 0 0 0 I I I I I I and simply test the effect of the intrusion as follows:

$$Y_T^* = \omega I_T$$

In this case, the intrusion equation contains only one independent variable, the intrusion variable. A statistically significant ω confirms an effect.

2. After we confirm an effect, to test for the form or nature of the effect, we again uses the scaling 0 0 0 0 1 1 1 1, but add Y_{T-1} as another independent variable. The effect of the intrusion then becomes as follows:

$$Y_T^* = \delta Y_{T-1} + \omega I_T$$

For this model, we interpret δ as the rate of the impact's persistence. If the effect is abrupt and constant, that is with a small δ, say in the range of ≤ 0.1, the effect erodes quickly and the abrupt, constant model, i.e., 0 0 0 0 1 1 1 1 probably holds. However, with a large δ, e.g., ≥ 0.90, the effect persists over a longer period, the gradual change, probably holds.[108] For a negative intrusion sign and a large δ, a gradual decline to zero reflects the effect. For a small δ and a positive ω, we have an abrupt intrusion, leveling off at a constant amount.

[108.] CD, 1979.

An Example

Let us suppose we have a time series representing the sales of our computer-consulting firm, StatSys, plotting its sales from January 1, 1998 to September 30, 2008 as figure 24.5.

Figure 24.5 Sales of StatSys

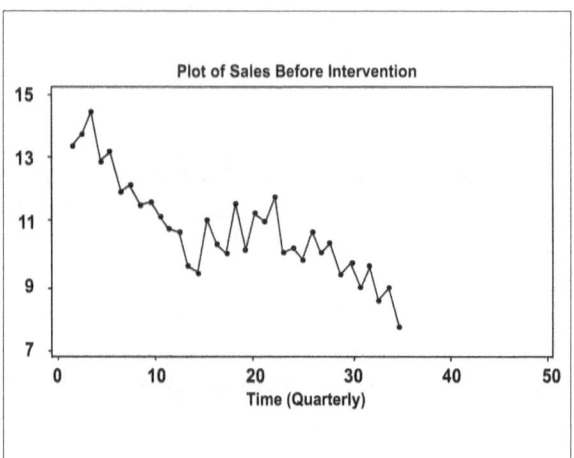

StatSys's management, desperate to turn the firm's fortunes around, hires a consulting firm to recommend a new direction. StatSys implements the consulting firm's recommendations immediately and subsequently, plots the sales again as figure 24.6.

Figure 24.6

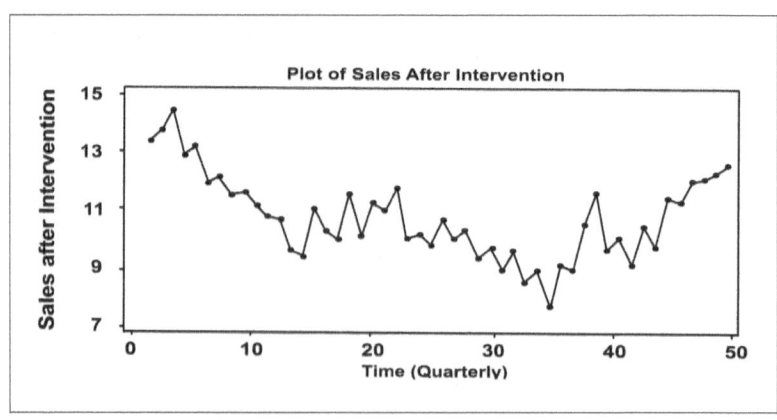

From figure 24.6 we see that the fortunes of StatSys improved after the intervention. The question, however, was the improvement caused by the consultant's change design or would the improvement occurred without the intervention? We can test this hypothesis by using an interrupted time series analysis.

Step 1: Check for Stationarity

From figure 24.6 we know the series must be differenced. Therefore, we difference once and plot the differenced sales time series as figure 24.7.

Figure 24.7 Plot of D1Sales after Intervention

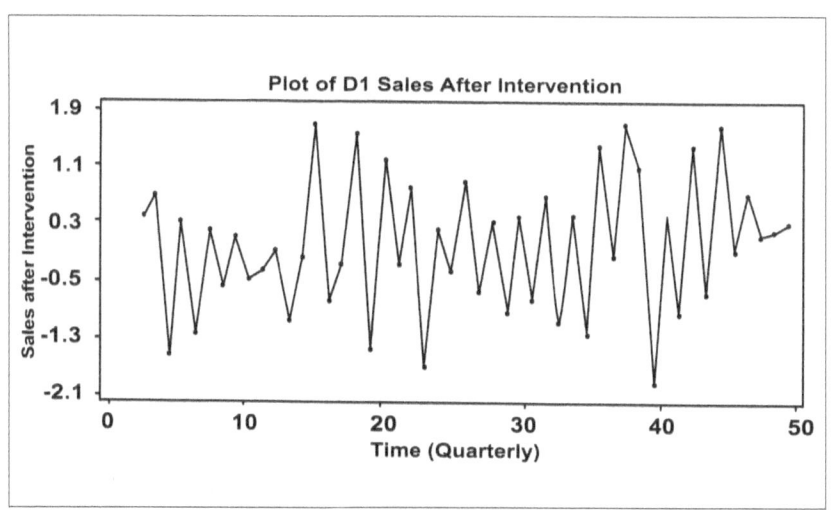

The series now looks stationary.

Step 2: Check the Autocorrelation and Partial Autocorrelation Plots of the Differenced Sales Series. Figure 24.8 plots the autocorrelations.

Figure 24.8 Autocorrelation Plot of D1Sales after Intervention

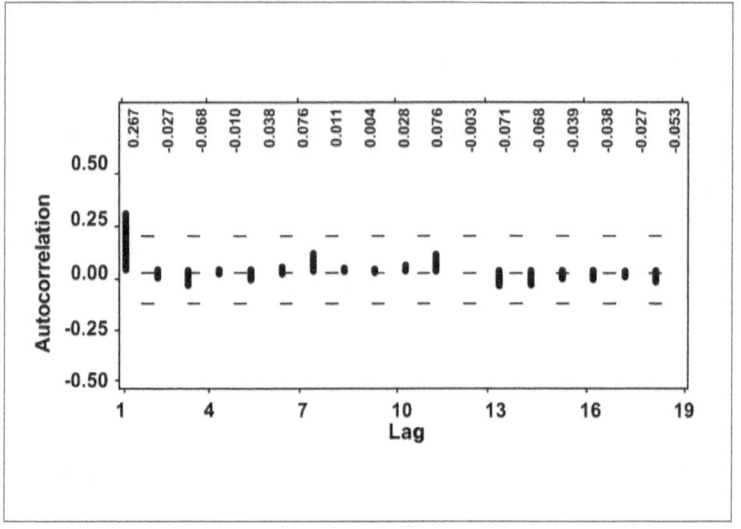

Figure 24.9 plots the partial autocorrelations.

Figure 24.9 Partial Autocorrelation Plot of D1 Sales after Intervention

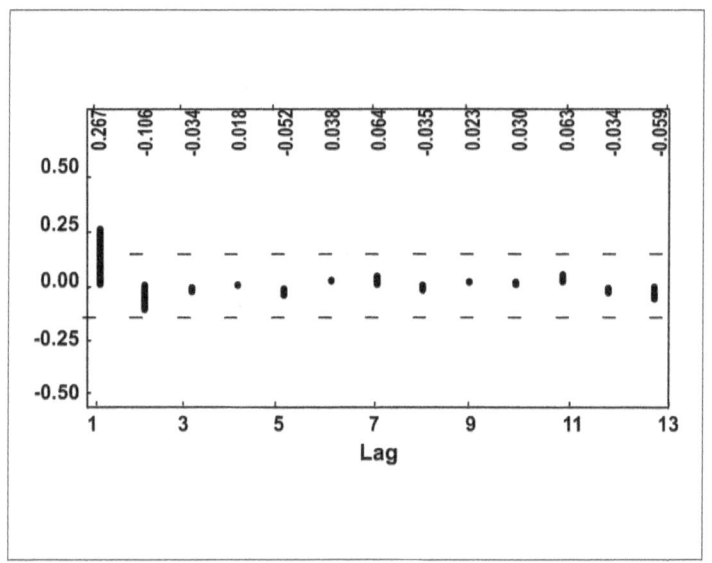

The two plots both spike, the first at regular MA1 and the second at a regular AR1. The monthly data look stationary; therefore they do not require seasonal differencing. So we define the tentative ARIMA model as a mixed PDQ (1,1,1), pdq (0,0,0).

Step 3: Use an ARIMA Program and Run This Model

In the tentative model (1,1,1) (0,0,0) the AR1 coefficient was significant; the MA1 coefficients was not, so we dropped MA1 and reran the model as (1,1,0) (0,0,0).

Table 24.2 Output of (1,1,0) (0,0,0) Model

Nonseasonal Differencing of Order 1
NOTE: No constant term in model

Term	Coefficient	Std Error	Coef/SE	P
AR 1	-0.50110	0.12595	-3.98	0.0001

Ljung-Box Portmanteau Lack-of-fit Diagnostics

LAG (DF) =	12(11)	24(23)	36(35)
Chi-Sq (P) =	5.29(0.9165)	18.56(0.7264)	30.97(0.6631)

This model produced a white noise residual, Y_T^*. The AR1 coefficient is statistically significant at $P < 0.0000$) and falls within the bounds of stationarity. The Portmanteau Lack-of-fit test shows no significance out to 48 lags and our residual check of confirms normality and homoscedasticity.

Step 4: Test the Intrusion Effect

We decided to test the effect as gradual and permanent. So we began with a gradual constant model where Y_T^* the white noise residual from the ARIMA = $Y_T^* = \omega I_T + Y_{T-1}$ with the intervention variable coded 0 0 0 0 1 1 1 1 or zero up to the intervention and 1 thereafter. Table 24.3 shows the result of that regression.

Table 24.3 Intervention as Gradual Constant

NOTE: Model Forced through origin

Predictor

Variables	Coefficient	Std Error	T	P	VIF
Grad Perm ω	0.59261	0.27864	2.13	0.0390	1.0
L1Sales δ	-0.01531	0.01304	-1.17	0.2466	1.0

R Squared	0.0891
F P	
2.20 0.1224	

The overall model was statistically insignificant; the intervention variable ω was significant, but the δ coefficient was not. Because δ, the test of gradual constant, was insignificant, we reran the intervention as an abrupt constant effect. Table 24.4 summarizes that regression.

Table 24.4 Intervention as Abrupt Constant

Variables	Coefficient	Std Error	T (Z)	P
Abrupt Perm ω	0.4649	0.23720	1.962	< 0.0500

R Squared	0.0612
F P	
3.84 <0.0500	

The positive ω coefficient was statistically significant suggesting a positive constant effect of about 0.465. Thus, our conclusion was that the advice of the consultants was sound; it arrested the decline in sales by the firm's redirection.

Two Variable Causal Analyses Using Cross Correlations

It is possible to enlarge our causal analysis to a situation in which we test the causal effect of one independent time series variable on a dependent time series variable. We call your attention to chapter 20 and our discussion of causal inference with cross-sectional data.

There we used an example of lung cancer and cigarette smoking. The method attempted to partial all the plausible alternative explanations for causes of lung cancer, e.g., heredity, diet, weight, etc., and tested the smoking variable for statistical significance. If this variable has the right sign and is statistically, we can tentatively conclude that cigarette smoking causes lung cancer.

With time series, we can, with two additional steps, do something similar without having to define, measure, and partial all possible confounders. The first step uses ARIMA and whitens the independent and dependent variables, eliminating the inherent time confounders. Correlating or regressing unwhitened time series may reflect only their internal, time driven dynamics.

The second step runs a cross correlation between the dependent and independent variables. For example, we may have a theory that hypothesizes a negative causal relationship between poverty and school achievement: as poverty declines, school achievement rises and as poverty increases, school achievement falls. Our theory also hypothesizes that the opposite is not the case. In table 24.5 and figure 24.10 we show the output of a cross correlation between whitened Poverty as an independent variable and whitened School Achievement as a dependent variable (nationwide annual data over a 40 year period).

Figure 24.10 Cross Correlation of Poverty and School Achievement

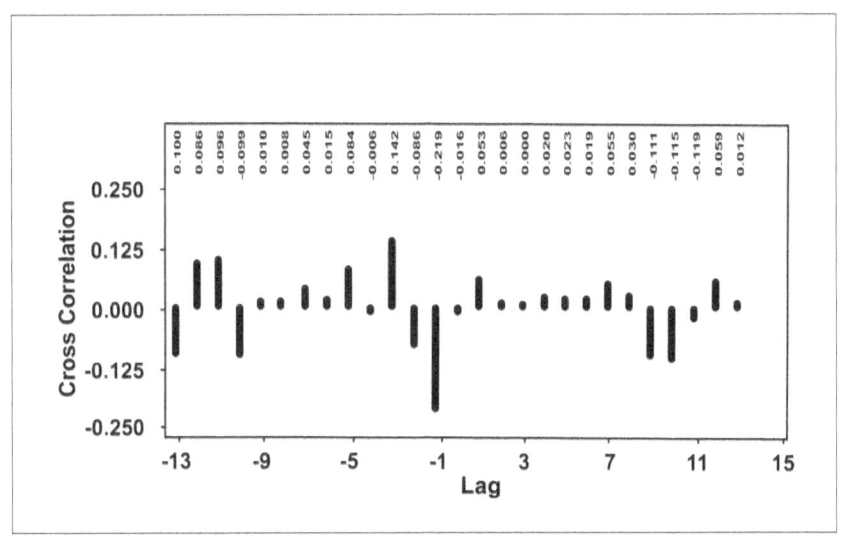

Table 24.5 Cross-Correlations of Poverty and School Achievement

Lag	Correlation	Lag	Correlation	Lag	Correlation	Lag	Correlation
-40	0.006075	-19	-0.185240	2	0.012355	23	-0.021837
-39	-0.171124	-18	0.069109	3	-0.000714	24	-0.000846
-38	-0.315951	-17	0.020724	4	0.018644	25	-0.006205
-37	0.107408	-16	-0.049407	5	0.025081	26	0.008784
-36	0.303876	-15	-0.051814	6	0.015453	27	-0.008009
-35	-0.110052	-14	0.038773	7	0.058021	28	-0.000743
-34	0.076312	-13	-0.098783	8	0.034674	29	-0.021354
-33	-0.032693	-12	0.089564	9	-0.116052	30	-0.000899
-32	-0.052269	-11	0.095460	10	-0.103237	31	0.010878
-31	-0.136975	-10	-0.098837	11	-0.048347	32	-0.027548
-30	0.057198	-9	0.009335	12	0.041777	33	0.013845
-29	0.169222	-8	0.009695	13	0.005545	34	-0.014927
-28	0.037969	-7	0.045039	14	0.040587	35	-0.011758
-27	-0.201768	-6	0.015233	15	0.051137	36	-0.022193
-26	0.048547	-5	0.084374	16	0.049556	37	0.009986
-25	0.046186	-4	-0.006125	17	-0.007586	38	0.013543
-24	0.014197	-3	0.142517	18	0.020446	39	0.020832
-23	-0.084184	-2	-0.083497	19	0.018487	40	0.014889
-22	0.015039	-1	-0.216108	20	-0.019242		
-21	0.152373	0	-0.014929	21	0.017029		
-20	-0.006654	1	0.054277	22	0.008444		

Figure 25.10, plots the lagged correlations of Poverty and School Achievement. The negative correlations represent lagged school achievement on poverty; the positive correlations show lagged poverty on school achievement. Verification of the hypothesis requires a significant, negative lag, that is, as poverty declines last year, achievement increases this year and the other way around.

From the graph and the cross-correlations of poverty and achievement, we see that poverty lags achievement, but, as our postulation suggests, school achievement does not lag poverty. Another interesting finding confirms the argument that educational achievement, while good for the individual, does not fix poverty. Certainly, this test, while not conclusive, represents a solid bit of causal evidence.

Structural and Reduced Form Equation
Path Models Using Time Series

In chapter 20, we explained the concept of causal analysis using cross-sectional data with structural and reduced form equations. This idea can be extended to time series by prewhitening and using lags of the independent variables. For example, in chapter 20 we estimated the causal effect of intelligence, time, training, and experience on wages, developing a path diagram showing how our hypotheses related to the causal effect of each independent variable. This path diagram also defined the structural equations necessary to run the model.

For a time series analysis, we would use the observations for variables across time, e.g., means for the dependent and independent variables. To run the analysis, we would whiten all the series variables, create appropriate lags for the independent variables e.g., experience, and run the model exactly as we ran the cross sectional data in chapter 20.

Econometric Models[109]

While we do not give an example, it is possible to forecast with more than one independent time series variable, To make these forecasts, do not whiten the variables because we are not interested in causal inference. Instead, we would run the unwhitened Y variable on unwhitened X variables, lagging the X variables as directed by our conceptual scheme. We can also lag the Y variable.

Summary

The critical concepts discussed in this chapter are the following:

1. ARIMA and the Intruded Time Series

    We use the intruded time series to assess the effects of an intrusion (X) on a white noise series. This quasi-experimental method treats political, social, economic, legal, business, advertising, and other purposeful, accidental, or natural intrusions as experimental variables. ARIMA methodology makes it possible to test the effects

---

[109.] PK, chapter 17.

of many things that would be difficult or impossible to test in a classical experimental or quasi-experimental design. For example, the impact of a hurricane on social services, the effect of legislation on educational drop-out rates, or increased advertising spending on sales. Figure 24.1 illustrates the classical intruded time series design.

Figure 24.1

$$O_1 \ O_2 \ O_3 \ O_4 \ O_5 \ O_6 \ X \ O_7 \ O_8 \ O_9 \ O_{10} \ O_{11} \ O_{12}$$

2. Abrupt Constant Change

Abrupt constant change, the simplest intervention effect, assumes that the time series changes at the intervention and remains in effect indefinitely. We scale the intrusion variable as 0 0 0 0 1 1 1 1 1. Figure 24.2 illustrates this intervention.

Figure 24.2 Abrupt Constant Change

```
1 1 1 1 1 1 1 1
        1
0 0 0 0 0 0 0 1
```

3. Abrupt or Gradual Temporary Change

The time series can be modeled as an abrupt temporary (pulse) change, assigning zeros to all observations before the intervention, one to the intervention, and zeros to all observations following the intervention. This model assumes the effect is abrupt, losing its power immediately [or quickly] after the intervention. Figure 24.3 shows this effect pattern

igure 24.3

```
    |
    ||
    |||
0 0 0 0 I I I I 0 0 0 0 0 0
```

4. Gradual Constant Change

We can model the intervention as a gradual, constant change. The coding for this model assigns zeros before the intervention, and, if we know how long it will take to achieve the full effect of the intervention, e.g., three months, one third (0.33) to the first period of the intervention, two thirds (0.67) to the second, and one (1) to the third with ones thereafter. Or if the analyst does not know how long it will take to achieve the full effect, zeros up to the intervention and ones thereafter. This model is often called a step function.

Figure 24.3 illustrates a gradual constant positive change.

Figure 24.3 Gradual Constant Change

```
          | | | | | | |
              |
     0 0 0 0 0 0 0 |
```

5. Cross Correlation

It is possible to enlarge our causal analysis to a situation in which we test the causal effect of one independent time series variable on a dependent time series variable. We begin this process by whitening the independent and dependent variables, eliminating the inherent time confounders. Then we run a cross correlation between the dependent and independent variables. Negative correlations represent lagged school achievement on poverty; positive correlations show lagged poverty on school achievement. Verification of the hypothesis requires a significant, negative lag, that is, as poverty declines last year, achievement increases this year and the other way around.

6. Structural and Reduced Form Equation Path Models Using Time Series

For causal inference using time series variables, we can use the values for variables across time, e.g., means for the dependent and independent variables. To run the analysis, we whiten all the series variables, create appropriate lags for the independent variables e.g., experience, and run the model exactly as the cross sectional data in chapter 20.

Appendix

Chapter 24

The data given below, a quarterly time series of a firm's output was recorded over 47 periods. At the 24th period, the firm's output was intruded by an event. Your task: Test the hypothesis that the intrusive event significantly affected the output of this enterprise.

Time	Series
1	2.37346629
2	2.471446308
3	2.338927621
4	2.397536699
5	2.362437096
6	2.384085402
7	2.367308576
8	2.438835255
9	2.535639561
10	2.365505355
11	2.443332676
12	2.717461524
13	2.785630819
14	2.68807759
15	2.753620079
16	2.705930736
17	2.75472521
18	2.818436009
19	2.757206626
20	2.896011798
21	2.740069053
22	2.761074974
23	2.891303123
24	2.855720937
25	2.84245185
26	2.899384527

27	2.942971084
28	2.928004291
29	2.97747298
30	2.93210317
31	2.860858108
32	2.933023454
33	2.991664944
34	2.981797242

PART VI

Multivariate Analysis

Introduction

In chapters 25 and 26 we introduce the expanded practice of multivariate data analysis. In previous discussions, we focused on univariate multiple regression, a technique using one dependent variable and two or more independent variables. In chapter 25, we begin a discussion of analyses with multiple dependent variables.

Chapter 25 develops the notion of factor analysis and principal component analysis. When measuring a set of variables, usually the variances of these variables overlap. We call this overlap shared common variance, or collinearity. Correlation coefficients measure the common variance between two variables. The part of variables that do not overlap represent unique variance specific to the variable including some random component, and perhaps some systematic variance. Sometimes, this shared variance is sparse; sometimes extensive. With extensive shared common variance, i.e., when the variables have correlation coefficients equal to or greater than 0.50, we can use factor analysis (FA) and principal component analysis (PCA) to extract from the data a smaller number of variables (factors or components) by synthesizing their overlapping variance.

Factor analysis, principally a theory testing technique, probes a set of variables for an underlying structure of unmeasurable constructs that lie within a set of variables. We give as an example the impossibility of directly measure teaching ability, an issue with which the Gates Foundation is currently grappling. Principal component analysis is a data reduction method that allows us to create useful linear combinations of shared variance as new variables for use in other analyses, e.g., multiple regressions.

Chapter 26 applies the concepts from chapter 25 to develop the notions of principal component correlation (PCA) and Jacob Cohen's powerful set regression and correlation (SRC).

CHAPTER 25

FACTOR AND PRINCIPAL COMPONENTS ANALYSIS

In previous discussions, we used the term univariate multiple regression, one dependent variable and two or more independent variables. Here, we begin a discussion of two techniques that form the basis for multivariate analyses with multiple dependent variables. This, of course, suggests that we cannot analyze multiple dependent variables by using a series of univariate multiple regressions treating each dependent variable separately. It should be apparent that such an approach tremendously increases the probability of a type I error.

The focus of this chapter is on Factor and Principal Components analyses (subsequently referred to as FA and PCA). These two methods use multiple variables without reference to their dependence or independence. While many people refer to these procedures synonymously (the two procedures do share many things), they are used for different purposes.

The Basic Idea Underlying FA and PCA

Factor and principal component analysis probe for an *underlying structure or structures* in a set of individual variables each of which purport to measure a fractional aspect (but not the whole) of some construct situated beneath the measured variables, e.g., the sociological notions of anomie or antinomy. This probing assumes that the variables in a data set have common or shared variance. If the variables have no shared variance, they are orthogonal, i.e., almost completely independent, making FA and PCA pointless.

To emphasize what we mean by orthogonal, figure 25.1 shows a picture of three completely orthogonal variables.

Figure 25.1 Orthogonal Variables

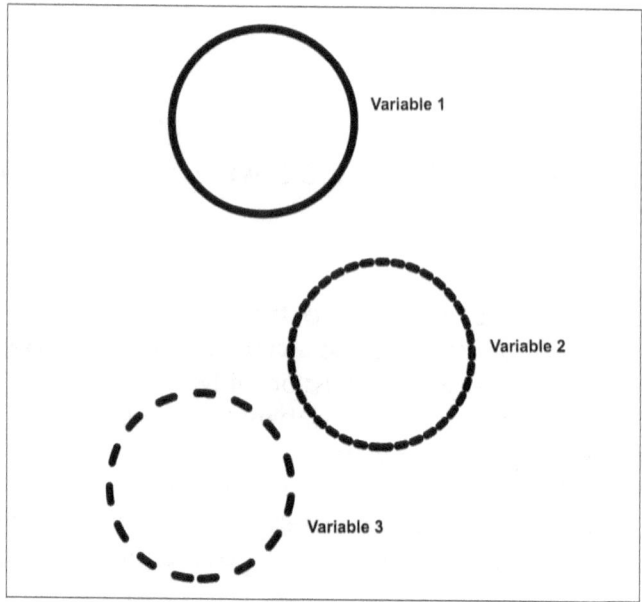

We drew the circles in figure 25.1 to represent three variables with variances having no shared common variance. To illustrate again what we mean by common variance, figure 25.2 depicts three variables with common variance that could conceivably be analyzed by FA or PCA.

Figure 25.2 Three Variables with Common Variance

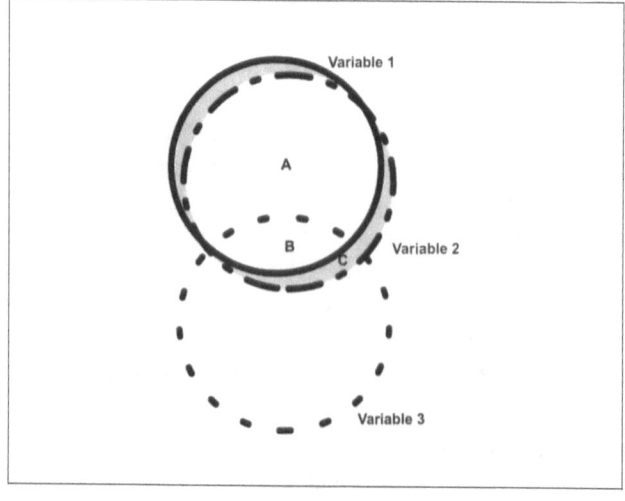

We drew the circles in figure 25.2 so that they would overlap. The overlapping represents the common variance among the three variables. Now, imagine that we can extract a new variable (factor or Component) that captures the greatest common variance among the three variables corresponding to areas A + B in figure 25.2. Then further imagine that we can extract a second factor or component that captures the next greatest common variance represented by area C. The areas of variables 1 and 2 that do not overlap represent their unique variance. The large blank space in variable 3 represents its unique variance.

The characteristic that defines and separates the FA from PCA lies in the variance identified and extracted. FA identifies only factors that represent *common variance*, ignoring the variance unique to each of the variables. PCA identifies and extracts components that contain all the variance in the data set: the common variance plus the unique variance.

Therefore, in figure 25.2, because FA identifies only the common variance, it would identify only two factors. PCA can extract as many components as there are variables in the data set or, as in this example, three.

FAs and PCAs usually begin as explorations based on hypotheses derived from theory, a confederacy of hunches, or simple speculation. Others are *confirmatory*, aimed at corroborating the findings of previous explorations. An example of an exploratory FA involves developing a measure of effective teaching at the university level. A soft theory (confederacy of hunches) might identify such variables as student evaluations, self evaluations, administrator evaluations, teaching experience, publications, citations of publications, each of which lays claim to capturing a fraction, but not all, of the underlying unmeasurable, effective teaching. In the parlance of these methods these variables are called manifest variables because they are perceived straightforwardly and measurable.

We note that FA, because it focuses only shared variance, cannot produce a mathematically true factor score because the factors it extracts do not include unique variance. However, some programs, e.g., NCSS and SPSS extract a pseudo factor score that, when used cautiously, can be useful in other analyses.

A Heuristic Example of Factor Analysis Using OLS

As a means of explanation and because it should now be familiar to our readers, we use OLS in our explanation of the workings of FA and PCA. While this analysis is not what your programs will use, for purposes of explanatory insight, OLS works.

Assume that you have been hired by our hypothetical firm StatSys, to search for an underlying structure of six manifest variables. These variables represent an attempt by management to evaluate employees. Table 25.1 shows the data set. The variables are given as Z scores.

Table 25.1 Five Variables as Surrogates for Employee Success

Intelligence	Creativity	Energy	Firm Loyalty	Longevity	Formal Training
0.2330	-0.8909	0.7476	-0.6820	0.1322	-2.1275
-0.0824	-0.1920	-0.7701	0.8848	-0.0413	0.8872
-0.8017	-1.2346	-1.0087	2.0624	2.1492	1.3766
0.2034	0.3465	0.5280	-0.9585	-0.4885	-0.2482
0.5680	1.4922	0.7285	0.8438	1.2912	1.8953
-0.2203	0.0142	-0.2260	-0.2417	-0.1873	0.2999
-0.9200	0.6100	-0.5983	-0.2212	0.4151	0.2705
0.7848	-0.2722	0.3085	0.8643	0.2325	-0.5810
0.2231	0.1059	-0.0733	1.1203	0.8623	0.4174
1.2085	0.7131	1.8644	-1.0609	-0.1143	0.1433
1.0903	-0.0775	1.0626	0.9155	1.0083	-0.2286
-0.6046	-0.4899	0.0317	-2.0440	-1.1456	-1.4130
-1.5999	-0.7993	-1.7914	-1.1428	-1.4468	-1.6283
-0.8313	-0.8107	-1.2474	0.5264	0.0409	-0.4342
0.1935	0.4496	-0.2356	0.9564	0.4516	0.9753
-1.1072	-0.9597	-1.2092	0.2499	0.2508	0.0944
2.2925	2.1109	1.4635	-0.8561	-2.3777	0.7306
-0.1415	-0.3868	-0.2260	-0.5898	-0.7166	-0.3265
1.3563	1.3547	1.2248	-1.1838	-1.5654	-0.0720
-0.1021	-0.5472	0.0031	0.5366	0.8623	0.2216

1.3465	1.8588	1.5112	0.8950	1.3278	1.9834
0.2132	0.9995	0.5567	-0.4567	0.1230	-0.2972
-1.0087	-1.1544	-0.6078	-1.2350	-0.6071	-0.1014
-0.3681	-0.5587	-0.3501	-0.0881	0.0318	-0.6202
-1.9251	-1.6815	-1.6864	-0.9380	-0.4885	-1.2172

The first step in an FA or PCA analysis constructs a correlation matrix. The correlation matrix expresses the relationships among the variables as Z scores[110] with a mean of zero and a variance of one. Table 25.2 shows that matrix.

Table 25.2 Correlation Matrix

	Intelligence	Creativity	Energy	Firm Loyalty	Longevity	Formal_ Training
Intelligence	1.000	0.826	0.929	0.111	-0.063	0.461
Creativity	0.826	1.000	0.811	-0.016	-0.100	0.536
Energy	0.929	0.811	1.000	-0.069	-0.029	0.382
Firm Loyalty	0.111	-0.016	-0.069	1.000	0.829	0.633
Longevity	-0.063	-0.100	-0.029	0.829	1.000	0.558
Formal Training	0.461	0.536	0.382	0.633	0.558	1.000

The matrix shows two identical triangles (lower left and upper right) depicting the correlations between the variables, for example the relationship between intelligence and creativity as measured by the coefficient of correlation equals 0.826. The diagonal of the matrix equals one, the correlation of each variable with itself.

A close examination of the table suggests that there are two factors underlying the variables in the matrix. The first factor seems to be represented by the three variables intelligence, creativity, and energy;

---

110. Remember, we calculate the correlation coefficient by dividing the covariance by the variances of the X and Y variables, a calculation that removes the unit of measure for both variables. To put the issue another way, the correlation coefficient is the covariance of two variables reexpressed as Z scores.

the second by the two variables firm loyalty and longevity. formal training seems to share variance with both sets.

Using a visual appraisal of the matrix, we proceeded as follows:[111]

1. First, we ran a simple bivariate regression between energy and intelligence, with energy as the dependent variable and saved the predicted values of energy, $(Y_{CEnergy} = B_1 Intelligence)$.[112] This regression captures the shared variance of energy and intelligence because the predicted values of energy represents a linear combination of the two. We began with that step because these two variables have the largest correlation.

2. Second, we ran a bivariate regression with the creativity as the dependent variable and the predicted values of energy as the independent variable and save the predicted values of Creativity, $Y_{CCreativity} = B_2 Y_{CEnergy}$. This step captures the common variance between the linear combination of energy and intelligence and creativity.

3. In this step we added the two predicted values $Y_{CEnergy}$ and $Y_{CCreativity}$ to realize the shared variance of the three variables: energy, intelligence, and creativity.

4. Then, we ran a bivariate regression between formal training and the variable realized in step 3, the common variance of energy, intelligence, and creativity and saved the predicted values of formal training, $Y_{CFT} = B_3 Common Variance_{ICE}$.

5. In step 5, we added the predicted values of *FT* to the common variance of energy, intelligence, and creativity to realize factor 1, a variable capturing the shared variance of energy, intelligence, creativity and formal training,

---

[111.] Our example contains only six variables making visual inspection easy. And we constructed the data set so that it would reflect two, certainly not more than three underlying factors or components. In practice, the procedures FA and PCA usually contain many more variables making visual inspection much more difficult.

[112.] Notice that we ran the regression without constant because all the variables were scaled as Z scores.

6. Following the same procedures as outlined in steps 1 through 5, we constructed Factor 2, the common variance between loyalty, longevity, and formal training.

7. To achieve orthogonality between the two factors, we ran a bivariate regression of Factor 2 on Factor 1, realized the residuals as a new variable, new Factor 2 with the effect of Factor 1 removed, $Y_{CFAC2} = B_6 + \epsilon$.

8. We now correlate all six variables with factors 1 and 2 to realize the correlation coefficients between each factor and the variables in the analysis. These correlation coefficients are called Factor Loadings that tell us how each variable is correlated with each factor. If we square the factor loadings, we obtain the eigenvalues, or the percentage of the factor explained by each manifest variable. Table 25.3 shows the factor loadings, eigenvalues, and communalties for each variable on factor1.

Table 25.3 Factor Loadings, Eigenvalues, and Communalties for Two Factors

	F1 Loadings	F1 Eigenvalues	F2 Loadings	F2 Eigenvalues	Communalties
Intelligence	0.96	0.922	-0.006	0.000	0.922
Creativity	0.937	0.878	0.018	0.000	0.879
Energy	0.895	0.801	-0.024	0.001	0.802
Firm Loyalty	0.066	0.004	0.867	0.751	0.755
Longevity	0.085	0.007	0.960	0.922	0.930
Firm Training	0.488	0.238	0.697	0.485	0.724
Eigenvalues	2.851			2.160	5.011
Explained Total Variance	47.5%			36.0%	83.5%
Explained Common Variance	56.89%			43.11%	100%

Summing the variable eigenvalues gives 2.851 and 2.160, the Factor eigenvalues, and dividing each by six, gives the percent of the total variance explained by the two factors. Dividing the two eigenvalues by 5.011 gives the percent of the common variance explained by the two factors. We obtained the Communalties summing the eigenvalues of the two factors giving us the percentage of common variance explained by each variable.

We can now run a multiple regression of Factor1 on all six variables to produce factor weights that we can use to produce a factor score for each worker. Thus, $Factor1 = B_1 Intelligence + B_2 Creativity + B_3 Energy + B_4 Loyality + B_5 Longevity + B_6 Formal$ training. By substituting each individual worker's score on the six variables, we obtain the factor scores for Factor 1. Notice again, we omitted the regression constant because all the variables are Z scores.

We emphasize that our computer programs use more sophisticated methods for FA and PCA. But our OLS example conveys the essential idea.

An Example of Factor Analysis Using a Computer Program[113]

For the remainder of this chapter, we use a computer program with a new hypothetical data set ($N = 100$) with eight variables. This program will give different (better) results than the OLS solution because it uses a different algorithm (Iterated Principal Axis) The hypothetical sample of 100 represents data taken from a population of neighborhoods in a large metropolitan area. Each observation represents a neighborhood. Most of the variables are indices ranging from 1 to 100 where a low score is poor and a high score good. The first variable, recreation measures the quality of city recreational facilities such as parks, jogging trails, athletic fields, and so forth. The second variable measures the quality of sanitation in the neighborhood: sewage. solid waste disposal and water run-off control. Variable 3, utilities, measures the quality of city-approved franchises providing electricity, natural gas, telephone, computer networks, and cable television. Variable 4, streets, measures road and street maintenance, efficient access to and egress from streets and vehicular movement on streets. The fifth variable measures the quality of police services. Variable 6, called stability, measures family stability and quality: Divorce and family violence. The seventh

---

113. For this example, we used the program NCSS.

variable, School, measures school quality. The last variable, wealth, measures after-tax annual income in dollars.

Our research interest lies in identifying one or more unmeasurable constructs, e.g., quality of life, by exploring the possibility of a structure or structures underlying these variables. So we classify our approach as exploratory.

The first step, an exploratory factor analysis, calculates a correlation matrix of the eight variables. Table 25.4 shows the correlation matrix.

Table 25.4 Correlation Matrix

Variables	Recreation	Sanitation	Utilities	Streets	Stability	Police	Schools	Wealth
Rec.	1.0000	0.5805	0.7069	0.6570	0.4338	0.2718	0.4089	0.3441
Sanitation	0.5805	1.0000	0.6866	0.8222	0.3171	0.2191	0.1765	0.1934
Utilities	0.7069	0.6866	1.0000	0.6031	0.4186	0.3521	0.4009	0.3570
Streets	0.6570	0.8222	0.6031	1.0000	0.3600	0.3956	0.3562	0.3438
Stability	0.4338	0.3171	0.4186	0.3600	1.0000	0.6142	0.6604	0.7511
Police	0.2718	0.2191	0.3521	0.3956	0.6142	1.0000	0.7895	0.7578
Schools	0.4089	0.1765	0.4009	0.3562	0.6604	0.7895	1.0000	0.8034
Wealth	0.3441	0.1934	0.3570	0.3438	0.7511	0.7578	0.8034	1.0000

Now, as you peruse the table, it is no longer easy to predict the factors by simply looking at the matrix.

The *second step* instructs the program to identify the number of factors inherent in the structure. Because the effort is exploratory, we issue a program command to extract the maximum number of factors allowed by the constraint of using only the shared variance of the eight variables. Even though we lacked a solid theory, we constructed the variables around what we believed were two underlying factors: the variable set consisting of the recreation, sanitation, utilities, and streets, tentatively called city services, and the set of variables measuring family issues, police protection, quality of schools, and family wealth, tentatively called familial concerns. So we were not completely flying blind.

When we ran the program, it identified five factors. Table 25.5 shows the factors and *factor loadings* produced by a five factor solution.

Table 25.5 Factors and Factor Loadings

Variables	Factor 1	Factor 2	Factor 3	Factor 4	Factor 5
Recreation	-0.7126	-0.3627	-0.3474	-0.1412	0.1476
Sanitation	-0.6572	0.3044	0.2097	0.1930	-0.0813
Utilities	-0.7300	-0.3652	-0.2523	0.0243	-0.2411
Streets	-0.7554	-0.4733	0.2994	-0.1525	0.1550
Stability	-0.7364	-0.6348	-0.0882	0.2624	0.1289
Police	-0.7281	0.4534	0.2053	-0.1105	-0.1503
Schools	-0.7674	0.4712	-0.0396	-0.1574	-0.0483
Wealth	-0.7616	0.5040	0.0179	0.1040	0.0771

These factor *loadings*, correlation coefficients between the variable and the factor, tell us how each variable is related to the identified factors.

The program then squared and summed the factor loadings for each factor producing *eigenvalues*[114]. Table 25.4 shows the result of these calculations.

Table 25.6 Squared Factor Loadings (Eigenvalues)

Variables	Factor 1	Factor 2	Factor 3	Factor 4	Factor 5	Communalities
Recreation	0.5078	0.1316	0.1207	0.0199	0.0218	0.8018
Sanitation	0.4319	0.0927	0.0440	0.0372	0.0066	0.6124
Utilities	0.5329	0.1334	0.0637	0.0006	0.0581	0.7887
Streets	0.5706	0.2240	0.0896	0.0233	0.0240	0.9315
Stability	0.5423	0.4030	0.0078	0.0689	0.0166	1.0386

[114.] The term *eigen* is a German word meaning "characteristic."

Police	0.5301	0.2056	0.0421	0.0122	0.0226	0.8126
Schools	0.5889	0.2220	0.0016	0.0248	0.0023	0.8396
Wealth	0.5800	0.2540	0.0003	0.0108	0.0059	0.8510
Eigenvalues 4.2846		1.6662	0.3698	0.1977	0.1580	6.6763

Squaring the factor loadings produced an R^2 value, the amount of the variance in the factor accounted for by that variable. The sums of those R^2s for each factor equal the eigenvalues for the factor, the amount of the common variance in the set of eight variables captured by that factor. The sums of the all the factor eigenvalues equal the total amount of common variance captured by the five factors. That total is 6.6763. Note: this value does not represent the total variance of the eight variables, just the common variance. The total variance is eight because each variable has been reexpressed as a Z score with a mean of zero and a variance of one. Thus, eight variables have a total variance of eight. The row communalities show the percentage of variation each variable shares with each factor. If we divide each factor's eigenvalue by 6.6763, we realize the percentages of common variance explained by each factor. If we sum the column and row totals, they both will equal 6.6763. Table 25.6 and 25.7 shows these calculations.

Table 25.7 Percentage of Total Common Variance Explained by Each Factor

	Factor 1	Factor 2	Factor 3	Factor 4	Factor 5	Total
Eigenvalues	4.2848	1.6662	0.3698	0.1977	0.1580	6.6763
Percentage of Total Common Variance Explained by Each Factor	64.2%	25.0%	5.5%	3.0%	2.3%	100.0%

Table 25.7 shows that two factors account for slightly over 89% of the total common variance.

In *step three* we determine the actual number of factors that we believe capture the concept. One device used for making that decision is the scree plot.[115] *Scree plots* use either eigenvalues or percentages as a guide for choosing the number of factors. Figure 26.3 shows a scree plot of the eigenvalues.

Figure 25.3 Scree Plot of Eigenvalues

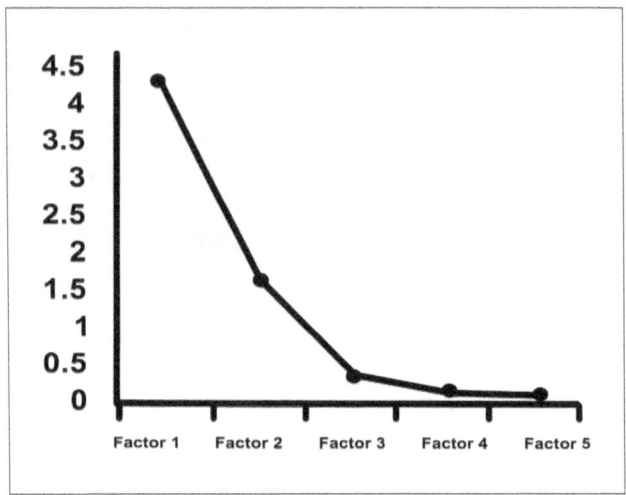

The scree plot shows that after two factors, all the other factors look like scree.[116] Another possible criterion identifies only the number of factors that have eigenvalues equal to or greater than one. This criterion is based on the assumption that a factor should explain at least as much variance as any single variable. In addition, from the outset, we built our variables to measure two different constructs. Using the scree plot, the eigenvalue criterion of one, and our original assumption concerning the variables, we decided two factors represent the best solution. So we instruct the program to rerun the analysis with two

---

115. The word *scree* means "an accumulation of rock debris at the base of a cliff, hill, or mountain slope." In this context, the mountain slope is the first two factors; the scree is the remaining factors.

116. There are other criteria for extracting factors. If the research effort is confirmatory, then we have working hypotheses as guides. Alternatively, we can use the percentage of common variance explained by the factor as the criterion. Some suggest that we retain enough factors to explain at least 70%. Finally some say retain only straightforwardly interpretable Factors. Our criteria, as you can see from the example, are pragmatic: we use all the information available.

factors. Table 25.7 shows the factor loadings and eigenvalues of the two-factor solution.

**Table 25.8 Unrotated Factor Loadings and
Eigenvalues for the Two-factor Solution**

Variables	Factor I Loadings	Factor I Eigenvalues	Factor II Loadings	Factor II Eigenvalues	Communalities
Recreation	0.686	0.471	0.352	0.123904	0.594904
Sanitation	0.639	0.408	0.272	0.073984	0.481984
Utilities	0.711	0.506	0.366	0.133956	0.639956
Streets	0.730	0.533	0.459	0.210681	0.743681
Stability	0.729	0.531	-0.626	0.391876	0.922876
Police	0.723	0.523	-0.416	0.173056	0.696056
Schools	0.776	0.602	-0.455	0.207025	0.809025
Wealth	0.776	0.602	-0.496	0.246016	0.848016
Eigenvalues		4.176		1.560	5.736

From table 25.5 we know that these two factors explain about 89% of the common variance among the eight variables or 100% of the common variance of the two factors. How do we interpret these factors?

Interpreting the Factors

We base our interpretation of the meaning of a factor by examining the factor loadings (the correlation of each variable with the Factor). Using table 26.7 to interpret the loadings has an air of ambiguity. On Factor 1, all eight of the variables have loadings equal to or greater than 0.69. Factor 2 has seven loadings equal to ± 0.30. This ambiguity results from the factor analytic's procedure requiring the first extracted factor to account for the greatest common variance with successive factors accounting for the greatest common variance in the residuals. Often, we cannot distinguish these aggregations in terms of their loadings on the first factor, because, as figure 25.5, shown below, Factor I falls between the two sets of factor loadings for the eight variables. To explain how we produced figure 25.5, we introduce a little high school trigonometry in figure 25.4.

Figure 25.4 Quadrant I of the Cartesian Plane

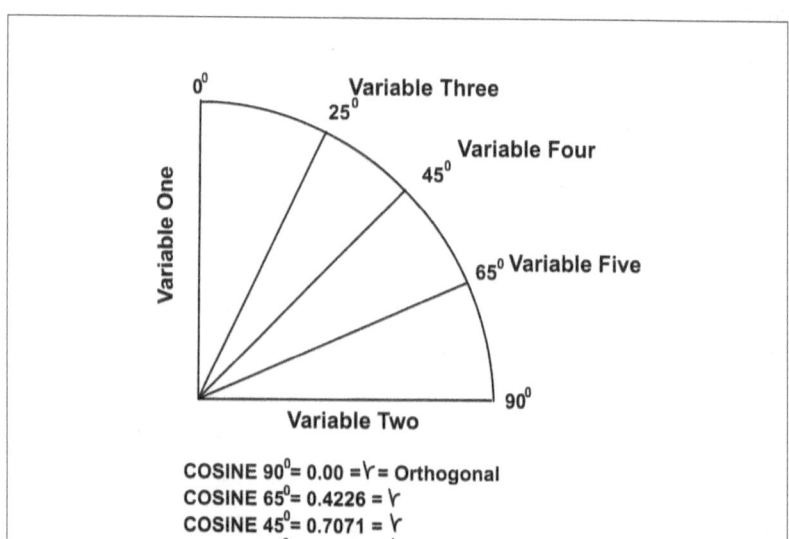

Figure 25.4 shows quadrant I, of the Cartesian plane, on which all values are positive because the X and Y axes are positive. The quarter circle, an arc, allows us to mark-off angles. Pretend that the X and Y axes represent two variables, One and Two with a 90 degree angle separating the two (we measure angles counterclockwise, i.e., from Y toward X). If we take the cosine of a 90 degree angle, zero, without proof, we assert that a cosine equal to zero is the correlation coefficient between the two variables. We repeat: the cosine of any angle formed between two variables equals the coefficient of correlation between the two variables. Thus, we see the meaning of the term orthogonal. Two variables at 90° angles have a correlation coefficient of zero.

Now, look at the other angles in figure 26.4. We can conceive another line with an angle of 65 degrees as variable 3. Taking the cosine of a 65 degree angle yields 0.4226, the correlation coefficient between variable 3 and variable 1. A fourth variable, with an angle of 45 degrees, has a cosine of 0.7071, the correlation coefficient for variable 4 and variable 1. Variable 5, with an angle 25 degrees has a cosine of 0.9063, the correlation coefficient between variable 5 and 1. Not incidentally, a zero angle means that a variable is on top of another variable and, with a cosine of zero, has a correlation coefficient of one. The two variables are perfectly correlated and surrogates for each other. Now, let's see how this discussion relates to a procedure called rotation used to clarify interpretation of the factors.

Figure 25.5 Two Unrotated Factors

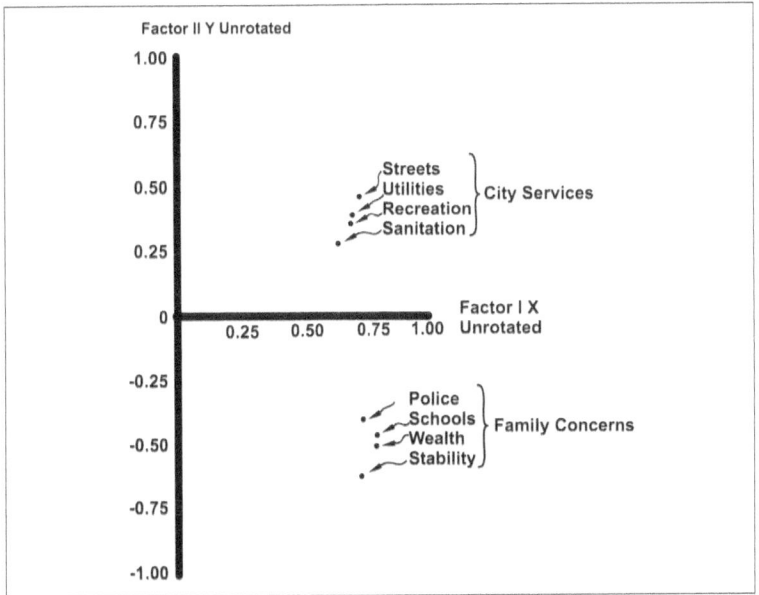

Figure 25.5 uses quadrants I and II[117] of the Cartesian plane and the variables recreation, sanitation, utilities, streets, stability, police, schools, and wealth. The X and Y axes represent unrotated factors I and II. Using the loadings from table 26, we plot the X and Y factor loadings for factor I in quadrant I, the loadings all plot as positive. The first plot, recreation, is $X = 0.686$ and $Y = 0.352$, both positive. We continue the plotting through the rest of the series for factor II, four of the loadings plot as positive and four plot as negative. In quadrant II, the coefficients plot positively for X, negatively for Y. On figure 26.5, we plotted the factor loadings for each variable. Note: because of the negative Y loadings on factor 2, the last four plots fall in quadrant II of the plane.

Given this visual, you can see the difficulty in interpreting these loadings. This ambiguity relates to the requirement that the first factor must capture the maximal amount of the common variance. In capturing

---

117. Using quadrant II allows negative observations because that part of the Y axis in quadrant II is negative.

the greatest amount common variance, Factor 1 falls between the two clusters. Figure 25.6 shows the solution to this ambiguity. In figure 25.6, we *literally rotated* clockwise the y—and x-axes (Factors I and II) as the arrows show, so that factors fit the loadings more closely.

Figure 25.6 Rotated Factors

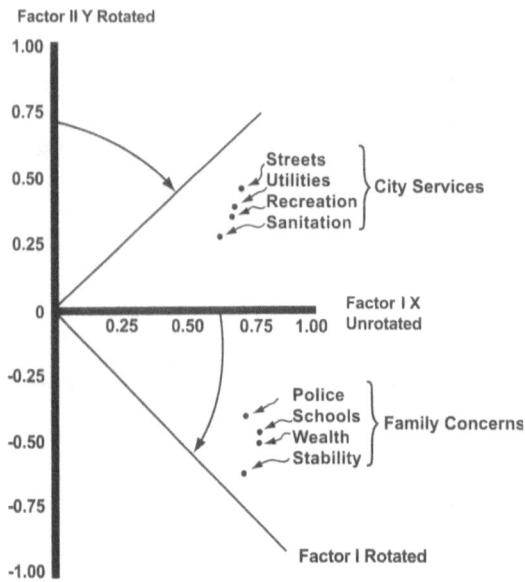

By measuring the new angles for the variable and taking the cosines of the angles, we produce a new set of factor loadings, summarized in table 25.9. .

Table 25.9 Rotated Factor Loading, Eigenvalues, and Percentages

Variables	Factor1 Loadings	Loadings Squared	Factor2 Loadings	Loadings Squared	Communalties
Recreation	0.27	0.07	0.72	0.52	0.59
Sanitation	0.05	0.00	0.89	0.80	0.80
Utilities	0.28	0.08	0.75	0.56	0.64
Streets	0.23	0.05	0.83	0.69	0.74
Stability	0.72	0.52	0.29	0.09	0.61
Police	0.81	0.66	0.18	0.03	0.69
Schools	0.88	0.77	0.19	0.04	0.81
Wealth	0.91	0.82	0.16	0.03	0.85
Eigenvalues		2.98		2.76	5.74

Two rotated factors did the job. Now, both factors have unambiguous loadings. The two rotated factors capture about 86% of the common variance equal to the amount captured by the unrotated factors. This fact emphasizes the notion that rotation is equivalent to a reexpression. We rotate to reduce ambiguity. The structure of the factors does not change when they are rotated.

In today's world, we no longer do the physical rotation; our computers do it with a mathematical algorithm.

Naming Factors

Now we are in a position that allows us to interpret and name the factors. The factors cluster around our original intent. Factor II clearly represents city services and Factor I familial concerns, just as we built the variables. Factor I shows a combination of family wealth, family stability, quality of police service, and school quality, the last two variables important to families. Factor II contains variables that directly measure city services: streets, utilities, sanitation, and recreation. Subsequently, we'll have more to say about this issue.

Because factor analysis extracts factors using only the common variance between variables (ignoring the unshared or unique variance), it cannot yield a true factor score and, therefore, cannot create true new variables representing the factors for further analysis. However, two of the programs we recommend, NCSS and SPSS, will produce pseudo factor scores that, if used cautiously, might yield useful information when used as dependent or independent variables in subsequent analyses.

Principal Component Analysis

Principal components, compared to factor analysis, extracts components using the total variance (common and unique) and produces real linear combinations of the variables analyzed. Because its components include all the variance, unlike FA, PCA can extract true component scores. All our recommended programs have Principal Component routines.

Principal Component Analyses extract components as new variables that show how much of the total variance each Principal Component explains. In other words, using PCA we calculate component equations just as we did in our OLS example, i.e., Principal Component $I = \beta_1 X_1 + \beta_2 X_2 + \ldots \beta_n X_n$ and use the β_s (called Eigenvectors) to calculate a PC that maximizes the explained variance in the X set. In other words, using the equation for PC I, we substitute the values of the variables for each individual in the equation which the produces component scores for each individual.

As a data reduction transformation, PCA has a different purpose than FA. PCA's central function deals with high multicollinearity in independent variables or reducing many variables to a manageable set. But we hope that you can see that PCA might also function as a follow-up to factor analysis using its findings to generate the true linear components.

Now, let's look at the PCA output of the same data set used in our discussion of FA, an eight variable analysis. Calculating the variable's correlation matrix also constitutes the first step in a principal component analysis.[118] Again, using the correlation matrix reexpresses the variables as Z scores making the total variance equal eight. Because it uses all the variance, PCA can extract as many components as variables in the analysis, with the first component capturing the maximal variance in the eight, the second, the next highest variance, and so on.

Table 25.8 shows the correlation matrix for the eight variables.

Table 25.10 Correlation Matrix, Phi, and Bartlett's Test

Variables	Recreation	Sanitation	Utilities	Streets	Stability	Police	School	Wealth
Recreation	1.000	0.581	0.707	0.657	0.434	0.272	0.409	0.344
Sanitation	0.581	1.000	0.687	0.822	0.317	0.219	0.177	0.193
Utilities	0.707	0.687	1.000	0.603	0.419	0.352	0.401	0.357
Streets	0.657	0.822	0.603	1.000	0.360	0.396	0.356	0.344
Stability	0.434	0.317	0.419	0.360	1.000	0.614	0.660	0.751
Police	0.272	0.219	0.352	0.396	0.614	1.000	0.790	0.758

---

118. We note here that, instead of using a correlation matrix, PCA might use a covariance matrix. Thus, instead of using Z scores, the PCA might use the raw score values of the matrix. This approach would result in raw score eigen vectors instead of Z scores.

| Schools | 0.409 | 0.177 | 0.401 | 0.356 | 0.660 | 0.790 | 1.000 | 0.803 |
| Wealth | 0.344 | 0.193 | 0.357 | 0.344 | 0.751 | 0.758 | 0.803 | 1.000 |

Phi = 0.531

Bartlett's

Test =

599.890

DF=28

Prob =
0.000

In table 25.8, we let the diagonal equal one showing our intent to use all the variance and introduce two new statistics: Phi and Bartlett. The Phi statistic measures how much common variance exists among the variables. A Phi statistic equal to 0.00 signals no correlation among the variables (no common variance) and neither FA nor PCA makes sense; a Phi statistic equal to 1.00 means perfect correlation among the variables and, therefore, any one variable could function as a perfect surrogate for the others. This statistic should minimally equal 0.50, but in many circumstances, even though the data do not achieve that level of common variance, a PCA or FA still might show an underlying structure. In our example, Phi exceeded 0.50.

Bartlett's tests the hypothesis that all correlations between the variables equal zero. If all correlations equal zero, we want to reject the null; failure to reject the null renders FA or PCA pointless. In our case, the P value, less than 0.000, forces rejecting of the null allowing proceeding with our PCA.

In the second step, we instruct the program to run the analysis with eight components, and, we note, in an exploratory effort, this is how you should approach the problem. Table 25.11 shows the unrotated component loadings for the eight variables, the eigenvalues, and percentages of explained variance for each component.

Table 25.11 Component Loadings, Eigenvalues, and Percentages

Variables	Comp 1	Comp 2	Comp 3	Comp 4	Comp 5	Comp 6	Comp 7
Recreation	-0.731	-0.409	-0.406	-0.128	-0.309	0.081	-0.075
Sanitation	-0.652	-0.642	0.267	0.173	0.116	-0.05	0.09
Utilities	-0.751	-0.413	-0.257	-0.146	0.407	-0.028	-0.034
Streets	-0.749	-0.48	0.345	-0.013	-0.226	-0.024	-0.018
Stability	-0.772	0.332	-0.168	0.485	0.01	0.151	0.061
Police	-0.748	0.476	0.281	-0.209	0.093	0.254	-0.126
Schools	-0.783	0.481	-0.035	-0.235	-0.058	-0.084	0.297
Wealth	-0.775	0.515	0.006	0.085	-0.023	-0.295	-0.193
Eigenvalues	4.455	1.813	0.53	0.41	0.338	0.192	0.161
Percent	55.70%	22.70%	6.60%	5.10%	4.20%	2.40%	2.00%

The eigenvalues tell us how much of the total variance of the eight variables each PC explains. For example, PC1 captures 4.455, about 56% of the total variance, 8; PC 2 about 23%; and PC 3 about 7%. If our criterion for retaining components was that the retained components should capture at least 80% of the variance in the set, we would extract three components.

Table 25.12 shows the component loadings, eigenvalues, and percent of total variation explained by the three component solution.

Table 25.12 Component Loadings, Eigen Values, and Percent of Total variation

Variables	PC 1	PC 2	PC 3
Recreation	-0.73	-0.41	-0.41
Sanitation	-0.65	-0.64	0.27
Utilities	-0.75	-0.41	-0.26
Streets	-0.75	-0.48	0.35
Stability	-0.77	0.33	-0.17
Police	-0.75	0.48	0.28

Schools	-0.78	0.48	-0.04
Wealth	-0.78	0.52	0.01
Eigenvalues	4.46	1.81	0.53
Percent	55.70%	22.70%	6.60%
Cum Percent	55.70%	78.40%	85.00%

Note: the decision to extract three components did not change the factor loadings, eigenvalues, or percents.

Now the question is, should we rotate the principal components?

Many who know argue against rotation with PCA, That claim centers on PCA as a data reduction transformation designed principally to deal with multicollinearity. While we empathize with this position, it does seem that, if we are using PCA as a follow-up to FA, rotation provides additional insight into the meaning of the PC. So we rotate to a three PC solution. Table 25.10 shows the factor loadings, eigenvalues, and percents of the rotated three PC solution.

**Table 25.13 Factor Loadings, Eigenvalues,
and Percents of the Three PC Rotated Solution**

Variables	PC 1	PC 2	PC 3
Recreation	-0.2035	-0.3724	-0.829
Sanitation	-0.0493	-0.8843	-0.3521
Utilities	-0.2285	-0.4754	0.723
Streets	-0.2384	-0.8824	-0.274
Stability	-0.7669	-0.0868	-0.3723
Police	-0.8908	-0.2606	0.0484
Schools	-0.8904	-0.0828	-0.2146
Wealth	-0.9122	-0.0833	-0.1655
Eigenvalues	3.159	2.0146	1.623
Percents	0.3949	0.2518	0.2029
Cum Percents	0.3949	0.6468	0.8497

Note: While the rotated loadings, eigenvalues, and percents changed, the percentage of the total variance captured by the three PCs did not. Comparing the rotated loadings with the unrotated, we can see that the first PC looks much like the first factor in the FA. It took two PCs, the second and third, to capture the notion of city services, combining sanitation and streets (PC 2) and recreation and utilities (PC 3).

Now, we do something that we did not do with FA; we instruct our program to calculate the component coefficients, the eigen vectors (betas). The component coefficients are the constants in an equation that allows us to create true variables representing the components that we can use in subsequent analyses. In other words, the factor scores are the sum of the betas × the values of that variable. We call these coefficients betas because the variables are in Z score format.

$$PC\ 1 = \beta \times Rec + \beta \times San + \beta \times Util + \beta \times Stre +$$
$$\beta \times Stab + \beta \times Pol + \beta \times Sch + \beta \times Wea \qquad \text{(Equation 25.1)}$$

After the rotation, Table 25.14 shows the component coefficients (betas).

Table 25.14 Component Coefficients

Variables	PC 1	PC 2	PC 3
Recreate	-0.35	-0.30	-0.56
Sanitation	-0.31	-0.48	0.37
Utilities	-0.36	-0.31	-0.35
Streets	-0.36	-0.36	0.47
Stability	-0.37	0.25	-0.23
Police	-0.35	0.35	0.39
Schools	-0.37	0.36	-0.05

These component coefficients are the betas[119] in equation 25.1. To calculate the factor scores for each observation, we plug the original values for each variable's observation and solve the equation for that

---

119. Betas, as opposed to regression coefficients containing the unit of measure of Y, are the regression coefficients of Z scores.

observation. This realizes the component score for that observation. Repeating that calculation through all the observations, we have a new variable, PC 1, labeled, if we wish, family concerns. Using the equation for PC 1 and two other equations for PC 2 and PC 3, our program now calculates and saves the three PCs. And you will find that they are orthogonal.

Now, if you calculate the PC scores for the unrotated solution, you will find that the rotated PC scores and unrotated PC scores are perfectly correlated. From that exercise, we see why some argue the pointlessness of PC rotation. If, however, we intend to use the PCs as dependent or independent variables in a subsequent regression, rotation gives a better bead on interpretation.

Finally, our program calculates the communalities in eight PC and three PC solutions table 25.15 shows the eight PC communalities.

Table 25.15 Communalties, Eigenvalues, and Percents in an Eight PC Solution

Variables	PC 1	PC 2	PC 3	PC 4	PC 5	PC 6	PC 7	PC 8	Communality
Recreation	0.01	0.14	0.73	0.03	0.08	0.00	0.01	0.00	1
Sanitation	0.00	0.82	0.02	0.02	0.09	0.00	0.00	0.05	1
Utilities	0.03	0.17	0.10	0.02	0.68	0.00	0.00	0.00	1
Streets	0.05	0.76	0.08	0.01	0.02	0.01	0.00	0.08	1
Stability	0.14	0.02	0.03	0.76	0.02	0.02	0.01	0.00	1
Police	0.91	0.02	0.00	0.05	0.01	0.01	0.00	0.00	1
Schools	0.49	0.00	0.04	0.08	0.02	0.04	0.33	0.00	1
Wealth	0.37	0.01	0.01	0.18	0.01	0.40	0.02	0.00	1
Eigenvalues	1.99	1.94	1.01	1.15	0.93	0.48	0.37	0.13	8
Percent of the Total Variance Explained	25%	24%	13%	14%	12%	6%	5%	2%	100%

The communalities are the squared values of the factor loadings, the coefficients of determination for each of the manifest[120] variables on their respective PCs. The coefficient of determination for each variable divided by the total variance of all eight variables gives the percentage of the total variance explained by that variable through its PC. The sum of each coefficient of determination for each PC equals the Eigenvalue for each PC. The sum of the coefficients of determination for each variable equals one, the total variance of each manifest variable. Table 25.16 shows the communalities for our three PC solution.

Table 25.16 Communalties, Eigenvalues, and Percents for a Three PC Solution

Variables	PC 1	PC 2	PC 3	Communality
Recreation	0.04	0.14	0.69	0.87
Sanitation	0.00	0.78	0.12	0.91
Utilities	0.05	0.23	0.52	0.80
Streets	0.06	0.78	0.08	0.91
Stability	0.59	0.01	0.14	0.73
Police	0.79	0.07	0.00	0.86
Schools	0.79	0.01	0.05	0.85
Wealth	0.83	0.01	0.03	0.87
Eigenvalues	3.16	2.01	1.62	6.8
Percent of the Total Variance Explained	39%	25%	20%	85%

The difference between tables 25.13 and table 25.14 reflects our decision to settle for a three PC solution that captured about 85% of the total variance in the eight variables.

[120.] The term *manifest variables* refers to measured variables as opposed to the PCs which are transformations.

Practical Uses of FA and PCA

The practical uses of FA and PCA depend on the purposes of the research effort. Exploring the possibility of a structure underlying a set of manifest variables, an exploratory FA is just the thing. An example is measuring the construct, teaching effectiveness. That process begins with the development of a tentative theory identifying potential manifest variables that are logical candidates for tapping into this unmeasurable construct, good teaching. We would then use an exploratory FA to search for an underlying structure. If the exploration found little or nothing, the effort would return to a reconstruction of the theory and a new sample. And with the new variable set, the exploration would continue. If exploration found what appeared to be a satisfactory underlying structure, the effort would turn into confirmatory replication in other settings and times. If results consistently confirmed the theory, then we could make a case for building an acceptable teaching effectiveness measure using indices or Principal Components of the confirmed variables.

We might add that the factors in factor analysis also can serve as reliability and validity coefficients for the manifest variables. High enough factor loadings give evidence of a variable's reliability, measuring something. The case for its validity lies in the context of the other variables in the factor analysis. If we start from theory and an FA provides confirmatory evidence, high factor loadings logically point to the validity of the variable in measuring the construct, in this case, effective teaching.

On the other hand, in doing multiple regressions with a set or part of a set of independent variables that are highly collinear, we might want to use PCA to transform the set into a smaller number of orthogonal PCs. Here we create PCs to reduce or eliminate damaging collinearity.

With reciprocal causation, we could use PCA to transform a set of X variables to one or more PCs as *instruments*. In the face of a problem of multiple dependent variables, we could use PCs to transform the multiple dependent variables into a single PC. In the next chapter, we discuss another use of PCA that more effectively deals with the problem of multiple dependent variables.

Summary

Chapter 25

The critical ideas discussed in this chapter are the following:

1. Factor analysis

    When measuring a set of variables, usually the variances of these variables overlap. We call this overlap shared common variance, or collinearity. Correlation coefficients measure the common variance between two or more variables. The parts of variables that do not overlap represent unique variance specific to the variable including random and perhaps some systematic variance. Sometimes this shared variance is sparse; sometimes extensive. With extensive common variance, i.e., when the variables have correlation coefficients equal to or greater than 0.50, we can use factor analysis (FA) to extract from the data a smaller number of variables (factors) by capturing their overlapping.

2. Factor loadings

    Factor loadings are the correlation coefficients between the factors created from variables and the variables themselves.

3. Eigenvalues

    Squaring the factor loadings, produces an R^2 value, the amount of the variance in the factor determined by that variable. The sums of these R^2s equals the eigenvalues, the amount of the shared common variance in a set of variables captured by that factor.

4. Factor rotation

    The factor analytic procedure requires the first factor to capture the maximal amount of the shared common variance. Also, the procedure requires the subsequent factors to capture the maximal amount of residual variance, causing Factor 1 and subsequent factors to fall between clusters, creating ambiguity in interpretation.

The solution to this ambiguity is factor rotation, a technique that makes the factors fit the loadings more closely. We note that the equivalence of factor rotation to variable reexpression; rotation does not change the factor structure.

5. Principal component analysis

Principal components, as contrasted with FA, extracts components using the total variance (common plus unique) and produces real linear combinations of the variables analyzed. Thus, unlike FA, PCA produces true component scores because its components include all the variance.

6. Practical uses of FA and PCA

The practical uses of FA and PCA depend on the purposes of the research effort. If we want to explore the possibility of a structure underlying a set of manifest variables, an exploratory FA is just the thing. If, in doing a multiple regression with a set or part of a set of collinear independent variables, we can use PCA to transform the set into a smaller number of orthogonal PCs. In this context, we create PCs to reduce or eliminate damaging multi collinearity.

Appendix

Chapter 25

The data set contains five variables. Your tasks are the following:

1. Name the variables as they might relate to your job or interests.
2. Run a factor analysis using OLS and explain your findings.
3. Using your statistical program, run a factor analysis and principal components analysis and explain your findings.
4. Use one of the variables as Y, the other four as Xs and check for multi collinearity. If multicollinearity is a problem, use PCA to solve it.

	X1	X2	X3	X4	X5
1	108.06	272.98	53.21	75.48	82.60
2	102.06	177.41	47.83	66.78	58.03
3	103.34	219.80	61.92	78.43	102.40

4	101.59	246.12	51.79	52.43	72.30
5	99.24	207.51	55.58	69.22	75.31
6	80.53	160.82	38.22	53.22	73.29
7	79.22	203.57	46.22	64.45	74.28
8	105.67	218.49	48.14	76.89	96.83
9	123.13	197.11	42.60	48.05	80.23
10	115.34	252.78	53.15	66.66	94.51
11	125.16	257.33	58.34	62.27	81.34
12	77.70	148.63	35.88	39.76	87.13
13	97.90	173.88	44.32	44.56	73.71
14	107.22	176.71	49.65	61.14	81.74
15	109.01	199.86	53.20	72.22	80.61
16	95.83	174.47	40.20	47.18	75.30
17	82.69	173.50	52.10	59.12	59.45
18	105.26	214.61	43.25	61.85	91.79
19	91.72	166.29	37.72	57.12	69.61
20	100.10	179.86	47.55	61.86	90.18
21	83.07	167.20	37.94	44.90	61.88
22	117.73	238.31	53.77	58.89	97.41
23	120.39	246.47	59.12	66.30	83.36
24	139.83	255.97	62.00	72.67	101.50
25	91.14	184.31	43.84	48.21	57.87
26	91.04	253.70	56.10	61.67	74.08
27	114.9	216.36	46.51	53.01	71.72
28	99.28	194.29	49.83	62.37	70.44
29	128.00	231.15	55.50	64.36	83.57
30	92.20	216.31	52.44	58.97	73.60
31	91.24	211.68	48.04	49.93	74.17
32	95.78	205.98	50.31	66.14	78.82
33	109.97	145.97	57.97	61.64	94.75
34	100.55	177.75	52.36	52.69	56.94
35	113.80	263.74	70.79	68.38	101.02
36	108.38	240.96	59.10	78.11	70.02
37	83.21	160.49	40.79	62.61	71.64
38	93.04	239.88	47.18	60.36	95.39
39	51.66	130.93	34.64	46.63	49.02
40	103.63	219.26	60.59	69.00	82.46
41	106.64	224.41	62.95	68.97	95.13
42	108.46	215.84	68.19	73.53	109.31
43	111.48	218.23	67.18	70.51	105.56
44	85.73	236.13	49.90	58.47	75.68
45	99.01	193.42	51.20	55.25	75.57

46	105.60	196.86	46.63	66.49	95.86
47	111.18	220.02	64.77	65.24	84.71
48	100.69	198.62	39.23	46.48	66.78
49	123.45	263.58	57.21	78.41	104.21
50	93.74	251.83	66.31	67.10	90.56
51	126.53	240.87	61.9	67.12	98.16
52	123.01	238.04	49.93	49.63	95.33
53	83.73	183.51	53.96	47.10	57.20
54	105.71	164.02	45.58	49.62	72.67
55	90.56	212.15	33.59	49.94	74.96
56	92.07	145.43	53.32	61.09	66.53
57	119.14	199.99	56.98	61.74	79.88
58	103.04	214.31	49.50	60.09	85.79
59	102.00	225.97	62.08	70.97	95.70
60	113.97	160.37	46.99	59.92	98.21

CHAPTER 26

CANONICAL CORRELATION AND JACOB COHEN'S SET CORRELATION

This chapter offers an introduction to regression analyses using multiple dependent variables putting to work the ideas discussed in chapter 25, with applications of principal component analysis at the center of our discussion.

Why do we need to treat multiple dependent variables in a separate chapter? In the face of, say, three dependent variables Y and five independent X variables, if we proceed to Y_1, Y_2, and Y_3 regressed three times on five X variables, this tack is inappropriate for two reasons. First, running three regressions one at a time will enormously increase the probability of a type I error, i.e., mistakenly identifying chance relationships as statistically significant. In this example, assuming three Y and five X variables, we must test the statistical significance of at least 15 regression coefficients at a 0.05 level of significance, making the probability of at least one type I error greater than 0.60. To reduce the type I threat we need a protected test, analogous to the F test, when running univariate multiple regressions. And second, running multiple univariate multiple regressions fails to take into account multicollinearity in the dependent set.

Canonical Correlation

In chapter 25 we discussed the techniques of factor and principal component analyses without reference to whether the variables used in the analyses were independent or dependent. In canonical correlation we introduce the use of principal components as dependent and independent variables.

Canonical correlation generally takes the following form:

$$Y_1 + Y_2 + \ldots Y_N = f(X_1 + X_2 + \ldots X_N). \qquad \text{(Equation 26.1)}$$

The strengths of canonical correlation analysis CCA are as follows: 1) it easily accommodates ordinal and nominal data, (2) largely overcomes the problem of type I error threats and collinear Y variables stemming from running a set of univariate regressions, and (3) because it uses principal components as the variables in the analysis is free from unreliability.

To explain how Canonical Components Analysis (CCA) works, let's assume that our hypothetical company, StatSys, has 74 retail outlets on which we have data on profits, sales, and inventory control. We also have seven indices measuring quality of the neighborhoods in which our outlets are located: crime, socioeconomic class, population size, alcohol/drug use, socioeconomic status, race-ethnic makeup, neighborhood stability, and family stability.

We have decided to use these variables to analyze the performance of our retail outlet managers. Because effective performance is a difficult construct to measure, we decided to use three manifest variables, inventory control, sales and profits, as measures of performance in our dependent variable set. In the same spirit, we chose five neighborhood variables as our independent variable set: crime, socio/economic status, race/ethnic concentration, population size, and neighborhood stability. The following is a step-by-step discussion of the conduct of a canonical correlation analysis using these data. We note that four of our recommended statistical packages have canonical routines: *OpenStat*, Systat *MyStat,* and *NCSS.* We will use some output from all four of these packages, plus some of our own hand calculations.

The method of Principal Components lies at the heart of canonical correlation analysis. That is, CCA extracts principal components from the X and Y sets and runs correlations on the X and Y principal components. It is important to understand that the extraction of canonical components is different from simple principle components (PCA) discussed in chapter 25. In simple Principal component analysis, the first PC (PC1) is extracted to capture the maximal amount of variance in the data set. In canonical principal component analysis, the process extracts *pairs of principal components from the Y and X sets,* canonical principal component Y1 and canonical principal component X1, canonical principle component Y2, canonical principal component X2, and canonical principal component Yn and canonical principal component Xn so *that the pairs* of canonical PCs are *maximally correlated.* The correlation between canonical principal component Y1 and the canonical principal component X1 is the highest, between CPC Y2 and CPC X2, the next highest, and so on. The maximal number of pairs that canonical analysis can extract is the smaller number of variables in the X and Y set. In our example, we have three variables in the Y set and five in the X set so the canonical analysis extracts three pairs.

Step 1 calculates the principal components of the Y and X sets. Table 26.1 shows the output of this first step. Remember that the extraction process is done either on a correlation matrix or the variables in their original metric.

Table 26.1 Canonical Correlations and Eigenvalues Between PC pairs

PC Pairs	Canonical Correlations	Canonical R Squares
PCY1 - PCX1	0.85	0.72
PCY2 - PCX2	0.26	0.068
PCY3 - PCX3	0.19	0.004

This table gives the correlation coefficients and the coefficients of determination for each pair of PCYs and PCXs. These values are symmetric for Xs and Ys. So PCY1 explains about 72% of the variance in PCX1 and vice versa.

The second step in CCA calculates the multivariate $MuR^2_{Y, X}$. This value is a generalization of the univariate MR^2. MuR^2, measures the generalized variance in the Y set accounted for by the X set. $MuR^2_{Y,X}$ is symmetric because it is also captures the generalized variance in the X set accounted for by the Y set.

We calculate $MuR^2_{Y, X}$ using the following equation:

$$MuR^2_{Y, X} = 1 - [(1 - R^2 PCY1) * (1 - R^2 PCY2) * (1 - R^2 PCY_N)] \quad \text{(Equation 26.2}_)$$

where N is the number of variables in the smaller of the two sets, in this case, R^2 equals R^2s from table 26.1.

Thus $MuR^2_{Y, X}$ = 1- [(1 - 0.780) × (1-0.567) × (1-0.0570)] = 1- (0.220 × 0.433) × 0.943) ≈ 0.79[121]

[121.] JCC

Thus, about 79% of the total variance of the Y pair sets is explained by the X pair sets and vice versa.

Step three tests the null hypothesis of no relationship between the Y and X pair sets, a specialized F test called Rao's F.

$MuR^2_{Y,X} = 0.79$

[122]Rao's F = 7.836[123]

$DF_{numerator} = 15$

$DF_{denominator} = 185.359$

$P < 0.000$

We note that NCSS, Systat, and MyStat make this test as part of their output.

We also note that NCSS, Systat, MyStat, and OpenStat give F or χ^2 tests for each of the pairs of canonical correlations making it unnecessary to hand calculate these tests. Table 26.2 summarizes this output.

Table 26.2

Canonical Pair	Canonical R	Canonical R Square	DF Numerator	DF Denominator	Probability of Type I Error
1	0.85	0.72	15.00	185.00	0.00
2	0.26	0.07	8.00	136.00	0.73
3	0.07	0.00	3.00	69.00	0.96

Only the first canonical pair has statistical significance.

---

122. For a more complete discussion of these concepts, see JCC.

$$F = \frac{(1 - R^2_{Y,X})}{(S-1)*(\frac{v}{v})}, \quad \textbf{where } S = \frac{(\sqrt{k_Y^2 * k_X^2} - 4)}{k_Y^2 + k_X^2}, \quad v = k_Y \times k_X, \quad v = ms + 1 - (\frac{v}{2}),$$

$m = (n - k_G) - \frac{k_Y + k_X + 3}{2}$, and k_G = the smaller of the number variables in X and Y sets.

123. Hand calculated and verified by SysStat.

NCSS, Systat, MyStat, and OpenStat, also provide tests of significance for each of the canonical pairs, plus four overall tests.[124]

Step 4 interprets the significant Y and X pairs, by correlating the actual Y and X variables against their respective principal components, a calculation of the component loadings.

Table 26.3 shows the component loadings between the Y manifest variables and component Y1 (the canonical Y principal component) and table 26.4 shows the component loadings of the X variables and component X1, the first X canonical principal component.

Table 26.3 Y Set Component Loadings on Y Component 1

Variables	Loadings
Sales	0.61
IC	0.72
Profit	1.00

Table 26.3 suggests that component Y1 is a generalized performance component.

[124.] This table is part of the OpenStat output.

Overall tests of significance:

Statistic	Approx. Stat.	Value	D.F.	Prob.
Wilkes Lambda	Chi-squared	94.2271	15	0.0000
Hotelling-Lawley	F Test	11.5407	15 194	0.0000
Pillai Trace	F Test	4.9013	15 204	0.0000
Roy's Largest Root	F Test	36.9114	5 71	0.0000

With respect to the Type I error, Pillai's Trace is the least conservative test, i.e., most likely to reject the null, followed by Wlikes's Lamda, Hotelling's Trace, and Roy's root in that order.

Table 26.4 X Set Variables Component Loadings on Component X1.

X Set Variables	Loadings
Crime	-0.73
Socioeconomic	0.51
Race ethnic	0.94
Population	0.72
Neighborhood Stability	0.09

Table 26.4 suggests that component $X1$ is a generalized quality of neighborhood component.

What do we conclude from this analysis? First, neighborhood quality is significantly related to performance. And we conclude that quality of neighborhood explains about 72% of the managerial performance. Of course, there are serious questions about this model specification because we did not include managerial variables.

Principal Component Regression

Another way to approach the problem of multiple dependent variables runs a canonical correlation analysis on the X and Y sets and makes a Rao's F test as an overall test of significance. If we find the F test significant, we return to the simple principal components routine, generate the number of X and Y components as described in chapter 25, and run a series of univariate multiple regressions on whatever number of canonical Y variables are generated. This is permissible because Rao's F test protects us from the type I error. Table 26.5 shows the eigenvalues of a simple PCA on the Y set.

Table 26.5 Eigenvalues of a Three PC Solution of the Y set

PC No.	Eigenvalue	Percent	Cumulative Percent
1	2.330	77.66	77.66
2	0.395	13.17	90.83
3	0.275	9.17	100

One principal component explains about 77% of the variance in the managerial performance set. So we extract one PC to represent the Y set. Then we repeat the PCA on the X set. Table 27.6 depicts the eigenvalues of that analysis.

Table 26.6 Eigenvalues and Percents of a Five Component Solution to a PC of the X Set

PC No.	Eigenvalue	Percent	Cumulative Percent
1	2.459	49.17	49.17
2	1.036	20.73	69.90
3	0.838	16.75	86.65
4	0.419	8.39	95.04
5	0.248	4.96	100.00

Using an Eigenvalue of one as the criterion for the number of factors, Table 26.6 suggests a two component extraction. Table 26.7 shows the component loadings on a two component solution.

Table 26.7 Unrotated Component Loadings of a Two Component Solution to a Simple PCA of the X Set

Variables	Comp. 1	Comp. 2
CRIME	0.87	-0.07
SE	-0.60	-0.32
RE	0.81	-0.07
POP	-0.83	0.04
NS	0.04	-0.96

X component one is a general neighborhood quality component while component two is a neighborhood stability factor. Using this information we run an OLS multiple regression where Y is PCY1 and the Xs are PCX1 and PCX2 or

$$PCY1_i = B_0 + B_1 PCX1 + B_2 PCX2 + \epsilon$$

Table 26.8 summarizes that calculation.

Table 26.8 Regression of PCY1 on PCX1 and PCX2

	DF	MR^2	F Ratio	Probability
Intercept	1			
Model	2.00	0.48	33.07	0.00
Error	72.00	0.52		
Total	74.00	1.00		

Independent Variables	Regression Coefficient	Standard Error	T Value	Probability
Intercept	0.00	0.08	0.00	1.00
General Quality of Neighborhood_ PCX1	0.69	0.09	8.11	0.00
Neighborhood Stability PCX2	0.05	0.09	0.62	0.54

The overall model, statistically significant at P less than 0.00 shows the general neighborhood quality component, PCX1 as statistically significant and positive. The neighborhood stability (PCX2) variable is not statistically significant.

An additional possibility in this arena runs a multiple regression with PCY1 as the dependent variable and the five raw score Xs as independent variables or

$$PCY1_i = B_0 + B_1X1 + B_2X2 + B_3X3 + B_4X4 + B_5X5 + \epsilon_i$$

Table 26.9 summarizes that regression.

Table 26.9 Regression of *PCY1* on All the Variables in the *X* Set

	DF	R²	F Ratio	Probability
Intercept	1.000			
Model	5.000	0.5724	18.475	0.000
Error	69.000	0.4276		

	Regression Coefficient	Standard Error	T Value	Probability
Independent Variables				
Intercept	-2.0679	1.4914	-1.387	0.1700
CRIME	0.3104	0.2527	1.228	0.2235
NS	-0.0866	0.0506	-1.714	0.0911
POP	-0.0246	0.0199	-1.236	0.2206
RE	0.1028	0.0182	5.661	0.0000
SE	0.0000	0.0000	0.724	0.4717

The overall model, statistically significant at *P* less than 0.000, shows the race-ethnic variable significant at *P* less than 0.0000 and neighborhood stability significant as a one tail test at *P* less than 0.1000.

To defeat the collinearity in the *X* set, we could develop a structural equation model by ranking the *X* variables from high to low importance in explaining performance. Table 27.10 shows that model.

Table 26.10 Structural Equation Model of Total Effects

	Variable	Incremental Regression Coefficient	Incremental Standard Error	Incremental T ratio	Incremental Probability	Incremental R Square	Total R Square
Step 1	Population	-0.09	0.0163	-5.536	0.0000	0.2957	0.2957
Step 2	Crime	0.7162	0.2889	2.479	0.0155	0.0554	0.3511
Step 3	Race ethnic	0.0949	0.0168	5.660	0.0000	0.2014	0.5525
Step 4	Socio economic	0.0000	0.0000	0.468	0.6410	0.0000	0.5502
Step 5	Neighborhood Stability	-0.0866	0.0506	-1.714	0.0911	0.0222	0.5724

The structural equation model seems more informative. The variables population, crime, and race-ethnic are statistically significant, with smaller populations increasing the probability of managerial effectiveness; lower crime and higher white-Anglo also predicting higher managerial performance. The socio-economic and neighborhood stability variables seem to have no effect.

Jacob Cohen's Set Correlation[125]

We believe Jacob Cohen's solution to the problem of regression with multiple dependent variables (multivariate analysis) is best. Cohen's approach requires the hardest thinking about model specification because of its use of structural equations. Significant amounts of multicollinearity, minimally requires specification of hierarchies for the X and Y variable sets.

Cohen's procedure requires running a canonical analysis using Rao's F to test the null hypothesis of no relationship between the Y and X

---

125. JCC, chapter 16, pp. 608-625.

sets, our protection against the increased threat of type I errors. A significant relationship between the joint variance of the two sets entitles running OLS multiple regressions of each Y variable on the independent variables in the X set. Multicollinearity forces us to deal with that problem in some form. If multicollinearity is low in both sets, as for example in a situation using three dependent and three independent variables, we can run three fully partialled regressions. If confronted by highly collinear independent variables and relatively orthogonal dependent variables, we can do the analysis using three structural models.

If Y and X sets are both highly collinear, then, we must develop structural equations for both the X and Y sets. This approach requires running three different structural models. The first model would regress the most important Y variable on the hierarchy of X variables. The second model partials the first Y variable from the second, and runs the residuals on the X hierarchy. The third model partials the first two Y variables from the last Y in the hierarchy and runs those residuals on the X hierarchy. This sounds a bit complicated, but Systat and MyStat have modules that do everything required.

To illustrate, we use the same data set from the previous discussion, three Y variables, and five X variables. To avoid the legitimate sniping about model specification, we simply call the three Y variables $Y1$, $Y2$, and $Y3$. In the same spirit we call the variables in the X set $X1$, $X2$, $X3$, $X4$, and $X5$. We use on both sets a structural equation model.

The first model regresses $Y1$ five times. The first is $Y1$ on $X1$. The second regresses $Y1$ on the residuals of $X2$ with $X1$ partialled. The third regresses $Y1$ on the residuals of $X3$ with $X1$ and $X2$ partialled. The fourth regresses Yi on the residuals of $X4$ with $X1$, $X2$, and $X3$ partialled. The last regresses $Y1$ on the residuals of $X5$ with $X1$, $X2$, and $X3$, and $X4$ partialled.

The second model regresses the residuals of $Y2$ with $Y1$ partialled on $X1$. Then the model regresses the same Y residuals on the residuals of $X2$ with $X1$ partialled. The analysis continues through the fifth X variable as described in model one.

The third model regresses the residuals of $Y3$ with $Y1$ and $Y2$ partialled on $X1$ and continues as described in models one and two.

Table 26.11 shows the first model output using Cohen's Set Correlation module.

Table 26.11 Model 1

Rao's F 9.369
$df = 12.0, 180.2$

Probability = 0.0000

$R^2 = 0.723$

		Model 1: Variable Y1				
	Independent Variable	Incremental Correlation Coefficient	Incremental Beta	Incremental Standard Error	Incremental T Ratio	Incremental Probability
Step 1	X1	-0.796	-0.796	0.071	-11.25	0.000
Step 2	X2 with X1 Partialled	-0.213	-0.213	0.067	-3.197	0.002
Step 3	X3 with X1 and X2 Partialled	0.136	0.136	0.065	2.08	0.041
Step 4	X4 with X1, X2, and X3 Partialled	0.057	0.057	0.065	0.871	0.387
Step 5	X5 with X1, X2, X3, and X4 Partialled	0.144	0.144	0.063	2.226	0.027

We note first the statistical significance of Rao's F allowing us to proceed to a series of OLS multiple regressions on each of the three ZY variables. Note the identity of the correlation coefficients and the betas. The reason for that is the analyses are run on the Z scores of the variables and, as we explained in an earlier discussion, the regression coefficients in a Z analysis exactly equal the correlation coefficients.

Table 26.11 shows four significant variables: X1, X2, X3, and X5. These effects are on the unpartialled Y1.

Table 26.12 gives the results of the second model: Y2 with Y1 partialled regressed on the hierarchical Xs.

Table 26.12 Model 2

Model 2:

Y2 with Y1
Partialled

	Independent Variable	Incremental Correlation Coefficient	Incremental Beta	Incremental Standard Error	Incremental T Ratio	Incremental Probability
Step 1	X1	0.006	0.006	0.071	0.09	0.928
Step 2	X2 with X1 Partialled	0.084	0.084	0.117	0.711	0.479
Step 3	X3 with X1 and X2 Partialled	-0.032	-0.032	0.118	-0.274	0.785
Step 4	X4 with X1, X2, and X3 Partialled	0.235	0.235	0.116	2.034	0.046
Step 5	X5 with X1, X2, X3, and X4 Partialled	0.021	0.021	0.116	0.182	0.856

In this model, only X4 shows statistical significance. Table 26.13 shows the results of the third model: Y3 with Y1 and Y2 partialled regressed on the hierarchical X set.

Table 26.13 Model 3

Model 3:
Y3 with Y1 and
Y2 Partialled

Independent Variable	Incremental Correlation Coefficient	Incremental Beta	Incremental Standard Error	Incremental T Ratio	Incremental Probability
Step 1					
X1	-0.007	-0.007	0.072	-0.101	0.919
Step 2					
X2 with X1 Partialled	-0.006	-0.006	0.119	-0.047	0.963
Step 3					
X3 with X1 and X2 Partialled	-0.044	-0.044	0.119	-0.374	0.709
Step 4					
X4 with X1, X2, and X3 Partialled	-0.076	-0.076	0.119	-0.639	0.525
Step 5					
X5 with X1, X2, X3, and X4 Partialled	0.033	0.033	0.12	0.272	0.786

There are no significant variables in this model.

We reiterate: Jacob Cohen's approach to analyzing problems with more than one dependent variable uses canonical analysis to test the hypothesis that the variance of the X set is significantly related to the variance of the Y set. This is equivalent to the protected F test in OLS multiple regression. In the canonical analysis, if Rao's F is statistically significant, we reject the null hypothesis of no relationship between the common variance of the X set and the common variance of the Y. The rejection of the null entitles us to proceed to a series of multiple regressions on the Y variables because we are protected from the type I error. The form of the model specification depends on the degree of multicollinearity in the two sets. If collinearity is relatively low, an analyst might opt for a series of regressions without building theoretical structural equation models as we did in this example. Whatever the case, Cohen's set correlation procedure, available in Systat, makes the procedure a cake-walk. O, if you do not have access to Systat, the procedure can be jerry-rigged from any of our recommended statistical programs.

Multinomial Logit

Earlier, we discussed a problem using a nominal dependent variable in which the nominal variable was restricted to two categories—for example, promoted/not promoted or won/lost, nominally coded one and zero. In this chapter, all our examples used ratio or interval hypothetical variables. Now, we address a circumstance in which the number of nominal categories exceeds two. This requires using an analytical technique that extends binary logit to multiple categories and provides a protected chi-square test equivalent to Rao's F. In the case of categorical dependent variables the test is chi-square. The interpretations are the same as a binary logit. Of the programs we recommend, only NCSS, Systat, MyStat, and SPSS have routines that do multinomial logit.

Summary

The critical ideas discussed in this chapter are the following:

1. Canonical correlation

    The method of Principal Components lies at the heart of canonical analysis, using principal components as dependent and independent variables. Canonical Analysis extracts Principal

Components from the X and Y sets and runs correlations on the X and Y PCs. The extraction of canonical principal components (CPCA) differs from simple principle components analysis (PCA) discussed in chapter 25. In simple principal component analysis, the first PC (PC1) captures the maximal amount of variance in the data set. In canonical principal component analysis (CPCA), the procedure extracts *pairs of principal components from the Y and X sets*, canonical PCY1 and canonical PCX1, . . . canonical PCYn and canonical PCXn, so that the pairs of canonical PCs are maximally correlated. The first canonical pair has the highest correlation; the second has the next highest, and so on.

2. Principal component regression

Regression with principal components first runs a Canonical analysis to make Rao's F test. If statistically significant, the analysis proceeds to extract principal components from the Y and X sets. Then the analysis uses OLS regressions of each Y principal component on the X set of principal components. This analysis works effectively with straightforward components.

3. Jacob Cohen's set correlation

Cohen's procedure requires running a canonical analysis using Rao's F to test the null hypothesis of no relationship between the Y and X sets, our protection against the increased threat of type I errors. A significant relationship between the joint variance of the two sets entitles running OLS multiple regressions of each Y variable on the independent variables in the X set. Multicollinearity forces us to deal with that problem in some form.

4. Multinomial logit

Multinomial logit addresses the problem of nominal dependent variables with more than two categories. The nominal variable equivalent of Cohen's procedure, Multinomial logit, uses a chi-square to test the null hypothesis of no relationship between the Y nominal set and any set of X variables. Interpreting the output of this procedure parallels the earlier discussion of Logit regression.

Appendix

Chapter 26

This data set contains three dependent variables labeled Y1, Y2, Y3; and five independent variables labeled X1, X2, X3, X4, and X5 with a sample size of 50.

Your tasks are the following: (1) Give names to the Y and X variables consistent with your interests or your job requirements, (2) run a canonical analysis using all the variables in the data set, (3) run OLS regressions on principal components extracted from the data set, (4) run Jacob Cohen's set correlation on the data, and (5) explain your findings and compare the three analyses.

ID	Y1	Y2	Y3	X1	X2	X3	X4	X5
1	52.43	40.39	-233.84	-36.90	87.77	24.33	17.38	65.02
2	89.91	51.12	188.69	83.01	91.89	32.07	20.08	74.30
3	124.53	61.32	607.78	194.90	107.75	37.43	26.03	84.48
4	103.97	45.18	88.29	56.42	76.42	33.41	20.89	69.70
5	105.20	43.32	103.45	52.21	75.04	24.83	21.07	66.59
6	86.93	51.69	186.33	69.07	99.52	31.69	22.10	77.94
7	73.75	45.41	22.79	23.20	81.92	24.47	16.41	64.69
8	118.71	44.36	318.42	87.28	96.60	39.16	26.24	76.16
9	92.22	55.03	271.21	102.60	102.66	32.02	24.18	80.46
10	103.79	46.32	187.09	85.14	93.32	32.76	20.69	67.77
11	101.54	57.37	412.57	119.30	102.20	38.06	25.87	92.72
12	103.47	58.14	428.68	139.10	106.46	31.25	20.95	68.93
13	117.19	43.56	225.37	80.95	95.17	33.29	23.25	74.10
14	9770%	34.69	21.02	20.43	87.69	30.28	20.77	79.16
15	121.22	56.84	557.70	158.50	114.04	34.03	25.10	85.91
16	106.92	58.24	403.94	147.30	133.18	38.01	24.39	98.57
17	106.64	49.75	270.55	89.25	61.73	30.53	17.33	72.45
18	104.67	46.92	185.66	68.84	106.35	37.03	21.09	67.78
19	94.57	42.38	39.41	28.29	60.65	28.16	13.83	52.10
20	99.2	54.42	333.21	103.40	82.12	28.19	19.62	71.68
21	108.23	48.06	296.46	92.90	91.58	38.53	20.72	85.47
22	107.5	61.10	548.33	152.00	107.83	35.15	20.94	94.78

23	82.26	44.45	32.98	20.30	85.37	27.92	20.98	74.84
24	77.58	35.15	-294.15	-16.19	84.50	25.14	22.38	49.10
25	88.97	36.18	-43.06	-0.61	79.77	26.71	19.91	71.85
26	116.69	59.21	544.61	185.90	116.02	35.13	20.93	78.77
27	71.93	42.40	-97.87	-5.34	87.38	24.36	17.14	69.02
28	113.46	61.88	526.45	175.10	88.01	29.00	20.63	71.19
29	114.2	58.74	534.88	170.90	96.72	32.56	24.27	87.27
30	91.73	55.73	250.99	99.69	110.55	26.95	18.61	64.57
31	102.28	49.43	281.65	107.30	76.32	36.11	20.46	59.86
32	101.15	50.12	235.90	101.60	95.41	26.43	17.24	72.05
33	82.9	49.54	56.40	53.22	84.49	32.50	20.83	56.66
34	61.13	42.13	-154.52	-12.88	69.30	24.72	16.41	66.37
35	75.44	48.67	36.19	31.77	82.46	19.85	18.88	76.84
36	97.59	38.15	102.43	34.81	93.41	30.18	18.90	64.19
37	82.07	45.42	81.51	28.37	70.04	22.97	17.22	44.26
38	89.48	47.55	187.31	51.89	98.92	26.95	19.05	58.85
39	119.68	58.98	518.70	151.70	87.38	37.96	19.26	76.87
40	82.98	47.51	61.74	31.34	76.24	29.21	20.71	70.37
41	83.28	49.12	33.85	48.42	94.45	25.52	23.15	66.48
42	93.91	53.51	325.28	96.96	103.23	29.85	19.29	84.58
43	90.37	40.10	-17.49	14.34	114.52	31.37	19.16	63.30
44	121.18	57.01	577.95	164.10	78.82	38.07	22.34	73.31
45	81.03	46.12	21.27	34.95	89.95	27.26	18.33	58.22
46	109.16	38.33	82.73	58.11	72.21	29.06	19.77	71.44
47	97.48	52.35	230.79	92.71	92.05	31.10	19.88	67.41
48	76.23	49.37	65.46	42.60	83.68	26.91	17.50	48.10
49	87.88	49.43	137.34	58.50	94.84	27.59	19.42	78.43
50	86.96	47.30	94.86	53.70	77.08	26.78	15.00	75.27

INDEX

A

Abrupt Constant Change, 422, 434, 489
absolute value, 417
alpha risk, 122–24, 126–30, 132, 136, 141, 143, 145, 177–78, 196
Analysis of variance, 135–37, 142, 144, 149, 193, 489
ARIMA (AutoRegressive Integrated Moving Average), 365, 401, 408, 410–12, 414–22, 429, 431, 433, 489
Arithmetic Mean, 8, 37–38, 40, 46, 51–52, 60–61, 75–76, 80, 192
autocorrelation, 223–24, 270–73, 370, 405, 412, 414
  detecting, 273
  first-order, 273–75, 284
  lagging and, 405
  plot for sales, 406–7, 412–15, 428
  positive, 223, 272
autonomous, 189, 197
autonomous values, 189, 197, 489
AutoRegressive Integrated Moving Average (ARIMA), 365, 401, 409–11, 420
average, 37, 46, 76
  moving, 365, 371–76, 382–83, 385–93, 398–99, 401, 408–11, 414–15, 418–20
  double, 388–90, 393, 398
  Holt's exponential, 393
  simple, 385–87
  single, 385, 387, 398
  single exponential, 391, 398
 single exponential weighted, 390

B

B_0, 187–91, 197–99, 209–10, 224–25, 294–95, 308–11, 320–21, 356–57, 388–89
B_1, 187, 189–90, 195–99, 205–6, 209–10, 214–15, 352–54, 388–89, 393–95
bar chart, 35, 46, 489
"best fit," 38–39, 46, 189, 338
beta risk, 120, 123, 129–30, 132, 489
 more on the idea of the, 129
bias, 99–100, 103, 105–6, 108, 221–23, 225, 270, 275, 344–46
 and mistake, 99–100, 105–6
bivariate, 153–54, 186, 188–90, 195–98, 201–5, 217, 225–26, 232–33, 446
bivariate regression, 153, 188
bivariate relationship, 153, 186, 210, 489
Box-Jenkins approach, 365, 489
Box-whisker diagrams, 67, 489

C

calendar, 368, 370, 373, 489
 effects, 370
 variation, 370, 373
Camp-Meidell inequality, 55, 71, 489
Canonical Components Analysis (CCA), 471
canonical correlation, 470–71, 475, 484
case for sampling, 100
categorizing data, 22, 25, 30. *See also* nominal data
causation, 5, 10, 155, 160–65, 224, 289, 343–44, 347, 494
 circularties, 10, 155, 160–65, 224, 289, 343–44, 347, 355, 361

distinction between tertium quid and, 160
reciprocal, 224, 361, 465
relationships and, 5, 10, 155, 160–65, 224, 289, 343–44, 347, 494
cause, 50, 160–61, 163
CCA (Canonical Components Analysis), 471–72, 489–90
centering, 234, 241, 278, 280, 286, 309, 311, 317–19, 372–73
central tendency, 5, 31, 36–37, 40, 42–43, 46, 48, 50, 74
Chebycheff's inequality, 56, 490
chi-square, 15, 135, 144, 147, 150, 334, 339, 417, 484–85
classical decomposition, 6, 367, 370, 373, 382–83, 418, 490
classical regression model, 189, 217, 490
classical time series analysis, 383
class interval, 33, 35, 490
coefficient, 174–76, 178–80, 192–93, 197–98, 205–9, 219–21, 258–62, 291–97, 454
  component, 462
  of correlation
    defined, 174
    measured, 174
    statistical inference for the, 176
  of determination, 138, 193, 198, 206–8, 214, 247, 464
  of kurtosis, 66–67, 98
  of nondetermination, 193, 198
  of skewness, 63, 70, 81, 83, 98
  of variation, 50, 56, 72, 104
  zero order correlation, 213
Cohen, Jacob, 15, 248, 250, 440
  set correlation, 470
collinearity, 225–26, 234, 241, 279–80, 286, 292, 343, 361, 465–67
communalities, 450–51, 453, 463–64, 490
concave, 217
confidence interval, 98, 112, 178, 194–95, 211, 387–88
confirmatory, 443, 452, 465, 490
confounders, 163, 213, 420–21, 431, 435, 490
consistency, 20, 219, 490
construct validity, 220–21
Continuous data, 21, 29, 490
continuous logit OLS, 327. See also proportion
convex, 217
correct model specification, 217, 222, 224, 229, 281, 286
  of independent variables, 217, 222, 224, 229, 281, 286
correlation
  coefficient, 176, 179–80, 183–84, 192, 219–20, 234, 283, 445, 454
  matrix, 202–4, 213, 247, 445, 449, 458, 472
  part-whole, 181
counts, 8, 144–45, 158, 203, 289, 327, 331–34, 340–41
covariance
  defined, 166
  measured, 167
covariates, 344, 490
criterion validity, 220–21, 490
Cronbach's alpha, 219, 490
cross correlation, 431, 435
cross-sectional data, 28–29, 160, 367, 430, 433, 490
curvilinear, 217, 231, 233, 235, 240–42, 277, 315, 334, 340
cycle, 367–70, 381
cyclical data, 367–68, 370–71, 377, 379–83, 408

D

data, 18–25, 28–35, 37–42, 44–48, 226–30, 244–50, 262–66, 302–6, 333–38
  cardinal, 22–25, 30, 198
  ordinal, 23–24
data set, 15, 19–20, 33–34, 47–48, 244–45, 258–59, 302–3, 305–6, 485–86
D1, 403–4, 414, 428
DD1, 403–4, 490
DDD1, 403, 490
Degrees of freedom, 112, 139–41, 149, 221, 246, 256, 282, 334, 339

departures from normality, 59
dependent variable, 154–55, 188–89,
    212–13, 230–31, 327–31, 334–38,
    340–41, 346–47, 350–51
deviation, 50–59, 70–72, 78–79, 104,
    106–12, 176–77, 194–95, 210–11,
    386–87
dfbetas, 258–59, 265, 491
DFFITS, 256–57, 259, 265, 491
diagonal, 203, 254, 259, 445, 459
Dichotomous Dependent Variables, 6,
    327, 491
differencing, 365, 401–4, 407–8, 413–17,
    420, 429, 491
direct effects, 355
discrepancy, 255–56, 260, 265–66, 491
discrete data, 21–22, 29, 45, 491
dispersion, 36, 50, 56, 60, 72, 74, 491
double moving average, 388–90, 393,
    398. See also under average
dummy coding, 296, 298–99, 303–6

E

effects coding, 291, 298–300, 303,
    305–6, 338
effects coefficient, 189–90, 491
e_i, 188–89, 191, 193, 202, 211, 408–9,
    491
eigenvalues, 447–48, 450–53, 456, 459–
    64, 466, 472, 475–76, 491
empirically estimating probability, 92
empirical sampling distribution of the
    mean, 106
endogenous, 351, 357, 491
equiprobability, 92–94, 491
error, 87–89, 98–100, 102–6, 110–13,
    115–17, 119–27, 209–11, 260–64,
    385–88
error term, 189, 198, 205, 275, 284, 357,
    491
estimating probability, 91–93
EV, 137–40, 142, 149, 191, 193, 208–9,
    491
exogenous, 186, 351, 357, 491
experimental, 92, 162–64, 343, 365,
    421, 433–34

explained variance, 139–40, 193, 208,
    458–59, 491
exploratory, 443, 449, 459, 465, 467,
    491
external standardized residual, 255,
    265

F

FA (factor analysis), 439, 441–46, 448,
    457–59, 461–62, 465–67
face validity, 220
factor analysis (FA), 439, 441–46, 448–
    49, 457–59, 461–62, 465–67, 491
factor loadings, 447, 450–51, 453, 455–
    56, 461, 464–66, 491
factors, 153–54, 365, 439, 441, 443,
    445–53, 455–57, 465–67, 470
first moment, 38, 60
forces producing a relationship, 159
form of relationships, 217
fourth Moment, 64, 66, 72
F ratio, 139–41, 143, 149, 209, 219, 295,
    302, 477–78
F ratios, sampling distribution of, 141,
    149
frequency distribution, 9, 32–35, 39, 46,
    48, 74, 85, 107, 125
    creating a, 32
functional form, 186–87

G

generalizability, 164, 262, 491
generalizable, 421
Geometric Mean, 74–78, 83, 491
goodness of fit, 386, 392–93
Gradual Constant Change, 423, 435,
    491
Gradual Temporary Change, 424, 434,
    491
Grand Mean, 137–38, 142, 298–99, 302,
    338, 491
Graphic Portrayal of Frequency
    Distributions, 35

H

Harmonic Mean, 74, 77–78, 83–84, 492
Hat Diagonals, 254–55, 259, 265, 492
heteroscedastic, 222, 275, 285, 333, 413, 492
heteroscedasticity, 5, 222, 270, 275–78, 285, 331, 333, 340, 413
    correcting for, 277
    definition of, 285
    patterns of, 276
    remediation of, 5, 222, 270, 275–78, 285, 331, 333, 340, 413
heuristic, 410, 444
histogram, 35–36, 46, 67, 69, 73, 107
Holt's exponential moving average, 393. See also under average
Holt's linear trend, 393
homoscedasticity, 222, 418–19, 429, 492
Hypothesis Testing, 5, 88, 119, 131, 135
Hypothesis Testing and Theory, 131, 492

I

importance of probability to inferential statistics, 97
independent variable, 213–14, 221–22, 254, 258–59, 264–65, 278–80, 291–92, 309–10, 355–56
indirect effects, 351–55, 361
inferential statistics, 5, 87, 89, 91, 97–98, 106, 208, 250, 492
influence, 39, 42, 54, 95, 163, 221, 256–57, 265, 390
Instat, 14, 492
instrument, 20, 22, 29, 163, 219, 358–60, 421
Integrated, 410
interaction, 278, 289, 308–24, 346
Intercept, 187, 206, 208, 233, 235, 260–61, 316, 332, 477–78
intruded time series, 365, 420–21, 433–34
invertibility, 417, 492
irregular variation, 368, 370–72, 374, 376–78, 382–83

K

K^2, 193–94, 198
kurtosis, 36, 63–64, 66–67, 72, 74, 79, 81–83, 98, 283
    coefficient of, 66–67, 98

L

lagged autocorrelation coefficients, 406
lags, 274–75, 284–85, 389, 405, 408, 411, 414–15, 417–18, 432–33
leptokurtic, 65, 67, 72, 109, 492
level of Significance, 123, 470
leverage, 254–57, 259–60, 265–66, 492
linear, 15, 187–88, 217–18, 232–33, 239–40, 285–86, 314, 393–95, 402–3
linearity, 217, 222, 229, 231, 238–42, 282, 327, 492
linear probability model, 337, 341
line chart, 35, 492
Ljung-Box Portmanteau Lack-of-fit Diagnostics, 416–17, 429
logarithmic transformations, 80–81, 492
logarithms, 33, 76, 84, 238, 242, 492
Logit, 327–30, 335–41, 343, 484–85
Logit regression, 330, 485, 492
lower whisker, 252
lowess, 231–33

M

m1, 38, 46, 60, 126, 388–89, 398
m2, 39, 46, 52, 56–57, 66–67, 71–72, 126, 388–89, 398
m3, 60–63, 72
m4, 66–67, 72
main effect, 309, 311–12, 316, 318, 321–23
maxima, 236–37, 492
Maximum Likelihood, 248, 333, 337–38, 341, 415–16, 492
maximum likelihood estimation (MLE), 248, 333, 415
mean difference, standard error of, 126
mean differences, sampling

distribution of, 125

Mean Substitution, 246–50, 263, 266, 493

mean substitution, with companion dummy variables, 248

measurement, 19–21, 23, 29, 52, 54–55, 99–100, 163, 219, 421

median, 37, 41–47, 61–62, 67–70, 113, 116–17, 132, 230, 252

mesokurtic, 64, 67, 72, 493

midrange, 37, 44–47, 493

missing data, 228–30, 244–49, 262–64, 286, 302–5, 493

Missing Value Imputation Using OLS Regression, 248

MLE (maximum likelihood estimation), 248, 333, 338–39, 343, 411, 415, 419, 492–93

mode, 37, 45–47, 55, 71, 493

Modeling Interventions, 422, 493

model specification, 5, 217, 222, 224, 229, 270, 281, 346, 479–80

moment, 28, 30, 38–39, 52, 60–61, 63–64, 66, 70–72, 422

monotonic, 230–32, 237–38, 240–42, 493

multicollinearity, 5, 225, 229, 270, 278–81, 285–86, 311, 479–80, 484–85

detecting, 279, 286

remediating, 279

multiple coefficient of determination, 208, 247

Multistage Area Random Samples, 102, 493

multivariate, 12, 186, 201, 249–50, 253–54, 264–65, 439, 441, 479

mutual causation, 289

MyStat, 13, 85, 471, 473–74, 480, 484, 493

N

n – 1, 143, 417

noise, 10, 39, 46–47, 52, 54–55, 153–54, 193, 411, 421–22

nominal data, 25, 30, 40, 58, 72, 144, 150, 302, 327

nominal variable coding, 6, 289, 291,

305, 493

nominal variables, 58–59, 153, 157–58, 171–72, 291, 293, 299–300, 304–5, 319

nonlinear, 233, 402

nonmonotonic, 230–31, 233, 241, 493

non-normal residuals, 224, 282–83, 286

detecting, 283

nonrecursive effect, 187

Nonseasonal Differencing, 416–17, 429, 493

non-stationarity, 408

normal curve, 53–54, 64, 74, 113, 116, 128

normality, 5, 52, 54–55, 59, 63–65, 67, 72–73, 108, 283

normal plot, 283

null hypothesis, 122, 124–27, 132, 139–41, 143–44, 146–47, 150–51, 210, 484–85

Numerical Limits, 93, 493

O

observation, 33, 41–45, 54, 68–70, 80–81, 224, 254–62, 264–66, 462–63

odds ratio, 97, 328–29, 339–40

one-and-two tailed test, more on the notion of, 128

one-tail test, 122, 124, 127–28, 132, 332, 334, 493

OpenStat, 13, 85, 281, 283, 385, 471, 473–74, 493

ordering, 22–23, 30, 362

ordinary least squares, 5, 186, 281, 286, 340, 408, 493

orthogonal, 280, 441–42, 454, 463, 465, 467, 480

outlier, 39, 62, 68, 76, 78, 226–28, 254–62, 264–66, 268

outliers, 39, 45–47, 62, 67–69, 76–77, 226–30, 250–62, 264–66, 268

P

pairwise deletion, 247, 263, 493

panel data, 29–30, 160, 493

parameter, 98–100, 111–12, 120, 127,

130, 135, 423

partial, 205, 207, 215, 275, 285–86, 357, 414–15, 428, 431

partial autocorrelation, 414–15, 428

partial autocorrelation plot, 414–15, 428

partialling, 194, 204–5, 222, 225, 233, 278, 293, 344, 410

path diagram, 350–52, 355–56, 358, 433, 494

path model, 355, 421

pdq, 411, 415–16, 429, 494

PDQ, 411, 415–16, 429

percentage, 17, 25–26, 30, 41, 53–56, 88–89, 144–45, 386–87, 451–52

percentile ranks, 17, 26–28, 30, 43, 335–36, 494

percentiles, deciles, and quartiles, 26

platykurtic, 65, 67, 72, 82, 109, 494

Poisson
  probability, 331, 333–34, 340, 494
  regression, 333–34

polygon, 35–36, 107

population, 31, 36–37, 59, 87–89, 98–116, 122–24, 133, 177–79, 208–10
  frame, 99, 178
  size, variance, sample error, and sample size, 103

powers, 238, 242

power transformations, 82–83, 238, 240, 242, 266, 494

principal components analysis, 6, 441, 467, 494

probability
  definition of, 91
  measures of, 93

Probit, 336–37, 494

proportions, 40–41, 59, 113–15, 117, 127, 143–44, 327–29, 334–35, 340–41
  theoretical sampling distribution of, 113

Q

Q statistic, 417

quadratic equation, 233–35, 241, 266,

277–78, 280, 378–79, 381

quadratic polynomial, 233, 494

R

R^2, 193, 203–5, 217–18, 227–28, 260–62, 276, 292, 310, 316

random
  error, 88, 99–100, 105–6, 108, 112, 116, 148, 210, 219
  samples, 101–6, 108, 112, 115, 133, 169–70, 300, 305, 315
    unrestricted, 101–2, 106, 111, 133, 150–51, 159
  walk, 402, 412

ranks, 6, 8, 17, 24, 26, 30, 327, 334–36, 340–41

rates, 20, 24–26, 29–30, 76–77, 182, 331, 421, 423, 425
  of change, 25, 30, 76

ratio variables, 22, 332

reciprocal, 77, 84, 238, 242
  causation, 224, 361, 465
    See also under causation
  transformations, 81–82, 84, 238, 242

recursive effect, 187

reduced-form equations, 343, 433

reexpression, 78–80, 84–85, 234, 241, 309, 457, 467, 494

reference group, 293–300, 302–5, 337–38, 345, 494

regression, 186–99, 201–6, 208–15, 221–35, 252–66, 275–82, 294–306, 327–41, 343–50
  coefficients, 209, 213–15, 217–19, 221–22, 224–26, 258–59, 275–76, 278–79, 297–99
  equations, 388–89, 393, 398
  line, 192–93, 197, 208, 218, 225–27, 232–33, 258, 275, 285
  principal component, 280–81, 475, 485
  ridge, 281, 286
  slope, 197–99, 214
  of the square root of counts, 332
  standard deviation of, 194, 198, 211

regular variation, 369–70, 382, 408, 411,

415, 429
relationship, 52–55, 144–45, 153–61,
    164–76, 181–83, 186–88, 230–34,
    237–42, 356–59
  between two nominal variables, 158
  negative, 169–70, 174
  positive, 156, 158, 168–69, 172, 174,
    192, 241
relationships, between nominal,
    cardinal, and interval variables,
    157
reliability, 20–21, 29, 219–20, 222, 421,
    465, 495
repeatability, 20, 219, 495
repeated trials, 94, 495
residuals, 193–94, 204–6, 221–24,
    231–33, 270–78, 282–86, 410–11,
    416–22, 480
  constant variance of, 222, 275
restricted random samples, 102
"robust," 39, 42, 47, 222, 260–61, 266,
    277
roots, 75–77, 80, 82, 238, 242, 331, 333,
    495
rotated factors, 456–57, 461–63
rotation, 455, 457, 461–63, 466–67, 495
Rstudent, 255–57, 259, 265, 495

S

samples, 31, 87–89, 98–117, 119–27,
    133–37, 139–40, 142–45, 177–80,
    195–96
  convenience, 99, 103, 105
  methods of choosing, 101
  purposive, 103
  types of, 105
sampling
  distribution of F ratios, 140
  distribution of R, 177
  purpose of, 100
SARIMA model, 411, 415
scatter plot, 230, 401, 495
seasonal
  exponential smooth, 396
  index, 373–74, 377–79, 389, 396
  variation, 369–70, 373, 376–78, 382,

387, 389–90, 396, 408
second moment, 39, 52, 56, 60, 63, 66,
    70–72
serial correlation, 223, 270–71, 495
Shapiro-Wilk (W) test, 283, 495
signal, 10, 39, 47, 52, 88–89, 193, 223–
    24, 254, 265
simple moving average, 385–87. See
    also under average
single exponential moving average,
    391, 398. See also average
single moving average, 385, 387, 398.
    See also average
skewed distributions, 35, 45, 56, 60–62,
    67, 72, 76, 80–82, 239–40
skewness, 36, 45, 60, 62–63, 67, 72,
    81–83, 242, 283
  coefficient of, 63, 70, 81, 83, 98
slope, 158, 172, 189, 197–99, 227, 258,
    314, 395–96, 452
sloped downward, 168
slope of the line, 189. See also under
    regression
smoothing, 365, 385, 390–91, 393,
    401–2, 418
spreadsheet, 15, 19, 47, 73, 85, 150–51,
    183–84, 198–99, 215
SPSS (Statistical Package for the Social
    Sciences), 12–13, 85, 281, 283,
    333, 419, 443, 457, 484
standard deviation, 50–59, 70–72, 78–
    79, 104, 106–12, 115–16, 176–77,
    194–95, 211
  of regression, 194, 198, 211
standard error, 108, 110–13, 115–17,
    123, 125–27, 177, 198, 210–11,
    386–87
  of B_1, 195, 198, 210
  of estimate, 211
  of the mean, 108, 110, 112, 115–16,
    123, 125–26
  of proportion difference, 127
  of proportions, 115, 127
  of R, 177
stationarity, 365, 401, 404, 408, 412–14,
    417–18, 427, 429, 496
statistic, 98–100, 104, 111, 117, 120, 127,

274, 417, 459
statistical
  packages, 12, 47–48, 85, 133, 408,
    411, 471, 495–96
  significance, 130–31, 133, 150,
    176–77, 282–83, 310–11, 318–20,
    406–7, 416–17
  more on the notion of, 130
Statistical Package for the Social
    Sciences (SPSS), 12–13, 85, 281,
    283, 333, 419, 443, 457, 495–96
statistics, 7–12, 14–15, 17–19, 36, 40–41,
    88–89, 97–100, 106, 142–43
Statistix (SX), 12–13, 27, 274, 281, 283,
    392–93, 396–97, 415–16, 418–19
stem-leaf plots, 69, 73, 85, 230, 496
straight line, 80, 171, 187–89, 217, 283
stratified random samples, 102. *See
    also under* random
structural equations, 347, 349–50, 352,
    355, 357, 361–62, 433, 479–80
sum of squares, 39, 50, 57, 137, 496
SX (Statistix), 12–13, 27, 274, 281, 392–
    93, 396–97, 415–16, 418–19, 496
S_{YC}, 198. *See also under* standard
    deviation

T

Tertium Quids, 159–60, 165, 490, 496
theoretical sampling distribution of the
    mean, 108–9
time series, 28–30, 270, 365, 367–73,
    381–83, 401–2, 410–13, 420–22,
    433–36
  analysis, 28, 270, 284, 365, 381, 420,
    433
  autocorrelation in, 275
tolerance, 92, 100, 105, 279, 281, 286,
    496
transformation, 80–84, 238, 240, 242,
    275, 277, 284–85, 327–29, 333–36
t ratio, 261, 311, 332
T ratio, 261, 311, 332
trend, 28, 365, 367–72, 376–83, 387–88,
    393, 396–98, 402, 408
two means, 80, 135, 144

difference between, 80, 135, 144
  testing among more than, 135
  testing the difference between, 125
2, 357, 359–60, 362
two stage least squares, 357, 362
two-tail test, 122, 129
Two-way Causation, 355, 496
type I error, 120–24, 130, 132, 136, 179–
    80, 209, 228, 470–71, 474–75
TYPE II error, 120–21, 130, 132, 220–21,
    234, 241, 247, 263, 496

U

uncorrelated residuals, 223, 229, 270,
    284
underlying structure, 280, 439, 441, 444,
    459, 465
unexplained variance (UV), 138–39,
    149, 193, 195, 208, 370
univariate forecasting, 365, 385, 408,
    411, 418, 496
unrotated factors, 453, 455, 457, 459,
    462–63, 476, 496
unweighted effects coding, 299, 305–6
unweighted grand mean, 298–99
upper whisker, 252
UV (unexplained variance), 137, 139–
    40, 142, 149, 191, 193, 208–9, 496

V

validity, 20–21, 29, 220–22, 421, 465,
    490–91, 496
variables, 50–62, 71–83, 211–15,
    219–22, 246–52, 291–96, 302–6,
    343–52, 445–52
  cardinal, 22, 48, 150, 157, 173, 183,
    259, 261, 304
  interval, 23, 157
variance, 56–59, 102–6, 135–40,
    277–81, 439–43, 445–49, 451–53,
    455–60, 466–67
variance inflation factor (VIF), 279
variation, 50–52, 56–57, 63, 72, 137–39,
    206–8, 368–73, 376–78, 381–83
Venn diagram, 193, 497

VIF (variance inflation factor), 279,
      281, 292, 296–97, 299, 310–11,
      320, 359–60, 497

W

weighted effects coding, 300, 303,
      305–6
weighted least squares, 277
weighted regression, 260, 277, 285, 497
white noise, 370, 411, 414, 417–18,
      421–22, 424, 429, 433

X

X variable, 172, 186

Y

YC, 188–94, 197–98, 205–6, 208, 211,
      276–77, 294–95, 308–12, 316
Y variable, 155

Z

Z score, 79, 129

REFERENCES

CC Croxton, Frederick .E. and Donald J. Cowden, *Applied General Statistics*, Second Edition, Prentice-Hall, Englewood Cliffs, NJ, 1960.

CD Cook, Thomas D. and Donald T. Campbell. *Quasi-Experimentation: Design & Analysis Issues for Fled Settings*. Houghton Mifflin Company, Boston, Massachusetts.1979

CS Campbell, Donald T. and Julian C. Stanley. *Experimental and Quasi-Experimental Design for Research*. Houghton Mifflin, Dallas, TX. 1963.

JCC Cohen, Jacob, Patricia Cohen, Stephen West, and Leona Aiken. *Applied Multiple Regression/Correlation Analysis for the Behavioral Sciences*, 3rd Edition, Lawrence Erlbaum, Mahwah, NJ, 2003.

KP Kennedy, Peter. *A Guide to Econometrics*, Fourth Edition, The MIT Press, Cambridge, Massachusetts, 2001.

KGO Kurnow, Frank, Gerald Glasser, and Frederick Ottman. *Statistics for Business Decisions*, Richard D. Irwin, Homewood, IL, 1959.

MCH McClearly, Richard and Richard A. Hay, Richard A. *Applied Time Series Analysis for the Social Sciences*, Sage Publications, Beverly Hills, 1980

ZC Zeller, Richard and Edward G. Carmines. *Measurement in the Social Sciences*. Cambridge University Press, London, 1980.

'35